Conformable Dynamic Equations on Time Scales

Conformable Dynamic Equations on Time Scales

Douglas R. Anderson
Concordia College

Svetlin G. Georgiev
Sorbonne University

CRC Press
Taylor & Francis Group
Boca Raton London New York

CRC Press is an imprint of the
Taylor & Francis Group, an **informa** business

A CHAPMAN & HALL BOOK

First edition published 2020
by CRC Press
6000 Broken Sound Parkway NW, Suite 300, Boca Raton, FL 33487-2742

and by CRC Press
2 Park Square, Milton Park, Abingdon, Oxon, OX14 4RN

© 2020 Douglas R. Anderson, Svetlin G. Georgiev

CRC Press is an imprint of Taylor & Francis Group, LLC

ISBN: 978-0-367-51701-4 (hbk)
ISBN: 978-0-367-52310-7 (pbk)
ISBN: 978-1-003-05740-6 (ebk)

**Visit the Taylor & Francis Web site at
http://www.taylorandfrancis.com**

**and the CRC Press Web site at
http://www.crcpress.com**

Contents

Preface

The concept of derivatives of non-integer order, known as fractional derivatives, first appeared in the letter between L'Hôpital and Leibniz in which the question of a half-order derivative was posed. Since then, many formulations of fractional derivatives have appeared. Recently, a new definition of fractional derivative, named "fractional conformable derivative", has been introduced. This new fractional derivative is compatible with the classical derivative and it has attracted attention in domains such as mechanics, electronics and anomalous diffusion.

This book is devoted to the qualitative theory of conformable dynamic equations on time scales. It summarizes the most recent contributions in this area, and vastly expands on them to create a comprehensive theory developed exclusively for this book. Except for a few sections in Chapter 1, the results here are presented for the first time. As a result, the book is intended for researchers who work on dynamic calculus on time scales and its applications. There are nine chapters in this book, all with new content not found in journals.

In Chapter 1 we introduce the concept of conformable dynamic differentiation and integration, and deduce some of the properties of these operations. Also in this chapter, we define basic elementary functions, such as conformable dynamic exponential, trigonometric, hyperbolic, and logarithmic functions, and we obtain a Taylor formula. Chapter 2 is devoted to first-order linear dynamic equations on time scales. In particular, we discuss conformable dynamic versions of Bernoulli, Riccati, and logistic equations. In Chapter 3, we investigate the structure of conformable dynamic systems with variable and constant coefficients. Chapter 4 deals with linear conformable inequalities such as Gronwall, Gamidov, Pachpatte, and Volterra-type inequalities. Chapter 5 is devoted to a Cauchy-type problem for some classes of nonlinear conformable dynamic equations. For this class, we prove the existence and uniqueness of solutions, and investigate the dependency of solutions on initial data. Lyapunov functions are also discussed, including some criteria for the boundedness and exponential stability of solutions. Higher-order linear dynamic equations with constant coefficients are investigated in Chapter 6. Chapter 7 deals with second-order conformable dynamic equations, and second-order self-adjoint conformable dynamic equations are investigated in Chapter 8. Chapter 9 is devoted to the Laplace transform, including its properties and applications.

This book is addressed to a wide audience of specialists such as mathematicians, physicists, engineers, and biologists. It can be used as a textbook at the graduate level and as a reference book for several disciplines.

The authors welcome any suggestions for improvement of the text.

Douglas R. Anderson, Svetlin G. Georgiev
Moorhead, Paris, August 2019

CHAPTER 1

Conformable Dynamic Calculus on Time Scales

1.1 INTRODUCTION

Motivated by a proportional-derivative (PD) controller from control theory, in this book we introduce a conformable dynamic calculus on time scales whose differential operator is modeled after a PD controller. This proportional derivative D^α of order $\alpha \in [0,1]$, where D^0 is the identity operator, and D^1 is the classical time scales delta derivative operator, will be used to construct a new calculus that is then explored extensively.

Let $\mathbb{T}$ be a time scale with forward jump operator and delta differentiation operator σ and Δ, respectively. Also, let $\alpha \in [0,1]$. Throughout this monograph we suppose

(A1) $k_0, k_1 : [0,1] \times \mathbb{T} \to [0,\infty)$ are continuous functions such that

$$\lim_{\alpha \to 0+} k_1(\alpha,t) = 1, \quad \lim_{\alpha \to 1-} k_1(\alpha,t) = 0, \quad t \in \mathbb{T},$$

$$\lim_{\alpha \to 0+} k_0(\alpha,t) = 0, \quad \lim_{\alpha \to 1-} k_0(\alpha,t) = 1, \quad t \in \mathbb{T},$$

$$k_1(\alpha,t) \neq 0, \quad \alpha \in [0,1), \quad t \in \mathbb{T}, \quad k_0(\alpha,t) \neq 0, \quad \alpha \in (0,1], \quad t \in \mathbb{T}.$$

Such functions k_1 and k_0 exist. For instance,

$$k_1(\alpha,t) = (1-\alpha)\left(1+t^2\right)^\alpha, \quad k_0(\alpha,t) = \alpha\left(1+t^2\right)^{1-\alpha}, \quad t \in \mathbb{T}, \quad \alpha \in [0,1],$$

$$k_1(\alpha,t) = (1-\alpha)|t|^\alpha, \quad k_0(\alpha,t) = \alpha|t|^{1-\alpha}, \quad t \in \mathbb{T}, \quad \alpha \in [0,1],$$

$$k_1(\alpha,t) = 1-\alpha, \quad k_0(\alpha,t) = \alpha, \quad t \in \mathbb{T}, \quad \alpha \in [0,1],$$

$$k_1(\alpha,t) = (1-\alpha)3^\alpha, \quad k_0(\alpha,t) = \alpha 3^{1-\alpha}, \quad t \in \mathbb{T}, \quad \alpha \in [0,1],$$

$$k_1(\alpha,t) \;=\; \cos\left(\alpha\frac{\pi}{2}\right)|t|^{\alpha}, \quad k_0(\alpha,t)=\sin\left(\alpha\frac{\pi}{2}\right)|t|^{1-\alpha}, \quad t\in\mathbb{T}, \quad \alpha\in[0,1],$$

satisfy $(A1)$.

1.2 CONFORMABLE DIFFERENTIATION

In control theory particularly, a proportional-derivative controller for controller output u at time t with two tuning parameters has the algorithm

$$u(t) = \kappa_p E(t) + \kappa_d \frac{d}{dt}E(t),$$

where κ_p is the proportional gain, κ_d is the derivative gain, and E is the error between the state variable and the process variable. This is the impetus for the definition and results to follow.

Definition 1.2.1 *Suppose that f is Δ-differentiable at some $t \in \mathbb{T}^{\kappa}$. The conformable Δ-derivative of f at t is defined by*

$$D^{\alpha}f(t) = k_1(\alpha,t)f(t) + k_0(\alpha,t)f^{\Delta}(t). \tag{1.1}$$

Here k_1 is a type of the proportional gain k_p, k_0 is a type of the derivative gain k_d, f is the error, and $D^{\alpha}f(t)$ is the controller output.

Remark 1.2.2 *We have*

$$D^0 f(t) = f(t), \quad D^1 f(t) = f^{\Delta}(t).$$

Remark 1.2.3 *Suppose that f is Δ-differentiable at $t \in \mathbb{T}^{\kappa}$.*

1. *Let $\alpha \in [0,1)$. Then*

$$k_1(\alpha,t)f(t) = D^{\alpha}f(t) - k_0(\alpha,t)f^{\Delta}(t),$$

   whereupon

$$f(t) = \frac{1}{k_1(\alpha,t)}D^{\alpha}f(t) - \frac{k_0(\alpha,t)}{k_1(\alpha,t)}f^{\Delta}(t).$$

2. *Let $\alpha \in (0,1]$. Then*

$$k_0(\alpha,t)f^{\Delta}(t) = D^{\alpha}f(t) - k_1(\alpha,t)f(t),$$

   whereupon

$$f^{\Delta}(t) = \frac{1}{k_0(\alpha,t)}D^{\alpha}f(t) - \frac{k_1(\alpha,t)}{k_0(\alpha,t)}f(t).$$

3. *Let $\alpha \in (0,1)$. Then*

$$D^\alpha f(t) = k_1(\alpha,t)\left(f^\sigma(t) - \mu(t)f^\Delta(t)\right) + k_0(\alpha,t)f^\Delta(t),$$

whereupon

$$
\begin{aligned}
k_1(\alpha,t)f^\sigma(t) &= D^\alpha f(t) + \mu(t)k_1(\alpha,t)f^\Delta(t) - k_0(\alpha,t)f^\Delta(t) \\[2mm]
&= D^\alpha f(t) + \left(\mu(t)k_1(\alpha,t) - k_0(\alpha,t)\right)f^\Delta(t),
\end{aligned}
$$

and

$$f^\sigma(t) = \frac{1}{k_1(\alpha,t)}D^\alpha f(t) + \left(\mu(t) - \frac{k_0(\alpha,t)}{k_1(\alpha,t)}\right)f^\Delta(t).$$

Example 1.2.4 *Let $\mathbb{T} = 2\mathbb{Z}$,*

$$k_1(\alpha,t) = (1-\alpha)t^{2\alpha}, \quad k_0(\alpha,t) = \alpha t^{2(1-\alpha)}, \quad \alpha \in [0,1], \quad t \in \mathbb{T},$$

and

$$f(t) = t^3 - t^2 + 2t + 3, \quad t \in \mathbb{T}.$$

In Fig. 1.1 is shown the function f.

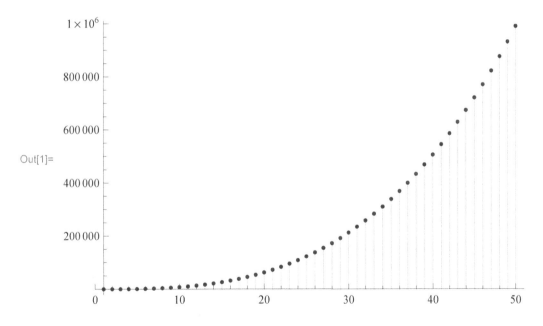

Out[1]=

Figure 1.1 $f(t)$ for $t \in [0, 100]$ in Example 1.2.4

We will find

$$D^{\frac{1}{2}}f(t), \quad t \in \mathbb{T}.$$

We have

$$\sigma(t) = t+2,$$

$$\begin{aligned}
f^\Delta(t) &= (\sigma(t))^2 + t\sigma(t) + t^2 - \sigma(t) - t + 2 \\
&= (t+2)^2 + t(t+2) + t^2 - t - 2 - t + 2 \\
&= t^2 + 4t + 4 + t^2 + 2t + t^2 - 2t \\
&= 3t^2 + 4t + 4, \quad t \in \mathbb{T}.
\end{aligned}$$

Fig. 1.2 shows the delta derivative f^Δ.

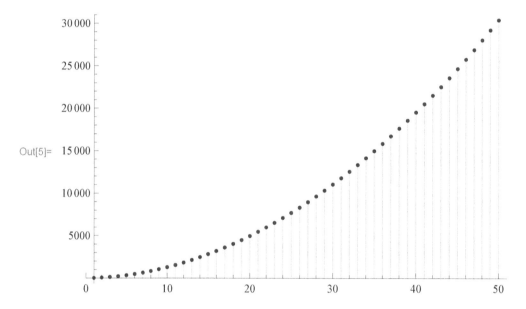

Out[5]=

Figure 1.2 $f^\Delta(t)$ for $t \in [0, 100]$ in Example 1.2.4

Next,

$$\begin{aligned}
D^{\frac{1}{2}} f(t) &= k_1\left(\frac{1}{2}, t\right) f(t) + k_0\left(\frac{1}{2}, t\right) f^\Delta(t) \\
&= \frac{1}{2} t \left(t^3 - t^2 + 2t + 3\right) + \frac{1}{2} t \left(3t^2 + 4t + 4\right) \\
&= \frac{1}{2} t \left(t^3 - t^2 + 2t + 3 + 3t^2 + 4t + 4\right)
\end{aligned}$$

$$= \frac{1}{2}t\left(t^3 + 2t^2 + 6t + 7\right), \quad t \in \mathbb{T}.$$

The Fig. 1.3 shows the conformable delta derivative $D^{\frac{1}{2}}f$.

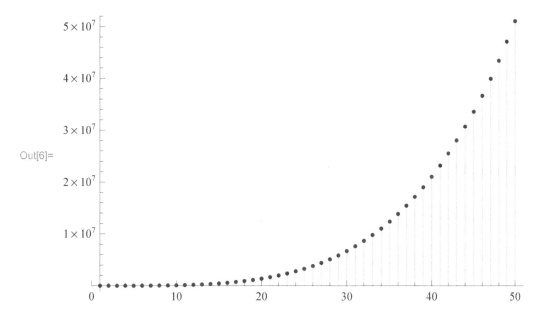

Out[6]=

Figure 1.3 $D^{\frac{1}{2}}f(t)$ for $t \in [0, 100]$ in Example 1.2.4

Example 1.2.5 *Let* $\mathbb{T} = 2^{\mathbb{N}_0}$,

$$k_1(\alpha, t) = (1-\alpha)t^{\alpha}, \quad k_0(\alpha, t) = \alpha t^{1-\alpha}, \quad \alpha \in [0,1], \quad t \in \mathbb{T},$$

$$f(t) = t^3 + t, \quad t \in \mathbb{T}.$$

We will find

$$D^{\frac{1}{4}}f(t), \quad t \in \mathbb{T}.$$

Here

$$\sigma(t) = 2t, \quad t \in \mathbb{T}.$$

Then

$$f^{\Delta}(t) = (\sigma(t))^2 + t\sigma(t) + t^2 + 1$$

$$= (2t)^2 + 2t^2 + t^2 + 1$$

$$= 4t^2 + 2t^2 + t^2 + 1$$

$$= 7t^2 + 1, \quad t \in \mathbb{T}.$$

Hence,

$$
\begin{aligned}
D^{\frac{1}{4}} f(t) &= k_1\left(\frac{1}{4},t\right) f(t) + k_0\left(\frac{1}{4},t\right) f^\Delta(t) \\
&= \frac{3}{4} t^{\frac{1}{4}}(t^3 + t) + \frac{1}{4} t^{\frac{3}{4}}(7t^2 + 1) \\
&= \frac{3}{4} t^{\frac{13}{4}} + \frac{3}{4} t^{\frac{5}{4}} + \frac{7}{4} t^{\frac{11}{4}} + \frac{1}{4} t^{\frac{3}{4}}, \quad t \in \mathbb{T}.
\end{aligned}
$$

Theorem 1.2.6 *Let f and g be Δ-differentiable at $t \in \mathbb{T}^\kappa$. Then*

1.
$$D^\alpha(f+g)(t) = D^\alpha f(t) + D^\alpha g(t),$$

2.
$$D^\alpha(af)(t) = aD^\alpha f(t)$$
for any $a \in \mathbb{R}$,

3.
$$
\begin{aligned}
D^\alpha(fg)(t) &= (D^\alpha f(t))g(t) + f^\sigma(t)(D^\alpha g(t)) \\
&\quad -k_1(\alpha,t)f^\sigma(t)g(t) \\
&= (D^\alpha f(t))g^\sigma(t) + f(t)(D^\alpha g(t)) \\
&\quad -k_1(\alpha,t)f(t)g^\sigma(t),
\end{aligned}
$$

4.
$$D^\alpha\left(\frac{f}{g}\right)(t) = \frac{g(t)D^\alpha f(t) - f(t)D^\alpha g(t)}{g(t)g^\sigma(t)} + k_1(\alpha,t)\frac{f(t)}{g(t)}$$
provided that $g(t)g^\sigma(t) \neq 0$.

Proof 1.2.7 *1. We have*

$$
\begin{aligned}
D^\alpha(f+g)(t) &= k_1(\alpha,t)(f+g)(t) + k_0(\alpha,t)(f+g)^\Delta(t) \\
&= k_1(\alpha,t)f(t) + k_0(\alpha,t)f^\Delta(t)
\end{aligned}
$$

$$+k_1(\alpha,t)g(t)+k_0(\alpha,t)g^\Delta(t)$$

$$=\ D^\alpha f(t)+D^\alpha g(t).$$

2. *We have*

$$D^\alpha(af)(t)\ =\ k_1(\alpha,t)(af)(t)+k_0(\alpha,t)(af)^\Delta(t)$$

$$=\ a\left(k_1(\alpha,t)f(t)+k_0(\alpha,t)f^\Delta(t)\right)$$

$$=\ aD^\alpha f(t).$$

3. *We have*

$$D^\alpha(fg)(t)\ =\ k_1(\alpha,t)(fg)(t)+k_0(\alpha,t)(fg)^\Delta(t)$$

$$=\ k_1(\alpha,t)f(t)g(t)+k_0(\alpha,t)f^\Delta(t)g(t)$$

$$+k_0(\alpha,t)f^\sigma(t)g^\Delta(t)$$

$$=\ \left(k_1(\alpha,t)f(t)+k_0(\alpha,t)f^\Delta(t)\right)g(t)$$

$$+\left(k_1(\alpha,t)g(t)+k_0(\alpha,t)g^\Delta(t)\right)f^\sigma(t)$$

$$-k_1(\alpha,t)f^\sigma(t)g(t)$$

$$=\ (D^\alpha f(t))\,g(t)+(D^\alpha g(t))\,f^\sigma(t)$$

$$-k_1(\alpha,t)f^\sigma(t)g(t)$$

$$=\ k_1(\alpha,t)f(t)g(t)+k_0(\alpha,t)f^\Delta(t)g^\sigma(t)$$

$$+k_0(\alpha,t)f(t)g^\Delta(t)$$

$$=\ \left(k_1(\alpha,t)g(t)+k_0(\alpha,t)g^\Delta(t)\right)f(t)$$

$$+\left(k_1(\alpha,t)f(t)+k_0(\alpha,t)f^\Delta(t)\right)g^\sigma(t)$$

$$-k_1(\alpha,t)f(t)g^{\sigma}(t)$$

$$= (D^{\alpha}f(t))g^{\sigma}(t)+f(t)D^{\alpha}g(t)$$

$$-k_1(\alpha,t)f(t)g^{\sigma}(t).$$

4. *We have*

$$D^{\alpha}\left(\frac{f}{g}\right)(t) = k_1(\alpha,t)\frac{f(t)}{g(t)}+k_0(\alpha,t)\left(\frac{f}{g}\right)^{\Delta}(t)$$

$$= k_1(\alpha,t)\frac{f(t)}{g(t)}+k_0(\alpha,t)\frac{f^{\Delta}(t)g(t)-f(t)g^{\Delta}(t)}{g(t)g^{\sigma}(t)}$$

$$= k_1(\alpha,t)\frac{f(t)}{g(t)}+k_0(\alpha,t)\frac{f^{\Delta}(t)}{g^{\sigma}(t)}$$

$$-\frac{f(t)}{g^{\sigma}(t)g(t)}k_0(\alpha,t)g^{\Delta}(t)$$

$$= k_1(\alpha,t)\frac{f(t)}{g(t)}$$

$$+\frac{1}{g^{\sigma}(t)}\left(k_1(\alpha,t)f(t)+k_0(\alpha,t)f^{\Delta}(t)\right)$$

$$-\frac{f(t)}{g(t)g^{\sigma}(t)}\left(k_1(\alpha,t)g(t)+k_0(\alpha,t)g^{\Delta}(t)\right)$$

$$= \frac{D^{\alpha}f(t)}{g^{\sigma}(t)}-\frac{f(t)}{g(t)g^{\sigma}(t)}D^{\alpha}g(t)+k_1(\alpha,t)\frac{f(t)}{g(t)}$$

$$= \frac{g(t)D^{\alpha}f(t)-f(t)D^{\alpha}g(t)}{g(t)g^{\sigma}(t)}+k_1(\alpha,t)\frac{f(t)}{g(t)}.$$

This completes the proof. □

Example 1.2.8 *Let* $\mathbb{T}=\mathbb{Z}$,

$$k_1(\alpha,t) = (1-\alpha)(1+t^2)^{\alpha}, \quad k_0(\alpha,t)=\alpha(1+t^2)^{1-\alpha}, \quad \alpha\in[0,1], \quad t\in\mathbb{T},$$

$$f(t) = t^2+t, \quad g(t)=t^3+t+1, \quad t\in\mathbb{T}.$$

We will find

$$D^{\frac{1}{2}}\left(\frac{f}{g}\right)(t), \quad t\in\mathbb{T}.$$

Here

$$\sigma(t)=t+1, \quad t\in\mathbb{T}.$$

Then

$$f^{\Delta}(t) \;=\; \sigma(t)+t+1$$

$$=\; t+1+t+1$$

$$=\; 2t+2,$$

$$g^{\Delta}(t) \;=\; (\sigma(t))^2+t\sigma(t)+t^2+1$$

$$=\; (t+1)^2+t(t+1)+t^2+1$$

$$=\; t^2+2t+1+t^2+t+t^2+1$$

$$=\; 3t^2+3t+2,$$

$$D^{\frac{1}{2}}f(t) \;=\; k_1\left(\frac{1}{2},t\right)f(t)+k_0\left(\frac{1}{2},t\right)f^{\Delta}(t)$$

$$=\; \frac{1}{2}\sqrt{1+t^2}(t^2+t)+\frac{1}{2}\sqrt{1+t^2}(2t+2)$$

$$=\; \frac{1}{2}\sqrt{1+t^2}(t^2+t+2t+2)$$

$$=\; \frac{1}{2}\sqrt{1+t^2}(t^2+3t+2),\cdot$$

$$D^{\frac{1}{2}}g(t) \;=\; k_1\left(\frac{1}{2},t\right)g(t)+k_0\left(\frac{1}{2},t\right)g^{\Delta}(t)$$

$$=\; \frac{1}{2}\sqrt{1+t^2}\left(t^3+t+1\right)+\frac{1}{2}\sqrt{1+t^2}\left(3t^2+3t+2\right)$$

$$=\; \frac{1}{2}\sqrt{1+t^2}\left(t^3+t+1+3t^2+3t+2\right)$$

$$=\; \frac{1}{2}\sqrt{1+t^2}\left(t^3+3t^2+4t+3\right),$$

$$g^{\sigma}(t) \;=\; (\sigma(t))^3+\sigma(t)+1$$

$$=\; (t+1)^3+t+1+1$$

$$= t^3 + 3t^2 + 3t + 1 + t + 2$$

$$= t^3 + 3t^2 + 4t + 3,$$

$$D^{\frac{1}{2}}\left(\frac{f}{g}\right)(t) = \frac{g(t)D^{\frac{1}{2}}f(t) - f(t)D^{\frac{1}{2}}g(t)}{g(t)g^{\sigma}(t)}$$

$$+ k_1\left(\frac{1}{2}, t\right)\frac{f(t)}{g(t)}$$

$$= \frac{(t^3 + t + 1)\frac{1}{2}\sqrt{1+t^2}(t^2 + 3t + 2)}{(t^3 + t + 1)(t^3 + 3t^2 + 4t + 3)}$$

$$- \frac{(t^2 + t)\frac{1}{2}\sqrt{1+t^2}(t^3 + 3t^2 + 4t + 3)}{(t^3 + t + 1)(t^3 + 3t^2 + 4t + 3)}$$

$$+ \frac{1}{2}\sqrt{1+t^2}\frac{t^2 + t}{t^3 + t + 1}$$

$$= \frac{\sqrt{1+t^2}}{2(t^3 + t + 1)(t^3 + 3t^2 + 4t + 3)}\Big(t^5 + 3t^4 + 2t^3 + t^3 + 3t^2 + 2t + t^2$$

$$+ 3t + 2 - t^5 - 3t^4 - 4t^3 - 3t^2 - t^4 - 3t^3 - 4t^2 - 3t\Big)$$

$$+ \frac{1}{2}\sqrt{1+t^2}\frac{t^2 + t}{t^3 + t + 1}$$

$$= \frac{\sqrt{1+t^2}\left(-t^4 - 4t^3 - 3t^2 + 2t + 2\right)}{2(t^3 + t + 1)(t^3 + 3t^2 + 4t + 3)}$$

$$+ \frac{1}{2}\sqrt{1+t^2}\frac{t^2 + t}{t^3 + t + 1}$$

$$= \frac{\sqrt{1+t^2}}{2(t^3 + t + 1)}\left(\frac{-t^4 - 4t^3 - 3t^2 + 2t + 2}{t^3 + 3t^2 + 4t + 3} + t^2 + t\right)$$

$$= \frac{\sqrt{1+t^2}}{2(t^3 + t + 1)(t^3 + 3t^2 + 4t + 3)}\Big(-t^4 - 4t^3 - 3t^2 + 2t + 2 + t^5$$

$$+ 3t^4 + 4t^3 + 3t^2 + t^4 + 3t^3 + 4t^2 + 3t\Big)$$

$$= \frac{\sqrt{1+t^2}\left(t^5 + 3t^4 + 3t^3 + 4t^2 + 5t + 2\right)}{2(t^3 + t + 1)(t^3 + 3t^2 + 4t + 3)}, \quad t \in \mathbb{T}.$$

Example 1.2.9 *Let* $\mathbb{T} = 2^{\mathbb{N}_0}$,

$$k_1(\alpha, t) = (1 - \alpha)t^{4\alpha}, \quad k_0(\alpha, t) = \alpha t^{4(1-\alpha)}, \quad \alpha \in [0, 1], \quad t \in \mathbb{T},$$

$$f(t) \;=\; t^2 - t, \quad g(t) = t^2 + 2t + 3, \quad t \in \mathbb{T}.$$

We will find

$$D^{\frac{1}{4}}(fg)(t) \quad and \quad D^{\frac{1}{2}}\left(\frac{f}{f+g}\right)(t), \quad t \in \mathbb{T}.$$

Here

$$\sigma(t) = 2t, \quad t \in \mathbb{T}.$$

We have

$$f^{\Delta}(t) \;=\; \sigma(t) + t - 1$$

$$=\; 2t + t - 1$$

$$=\; 3t - 1,$$

$$g^{\Delta}(t) \;=\; \sigma(t) + t + 2$$

$$=\; 2t + t + 2$$

$$=\; 3t + 2,$$

$$f^{\sigma}(t) \;=\; (\sigma(t))^2 - \sigma(t)$$

$$=\; (2t)^2 - 2t$$

$$=\; 4t^2 - 2t,$$

$$g^{\sigma}(t) \;=\; (\sigma(t))^2 + 2\sigma(t) + 3$$

$$=\; (2t)^2 + 2(2t) + 3$$

$$=\; 4t^2 + 4t + 3,$$

$$D^{\frac{1}{4}}f(t) \;=\; k_1\left(\frac{1}{4},t\right)f(t) + k_0\left(\frac{1}{4},t\right)f^{\Delta}(t)$$

$$=\; \frac{3}{4}t\left(t^2 - t\right) + \frac{1}{4}t^3(3t - 1)$$

$$= \frac{1}{4}t^2(3t - 3 + 3t^2 - t)$$

$$= \frac{1}{4}t^2(3t^2 + 2t - 3),$$

$$D^{\frac{1}{4}}g(t) = k_1\left(\frac{1}{4}, t\right)g(t) + k_0\left(\frac{1}{4}, t\right)g^\Delta(t)$$

$$= \frac{3}{4}t\left(t^2 + 2t + 3\right) + \frac{1}{4}t^3(3t + 2)$$

$$= \frac{1}{4}t\left(3t^2 + 6t + 9 + 3t^3 + 2t^2\right)$$

$$= \frac{1}{4}t\left(3t^3 + 5t^2 + 6t + 9\right),$$

$$D^{\frac{1}{4}}(fg)(t) = \left(D^{\frac{1}{4}}f(t)\right)g(t) + f^\sigma(t)\left(D^{\frac{1}{4}}g(t)\right)$$

$$\qquad - k_1\left(\frac{1}{4}, t\right)f^\sigma(t)g(t)$$

$$= \frac{1}{4}t^2\left(3t^2 + 2t - 3\right)\left(t^2 + 2t + 3\right)$$

$$\qquad + \left(4t^2 - 2t\right)\frac{1}{4}t\left(3t^3 + 5t^2 + 6t + 9\right)$$

$$\qquad - \frac{3}{4}t\left(4t^2 - 2t\right)\left(t^2 + 2t + 3\right)$$

$$= \frac{1}{4}t^2\left(3t^2 + 2t - 3\right)\left(t^2 + 2t + 3\right)$$

$$\qquad + \frac{1}{4}t^2(4t - 2)\left(3t^3 + 5t^2 + 6t + 9\right)$$

$$\qquad - \frac{1}{4}t^2(12t - 6)\left(t^2 + 2t + 3\right)$$

$$= \frac{1}{4}t^2\Bigg(3t^4 + 6t^3 + 9t^2 + 2t^3 + 4t^2 + 6t - 3t^2 - 6t - 9$$

$$\qquad + 12t^4 + 20t^3 + 24t^2 + 36t - 6t^3 - 10t^2 - 12t - 18$$

$$\qquad - 12t^3 - 24t^2 - 36t + 6t^2 + 12t + 18\Bigg)$$

$$= \frac{1}{4}t^2\left(15t^4 + 10t^3 + 6t^2 - 9\right),$$

$$D^{\frac{1}{2}}f(t) = k_1\left(\frac{1}{2}, t\right)f(t) + k_0\left(\frac{1}{2}, t\right)f^{\Delta}(t)$$

$$= \frac{1}{2}t^2(t^2 - t) + \frac{1}{2}t^2(3t - 1)$$

$$= \frac{1}{2}t^2(t^2 - t + 3t - 1)$$

$$= \frac{1}{2}t^2(t^2 + 2t - 1)$$

$$= \frac{1}{2}t^4 + t^3 - \frac{1}{2}t^2,$$

$$D^{\frac{1}{2}}g(t) = \frac{1}{2}t^2 g(t) + \frac{1}{2}t^2 g^{\Delta}(t)$$

$$= \frac{1}{2}t^2(t^2 + 2t + 3) + \frac{1}{2}t^2(3t + 2)$$

$$= \frac{1}{2}t^2(t^2 + 5t + 5),$$

$$f^{\sigma}(t) + g^{\sigma}(t) = (\sigma(t))^2 - \sigma(t) + (\sigma(t))^2 + 2\sigma(t) + 3$$

$$= 2(\sigma(t))^2 + \sigma(t) + 3$$

$$= 8t^2 + 2t + 3,$$

$$D^{\frac{1}{2}}\left(\frac{f}{f+g}\right)(t) = \frac{(f(t) + g(t))D^{\frac{1}{2}}f(t) - f(t)\left(D^{\frac{1}{2}}f(t) + D^{\frac{1}{2}}g(t)\right)}{(f(t) + g(t))(f^{\sigma}(t) + g^{\sigma}(t))}$$

$$+ k_1\left(\frac{1}{2}, t\right)\frac{f(t)}{f(t) + g(t)}$$

$$= \frac{1}{(2t^2 + t + 3)(8t^2 + 2t + 3)}\left((2t^2 + t + 3)\left(\frac{1}{2}t^4 + t^3 - \frac{1}{2}t^2\right)\right.$$

$$\left. - (t^2 - t)\left(\frac{1}{2}t^4 + t^3 - \frac{1}{2}t^2 + \frac{1}{2}t^4 + \frac{5}{2}t^3 + \frac{5}{2}t^2\right)\right)$$

$$+ \frac{1}{2}t^2\frac{t^2 - t}{2t^2 + t + 3}$$

$$= \frac{1}{(2t^2+t+3)(8t^2+2t+3)}\left(t^6+2t^5-t^4+\frac{1}{2}t^5+t^4-\frac{1}{2}t^3+\frac{3}{2}t^4\right.$$

$$\left.+3t^3-\frac{3}{2}t^2-(t^2-t)\left(t^4+\frac{7}{2}t^3+2t^2\right)\right)$$

$$+\frac{t^4-t^3}{2(2t^2+t+3)}$$

$$= \frac{1}{(2t^2+t+3)(8t^2+2t+3)}\left(t^6+\frac{5}{2}t^5+\frac{3}{2}t^4+\frac{5}{2}t^3-\frac{3}{2}t^2-t^6\right.$$

$$\left.-\frac{7}{2}t^5-2t^4+t^5+\frac{7}{2}t^4+2t^3\right)$$

$$+\frac{t^4-t^3}{2(2t^2+t+3)}$$

$$= \frac{3t^4+\frac{9}{2}t^3-\frac{3}{2}t^2}{(2t^2+t+3)(8t^2+2t+3)}+\frac{t^4-t^3}{2(2t^2+t+3)}$$

$$= \frac{6t^4+9t^3-3t^2+(8t^2+2t+3)(t^4-t^3)}{2(2t^2+t+3)(8t^2+2t+3)}$$

$$= \frac{6t^4+9t^3-3t^2+8t^6-8t^5+2t^5-2t^4+3t^4-3t^3}{2(2t^2+t+3)(8t^2+2t+3)}$$

$$= \frac{8t^6-6t^5+7t^4+6t^3-3t^2}{2(2t^2+t+3)(8t^2+2t+3)}, \quad t \in \mathbb{T}.$$

Example 1.2.10 *Let* $\mathbb{T}=2^{\mathbb{N}_0}$,

$$k_1(\alpha,t) = (1-\alpha)t^{4\alpha}, \quad k_0(\alpha,t)=\alpha t^{4(1-\alpha)}, \quad \alpha \in [0,1], \quad t \in \mathbb{T},$$

$$f(t) = t, \quad t \in \mathbb{T}.$$

Then

$$\sigma(t) = 2t,$$

$$f^{\Delta}(t) = 1,$$

$$D^{\frac{1}{2}}f(t) = k_1\left(\frac{1}{2},t\right)f(t)+k_0\left(\frac{1}{2},t\right)f^{\Delta}(t)$$

$$= \frac{1}{2}t^3 + \frac{1}{2}t^2$$

$$= \frac{1}{2}(t^3 + t^2),$$

$$\left(D^{\frac{1}{2}}f\right)^{\Delta}(t) = \frac{1}{2}\left((\sigma(t))^2 + t\sigma(t) + t^2 + \sigma(t) + t\right)$$

$$= \frac{1}{2}(4t^2 + 2t^2 + t^2 + 2t + t)$$

$$= \frac{1}{2}(7t^2 + 3t),$$

$$D^{\frac{1}{4}}\left(D^{\frac{1}{2}}f\right)(t) = k_1\left(\frac{1}{4}, t\right)\left(D^{\frac{1}{2}}f\right)(t) + k_0\left(\frac{1}{4}, t\right)\left(D^{\frac{1}{2}}f\right)^{\Delta}(t)$$

$$= \frac{3}{4}t\left(\frac{1}{2}(t^3 + t^2)\right) + \frac{1}{4}t^3\left(\frac{1}{2}(7t^2 + 3t)\right)$$

$$= \frac{3}{8}(t^4 + t^3) + \frac{1}{8}(7t^5 + 3t^4)$$

$$= \frac{1}{8}\left(7t^5 + 3t^4 + 3t^4 + 3t^3\right)$$

$$= \frac{1}{8}\left(7t^5 + 6t^4 + 3t^3\right),$$

$$D^{\frac{1}{4}}f(t) = k_1\left(\frac{1}{4}, t\right)f(t) + k_0\left(\frac{1}{4}, t\right)f^{\Delta}(t)$$

$$= \frac{3}{4}t^2 + \frac{1}{4}t^3$$

$$= \frac{1}{4}(t^3 + 3t^2),$$

$$\left(D^{\frac{1}{4}}f\right)^{\Delta}(t) = \frac{1}{4}\left((\sigma(t))^2 + t\sigma(t) + t^2 + 3\sigma(t) + 3t\right)$$

$$= \frac{1}{4}\left(4t^2 + 2t^2 + t^2 + 6t + 3t\right)$$

$$= \frac{1}{4}(7t^2 + 9t),$$

$$D^{\frac{1}{2}}\left(D^{\frac{1}{4}}f\right)(t) = k_1\left(\frac{1}{2}, t\right)D^{\frac{1}{4}}f(t) + k_0\left(\frac{1}{2}, t\right)\left(D^{\frac{1}{4}}f\right)^{\Delta}(t)$$

$$= \frac{1}{2}t^2\left(\frac{1}{4}(t^3+3t^2)\right) + \frac{1}{2}t^2\left(\frac{1}{4}(7t^2+9t)\right)$$

$$= \frac{1}{8}t^2\left(t^3+3t^2+7t^2+9t\right)$$

$$= \frac{1}{8}t^2\left(t^3+10t^2+9t\right), \quad t \in \mathbb{T}.$$

Therefore

$$D^{\frac{1}{4}}\left(D^{\frac{1}{2}}f\right)(t) \neq D^{\frac{1}{2}}\left(D^{\frac{1}{4}}f\right)(t), \quad t \in \mathbb{T}.$$

Remark 1.2.11 *Assume* $\alpha, \beta \in [0,1]$, k_1, *and* k_0 *are* Δ-*differentiable at* $t \in \mathbb{T}^{\kappa^2}$, *and* $f : \mathbb{T} \to \mathbb{R}$ *is twice* Δ-*differentiable at* $t \in \mathbb{T}^{\kappa^2}$. *Then, in the general case, we have*

$$D^\alpha\left(D^\beta f\right)(t) \neq D^\beta\left(D^\alpha f\right)(t).$$

Definition 1.2.12 *Let* k_1, k_0 *be* $n-1$-*times* Δ-*differentiable at* $t \in \mathbb{T}^{\kappa^{n-1}}$, $f : \mathbb{T} \to \mathbb{R}$ *be* n-*times* Δ-*differentiable at* $t \in \mathbb{T}^{\kappa^n}$, $n \in \mathbb{N}$. *Then we define*

$$(D^\alpha)^n f(t) = \underbrace{D^\alpha\left(D^\alpha\left(\ldots\left(D^\alpha f\right)\ldots\right)\right)}_{n}(t), \quad t \in \mathbb{T}^{\kappa^n}.$$

Remark 1.2.13 *Note that in the general case, we have*

$$(D^\alpha)^n f(t) \neq D^{n\alpha} f(t), \quad t \in \mathbb{T}^{\kappa^n},$$

if $n\alpha \in (0,1]$.

Example 1.2.14 *Let* $\mathbb{T} = 2^{\mathbb{N}_0}$,

$$k_1(\alpha,t) = (1-\alpha)t^{4\alpha}, \quad k_0(\alpha,t) = \alpha t^{4(1-\alpha)}, \quad \alpha \in [0,1], \quad t \in \mathbb{T},$$

$$f(t) = t, \quad t \in \mathbb{T}.$$

We have

$$\sigma(t) = 2t, \quad t \in \mathbb{T}.$$

By the previous example, we have

$$f^\Delta(t) = 1,$$

$$D^{\frac{1}{4}}f(t) = \frac{3}{4}t^2 + \frac{1}{4}t^3,$$

$$\left(D^{\frac{1}{4}}f\right)^\Delta(t) = \frac{9}{4}t + \frac{7}{4}t^2,$$

$$D^{\frac{1}{2}} f(t) = \frac{1}{2} t^3 + \frac{1}{2} t^2, \quad t \in \mathbb{T}.$$

Then

$$
\begin{aligned}
D^{\frac{1}{4}} \left(D^{\frac{1}{4}} f \right)(t) &= k_1 \left(\frac{1}{4}, t \right) D^{\frac{1}{4}} f(t) + k_0 \left(\frac{1}{4}, t \right) \left(D^{\frac{1}{4}} f \right)^{\Delta}(t) \\
&= \frac{3}{4} t \left(\frac{3}{4} t^2 + \frac{1}{4} t^3 \right) + \frac{1}{4} t^3 \left(\frac{9}{4} t + \frac{7}{4} t^2 \right) \\
&= \frac{1}{16} t^3 \left(3(3+t) + 9t + 7t^2 \right) \\
&= \frac{1}{16} t^3 \left(9 + 3t + 9t + 7t^2 \right) \\
&= \frac{1}{16} t^3 \left(9 + 12t + 7t^2 \right), \quad t \in \mathbb{T}.
\end{aligned}
$$

Consequently,

$$D^{\frac{1}{4}} \left(D^{\frac{1}{4}} f \right)(t) \neq D^{\frac{1}{2}} f(t), \quad t \in \mathbb{T}.$$

1.3 CONFORMABLE REGRESSIVE FUNCTIONS

Definition 1.3.1 *We say that a function $f : \mathbb{T} \to \mathbb{R}$ is a conformable regressive function if*

$$k_0(\alpha, t) - \mu(t) k_1(\alpha, t) \neq 0$$

and

$$k_0(\alpha, t) + \mu(t) \left(f(t) - k_1(\alpha, t) \right) \neq 0$$

for any $\alpha \in (0, 1]$ and for any $t \in \mathbb{T}$. The set of all conformable regressive functions on $\mathbb{T}$ will be denoted by $\mathscr{R}_c$.

Definition 1.3.2 *For $f, g \in \mathscr{R}_c$, we define "conformable circle plus" $\oplus_c$ as follows:*

$$(f \oplus_c g)(t) = \frac{(f(t) + g(t) - k_1(\alpha, t)) k_0(\alpha, t) + \mu(t) (f(t) - k_1(\alpha, t)) (g(t) - k_1(\alpha, t))}{k_0(\alpha, t)},$$

$t \in \mathbb{T}$, $\alpha \in (0, 1]$.

Remark 1.3.3 *When $\alpha = 1$, we have*

$$\mathscr{R}_c = \mathscr{R} \quad and \quad \oplus_c = \oplus.$$

Theorem 1.3.4 *We have that $(\mathscr{R}_c, \oplus_c)$ is an Abelian group.*

Proof 1.3.5 *Let $f, g, h \in \mathscr{R}_c$ be arbitrarily chosen. Then*

$$k_0 + \mu \left((f \oplus_c g) - k_1 \right) = k_0 + \frac{\mu}{k_0} \left((f + g - k_1) k_0 + \mu (f - k_1)(g - k_1) - k_1 k_0 \right)$$

$$= \frac{1}{k_0} \left(k_0^2 + \mu \left((f - k_1)(k_0 + \mu(g - k_1)) + g k_0 - k_1 k_0 \right) \right)$$

$$= \frac{1}{k_0} \left(k_0 (k_0 + \mu(g - k_1)) + \mu(f - k_1)(k_0 + \mu(g - k_1)) \right)$$

$$= \frac{1}{k_0} \left(k_0 + \mu(f - k_1) \right) \left(k_0 + \mu(g - k_1) \right)$$

$$\neq 0 \quad on \quad \mathbb{T},$$

i.e.,

$$f \oplus_c g \in \mathscr{R}_c.$$

Also,

$$k_0 + \mu \left(-\frac{k_0(f - k_1)}{k_0 + \mu(f - k_1)} + k_1 - k_1 \right)$$

$$= k_0 - \frac{\mu k_0 (f - k_1)}{k_0 + \mu(f - k_1)}$$

$$= \frac{k_0^2 + \mu k_0 (f - k_1) - \mu k_0 (f - k_1)}{k_0 + \mu(f - k_1)}$$

$$= \frac{k_0^2}{k_0 + \mu(f - k_1)}$$

$$\neq 0 \quad on \quad \mathbb{T},$$

i.e.,

$$-\frac{k_0(f - k_1)}{k_0 + \mu(f - k_1)} + k_1 \in \mathscr{R}_c.$$

We have

$$f \oplus_c \left(-\frac{k_0(f - k_1)}{k_0 + \mu(f - k_1)} + k_1 \right)$$

$$= \frac{1}{k_0} \left(\left(f - \frac{k_0(f - k_1)}{k_0 + \mu(f - k_1)} + k_1 - k_1 \right) k_0 \right.$$

$$\left. + \mu \left(-\frac{k_0(f - k_1)}{k_0 + \mu(f - k_1)} + k_1 - k_1 \right) (f - k_1) \right)$$

$$= \frac{1}{k_0}\left(\left(f - \frac{k_0(f-k_1)}{k_0 + \mu(f-k_1)}\right)k_0 - \mu\frac{k_0(f-k_1)}{k_0 + \mu(f-k_1)}(f-k_1)\right)$$

$$= \frac{1}{k_0}\left(fk_0 - \frac{k_0^2(f-k_1)}{k_0 + \mu(f-k_1)} - \frac{\mu k_0(f-k_1)^2}{k_0 + \mu(f-k_1)}\right)$$

$$= \frac{1}{k_0}\left(fk_0 - \frac{k_0(f-k_1)(k_0 + \mu(f-k_1))}{k_0 + \mu(f-k_1)}\right)$$

$$= \frac{1}{k_0}\left(fk_0 - k_0f + k_0k_1\right)$$

$$= k_1,$$

i.e., the conformable addition inverse of f is

$$-\frac{k_0(f-k_1)}{k_0 + \mu(f-k_1)} + k_1.$$

Next,

$$k_0 + \mu(k_1 - k_1) = k_0$$

$$\neq 0, \quad for \quad \alpha \in (0,1] \quad and \quad t \in \mathbb{T},$$

i.e., $k_1 \in \mathscr{R}_c$, and

$$f \oplus_c k_1 = \frac{(f + k_1 - k_1)k_0 + \mu(f - k_1)(k_1 - k_1)}{k_0}$$

$$= \frac{fk_0}{k_0}$$

$$= f$$

$$= \frac{(k_1 + f - k_1)k_0 + \mu(k_1 - k_1)(f - k_1)}{k_0}$$

$$= k_1 \oplus_c f,$$

i.e., k_1 is the conformable additive identity for $\oplus_c$. Next,

$$(f \oplus_c g) \oplus_c h$$

$$= ((f \oplus_c g) + h - k_1) + \frac{\mu}{k_0}(f \oplus_c g - k_1)(h - k_1)$$

$$= \left(f + g - k_1 + \frac{\mu}{k_0}(f - k_1)(g - k_1) + h - k_1\right)$$

$$+\frac{\mu}{k_0}\left(f+g-k_1+\frac{\mu}{k_0}(f-k_1)(g-k_1)-k_1\right)(h-k_1)$$

$$= f+g+h-2k_1+\frac{\mu}{k_0}(f-k_1)(g-k_1)$$

$$+\frac{\mu}{k_0}(f-k_1)(h-k_1)+\frac{\mu}{k_0}(g-k_1)(h-k_1)$$

$$+\frac{\mu^2}{k_0^2}(f-k_1)(g-k_1)(h-k_1) \quad on \quad \mathbb{T},$$

and

$$f\oplus_c(g\oplus_c h)$$

$$= (f+(g\oplus_c h)-k_1)+\frac{\mu}{k_0}(f-k_1)(g\oplus_c h-k_1)$$

$$= \left(f+g+h-k_1+\frac{\mu}{k_0}(g-k_1)(h-k_1)-k_1\right)$$

$$+\frac{\mu}{k_0}(f-k_1)\left(g+h-k_1+\frac{\mu}{k_0}(g-k_1)(h-k_1)-k_1\right)$$

$$= f+g+h-2k_1+\frac{\mu}{k_0}(g-k_1)(h-k_1)+\frac{\mu}{k_0}(g-k_1)(f-k_1)$$

$$+\frac{\mu}{k_0}(f-k_1)(h-k_1)+\frac{\mu^2}{k_0^2}(f-k_1)(g-k_1)(h-k_1) \quad on \quad \mathbb{T}.$$

Consequently,

$$(f\oplus_c g)\oplus_c h = f\oplus_c(g\oplus_c h) \quad on \quad \mathbb{T}.$$

Also,

$$f\oplus_c g = (f+g-k_1)+\frac{\mu}{k_0}(f-k_1)(g-k_1)$$

$$= (g+f-k_1)+\frac{\mu}{k_0}(g-k_1)(f-k_1)$$

$$= g\oplus_c f \quad on \quad \mathbb{T}.$$

This completes the proof. □

Definition 1.3.6 *Let $f\in\mathcal{R}_c$. We define the conformable addition inverse of f under the operation $\ominus_c$ as follows*

$$\ominus_c f = -\frac{k_0(f-k_1)}{k_0+\mu(f-k_1)}+k_1.$$

For $f \in \mathscr{R}_c$, we have

$$
\begin{aligned}
\ominus_c(\ominus_c f) &= -\frac{k_0(\ominus_c f - k_1)}{k_0 + \mu(\ominus_c f - k_1)} + k_1 \\[2mm]
&= -\frac{k_0\left(-\frac{k_0(f-k_1)}{k_0+\mu(f-k_1)} + k_1 - k_1\right)}{k_0 + \mu\left(-\frac{k_0(f-k_1)}{k_0+\mu(f-k_1)} + k_1 - k_1\right)} + k_1 \\[2mm]
&= \frac{k_0^2(f-k_1)}{k_0^2 + \mu k_0(f-k_1) - \mu k_0(f-k_1)} + k_1 \\[2mm]
&= f - k_1 + k_1 \\[4mm]
&= f.
\end{aligned}
$$

Definition 1.3.7 *Let $f, g \in \mathscr{R}_c$. We define "conformable circle minus" subtraction $\ominus_c$ as follows:*

$$
f \ominus_c g = f \oplus_c (\ominus_c g).
$$

For $f, g \in \mathscr{R}_c$, we have

$$
\begin{aligned}
f \ominus_c g &= f \oplus_c (\ominus_c g) \\[2mm]
&= f + (\ominus_c g) - k_1 + \mu \frac{(f-k_1)(\ominus_c g - k_1)}{k_0} \\[2mm]
&= f - \frac{k_0(g-k_1)}{k_0 + \mu(g-k_1)} + k_1 - k_1 \\[2mm]
&\quad + \frac{\mu(f-k_1)\left(-\frac{k_0(g-k_1)}{k_0+\mu(g-k_1)} + k_1 - k_1\right)}{k_0} \\[2mm]
&= f - \frac{k_0(g-k_1)}{k_0 + \mu(g-k_1)} - \mu \frac{k_0(f-k_1)(g-k_1)}{k_0(k_0 + \mu(g-k_1))} \\[2mm]
&= f - \frac{k_0(g-k_1)}{k_0 + \mu(g-k_1)} - \mu \frac{(f-k_1)(g-k_1)}{k_0 + \mu(g-k_1)} \\[2mm]
&= \frac{fk_0 + \mu f(g-k_1) - (k_0 + \mu(f-k_1))(g-k_1)}{k_0 + \mu(g-k_1)} \\[2mm]
&= \frac{fk_0 + (g-k_1)(\mu f - k_0 - \mu f + \mu k_1)}{k_0 + \mu(g-k_1)} \\[2mm]
&= \frac{fk_0 - (g-k_1)(k_0 - \mu k_1)}{k_0 + \mu(g-k_1)}
\end{aligned}
$$

$$= \frac{fk_0 - gk_0 + \mu g k_1 + k_0 k_1 - \mu k_1^2}{k_0 + \mu(g - k_1)}$$

$$= \frac{(f - g)k_0 + k_1(k_0 + \mu(g - k_1))}{k_0 + \mu(g - k_1)} \quad on \quad \mathbb{T},$$

and

$$(f \ominus_c g) - k_1 = \frac{(f - g)k_0}{k_0 + \mu(g - k_1)} + k_1 - k_1$$

$$= \frac{(f - g)k_0}{k_0 + \mu(g - k_1)} \quad on \quad \mathbb{T},$$

and

$$k_0 + \mu((f \ominus_c g) - k_1) = k_0 + \mu \frac{k_0(f - g)}{k_0 + \mu(g - k_1)}$$

$$= k_0 \left(1 + \frac{\mu f - \mu g}{k_0 + \mu(g - k_1)}\right)$$

$$= k_0 \frac{k_0 + \mu g - \mu k_1 + \mu f - \mu g}{k_0 + \mu(g - k_1)}$$

$$= k_0 \frac{k_0 + \mu(f - k_1)}{k_0 + \mu(g - k_1)}$$

$$\neq 0 \quad on \quad \mathbb{T},$$

i.e., $f \ominus_c g \in \mathcal{R}_c$.

Definition 1.3.8 *Let* $f \in \mathcal{R}_c$. *The conformable generalized square of* f *is defined as follows*

$$f^{\textcircled{2}} = -f(\ominus_c f).$$

For $f \in \mathcal{R}_c$, we have

$$f^{\textcircled{2}} = -f(\ominus_c f)$$

$$= -f\left(-\frac{k_0(f - k_1)}{k_0 + \mu(f - k_1)} + k_1\right)$$

$$= f\left(\frac{k_0(f - k_1)}{k_0 + \mu(f - k_1)} - k_1\right)$$

$$= f \frac{k_0(f - k_1) - k_0 k_1 - \mu k_1(f - k_1)}{k_0 + \mu(f - k_1)}$$

$$= f \frac{(k_0 - \mu k_1)(f - k_1) - k_0 k_1}{k_0 + \mu(f - k_1)}, \quad t \in \mathbb{T}^\kappa.$$

Theorem 1.3.9 *Let $f \in \mathscr{R}_c$. Then*

$$(\ominus_c f)^{②} = f^{②}.$$

Proof 1.3.10 *We have*

$$
\begin{aligned}
(\ominus_c f)^{②} &= -(\ominus_c f)(\ominus_c(\ominus_c f)) \\[2mm]
&= -f(\ominus_c f) \\[2mm]
&= f^{②}.
\end{aligned}
$$

This completes the proof. □

1.4   THE CONFORMABLE EXPONENTIAL FUNCTION

Definition 1.4.1 *Suppose that $\alpha \in (0,1]$, $p \in \mathscr{R}_c$. For $t, t_0 \in \mathbb{T}$, we define the conformable exponential function as follows*

$$E_p(t, t_0) = e_{\frac{p - k_1}{k_0}}(t, t_0).$$

We have

$$
\begin{aligned}
D^{\alpha} E_p(t, t_0) &= k_1(\alpha, t) e_{\frac{p-k_1}{k_0}}(t, t_0) + k_0(\alpha, t) e^{\Delta}_{\frac{p-k_1}{k_0}}(t, t_0) \\[2mm]
&= k_1(\alpha, t) e_{\frac{p-k_1}{k_0}}(t, t_0) + k_0(\alpha, t) \frac{p(t) - k_1(\alpha, t)}{k_0(\alpha, t)} e_{\frac{p-k_1}{k_0}}(t, t_0) \\[2mm]
&= p(t) e_{\frac{p-k_1}{k_0}}(t, t_0) \\[2mm]
&= p(t) E_p(t, t_0), \quad t \in \mathbb{T}^{\kappa}.
\end{aligned}
$$

Below we will list some of the properties of the conformable exponential function.

Theorem 1.4.2 (Semigroup Property) *The exponential function defined above in Definition 1.8 satisfies*

$$E_p(t, s) E_p(s, r) = E_p(t, r), \quad t, s, r \in \mathbb{T}.$$

Proof 1.4.3 *We have*

$$
\begin{aligned}
E_p(t, s) E_p(s, r) &= e_{\frac{p-k_1}{k_0}}(t, s) e_{\frac{p-k_1}{k_0}}(s, r) \\[2mm]
&= e_{\frac{p-k_1}{k_0}}(t, r) \\[2mm]
&= E_p(t, r), \quad t, s, r \in \mathbb{T}.
\end{aligned}
$$

This completes the proof. □

Theorem 1.4.4 *We have*

$$E_{k_1}(t,t_0) = 1, \quad E_p(t,t) = 1, \quad t,t_0 \in \mathbb{T}.$$

Proof 1.4.5 *We have*

$$E_{k_1}(t,t_0) \;=\; e_0(t,t_0) = 1$$

and

$$E_p(t,t) \;=\; e_{\frac{p-k_1}{k_0}}(t,t) = 1, \quad t,t_0 \in \mathbb{T}.$$

This completes the proof. □

Theorem 1.4.6 *We have*

$$E_p(\sigma(t),t_0) = \left(1 + \mu(t)\frac{p(t) - k_1(\alpha,t)}{k_0(\alpha,t)}\right) E_p(t,t_0), \quad t,t_0 \in \mathbb{T}.$$

Proof 1.4.7 *Using the definition of the exponential function,*

$$
\begin{aligned}
E_p(\sigma(t),t_0) &= e_{\frac{p-k_1}{k_0}}(\sigma(t),t_0) \\[2mm]
&= \left(1 + \mu(t)\frac{p(t) - k_1(\alpha,t)}{k_0(\alpha,t)}\right) e_{\frac{p-k_1}{k_0}}(t,t_0) \\[2mm]
&= \left(1 + \mu(t)\frac{p(t) - k_1(\alpha,t)}{k_0(\alpha,t)}\right) E_p(t,t_0), \quad t,t_0 \in \mathbb{T}.
\end{aligned}
$$

This completes the proof. □

Theorem 1.4.8 *Let* $p \in \mathscr{R}_c$. *Then*

$$E_p(t,t_0) = E_{\ominus_c p}(t_0,t), \quad \alpha \in (0,1], \quad t_0,t \in \mathbb{T}.$$

Proof 1.4.9 *We have*

$$
\begin{aligned}
E_p(t,t_0) &= e_{\frac{p-k_1}{k_0}}(t,t_0) \\[2mm]
&= e_{\ominus\left(\frac{p-k_1}{k_0}\right)}(t_0,t), \quad \alpha \in (0,1], \quad t_0,t \in \mathbb{T}.
\end{aligned}
$$

Note that

$$
\begin{aligned}
\ominus\left(\frac{p-k_1}{k_0}\right)(t) &= -\frac{\dfrac{p(t)-k_1(\alpha,t)}{k_0(\alpha,t)}}{1 + \mu(t)\dfrac{p(t)-k_1(\alpha,t)}{k_0(\alpha,t)}} \\[4mm]
&= -\frac{p(t)-k_1(\alpha,t)}{k_0(\alpha,t) + \mu(t)\left(p(t)-k_1(\alpha,t)\right)}, \quad \alpha \in (0,1], \quad t \in \mathbb{T}.
\end{aligned}
$$

Next,

$$\frac{(\ominus_c p)(t) - k_1(\alpha,t)}{k_0(\alpha,t)} = \frac{1}{k_0(\alpha,t)} \left(-\frac{k_0(\alpha,t)(p(t) - k_1(\alpha,t))}{k_0(\alpha,t) + \mu(t)(p(t) - k_1(\alpha,t))} - k_1(\alpha,t) + k_1(\alpha,t) \right)$$

$$= -\frac{p(t) - k_1(\alpha,t)}{k_0(\alpha,t) + \mu(t)(p(t) - k_1(\alpha,t))}, \quad \alpha \in (0,1], \quad t \in \mathbb{T}.$$

Therefore

$$E_p(t,t_0) = E_{\ominus_c p}(t_0,t), \quad t,t_0 \in \mathbb{T}.$$

This completes the proof. □

Theorem 1.4.10 *Let* $f,g \in \mathscr{R}_c$. *Then*

$$E_f(t,t_0)E_g(t,t_0) = E_{f \oplus_c g}(t,t_0), \quad \alpha \in (0,1], \quad t,t_0 \in \mathbb{T}.$$

Proof 1.4.11 *We have*

$$E_f(t,t_0)E_g(t,t_0) = e_{\frac{f-k_1}{k_0}}(t,t_0)e_{\frac{g-k_1}{k_0}}(t,t_0)$$

$$= e_{\left(\frac{f-k_1}{k_0}\right) \oplus \left(\frac{g-k_1}{k_0}\right)}(t,t_0), \quad \alpha \in (0,1], \quad t,t_0 \in \mathbb{T}.$$

Note that

$$\left(\left(\frac{f-k_1}{k_0}\right) \oplus \left(\frac{g-k_1}{k_0}\right) \right)(t) = \frac{f(t) - k_1(\alpha,t)}{k_0(\alpha,t)} + \frac{g(t) - k_1(\alpha,t)}{k_0(\alpha,t)}$$

$$+ \mu(t)\frac{(f(t) - k_1(\alpha,t))(g(t) - k_1(\alpha,t))}{(k_0(\alpha,t))^2}, \quad \alpha \in (0,1], \quad t \in \mathbb{T},$$

and

$$\frac{(f \oplus_c g)(t) - k_1(\alpha,t)}{k_0(\alpha,t)} = \frac{1}{k_0(\alpha,t)} \left(f(t) + g(t) - k_1(\alpha,t) \right.$$

$$\left. + \frac{\mu(t)}{k_0(\alpha,t)}(f(t) - k_1(\alpha,t))(g(t) - k_1(\alpha,t)) - k_1(\alpha,t) \right)$$

$$= \frac{f(t) - k_1(\alpha,t)}{k_0(\alpha,t)} + \frac{g(t) - k_1(\alpha,t)}{k_0(\alpha,t)}$$

$$+ \mu(t)\frac{(f(t) - k_1(\alpha,t))(g(t) - k_1(\alpha,t))}{(k_0(\alpha,t))^2},$$

$\alpha \in (0,1], t \in \mathbb{T}$. *Therefore*

$$\left(\left(\frac{f-k_1}{k_0}\right) \oplus \left(\frac{g-k_1}{k_0}\right) \right)(t) = \frac{(f \oplus_c g)(t) - k_1(\alpha,t)}{k_0(\alpha,t)}, \quad \alpha \in (0,1], \quad t \in \mathbb{T}.$$

Hence,

$$E_f(t,t_0)E_g(t,t_0) = E_{f \oplus_c g}(t,t_0), \quad \alpha \in (0,1], \quad t,t_0 \in \mathbb{T}.$$

This completes the proof. □

Theorem 1.4.12 *Let* $f, g \in \mathscr{R}_c$. *Then*

$$\frac{E_f(t, t_0)}{E_g(t, t_0)} = E_{f \ominus_c g}(t, t_0), \quad \alpha \in (0, 1], \quad t, t_0 \in \mathbb{T}.$$

Proof 1.4.13 *Using Theorem 1.4.8 and Theorem 1.4.10, we get*

$$\frac{E_f(t, t_0)}{E_g(t, t_0)} = E_f(t, t_0) E_{\ominus_c g}(t, t_0)$$

$$= E_{f \ominus_c g}(t, t_0), \quad \alpha \in (0, 1], \quad t, t_0 \in \mathbb{T}.$$

This completes the proof. □

1.5 CONFORMABLE HYPERBOLIC AND TRIGONOMETRIC FUNCTIONS

Definition 1.5.1 *Let* $\pm f \in \mathscr{R}_c$. *Define the conformable hyperbolic functions $Cosh_f$ and $Sinh_f$ by*

$$Cosh_f = \frac{E_f + E_{-f}}{2} \quad and \quad Sinh_f = \frac{E_f - E_{-f}}{2}.$$

Theorem 1.5.2 *Let* $\pm f \in \mathscr{R}_c$. *Then*

$$D^\alpha Cosh_f = f Sinh_f, \quad D^\alpha Sinh_f = f Cosh_f,$$

$$Cosh_f^2 - Sinh_f^2 = E_g,$$

where

$$g = f \oplus_c (-f)$$

$$= -k_1 - \frac{\mu}{k_0} \left(f^2 - k_1^2 \right).$$

Proof 1.5.3 *We have*

$$D^\alpha Cosh_f = D^\alpha \left(\frac{E_f + E_{-f}}{2} \right)$$

$$= \frac{D^\alpha E_f + D^\alpha E_{-f}}{2}$$

$$= \frac{f E_f - f E_{-f}}{2}$$

$$= f\frac{E_f - E_{-f}}{2}$$

$$= f\,Sinh_f,$$

$$D^\alpha Sinh_f = D^\alpha\left(\frac{E_f - E_{-f}}{2}\right)$$

$$= \frac{D^\alpha E_f - D^\alpha E_{-f}}{2}$$

$$= \frac{fE_f + fE_{-f}}{2}$$

$$= f\frac{E_f + E_{-f}}{2}$$

$$= f\,Cosh_f,$$

$$Cosh_f^2 - Sinh_f^2 = \left(\frac{E_f + E_{-f}}{2}\right)^2 - \left(\frac{E_f - E_{-f}}{2}\right)^2$$

$$= \frac{1}{4}\left(E_f^2 + 2E_f E_{-f} + E_{-f}^2\right.$$

$$\left. -E_f^2 + 2E_f E_{-f} - E_{-f}^2\right)$$

$$= E_f E_{-f}$$

$$= E_{f\oplus_c(-f)}.$$

Note that

$$f\oplus_c(-f) = \frac{(f - f - k_1)k_0 + \mu(f - k_1)(-f - k_1)}{k_0}$$

$$= -\frac{k_0 k_1 + \mu(f - k_1)(f + k_1)}{k_0}$$

$$= -k_1 - \frac{\mu}{k_0}\left(f^2 - k_1^2\right)$$

$$= g.$$

Therefore

$$Cosh_f^2 - Sinh_f^2 = E_g.$$

This completes the proof. □

Definition 1.5.4 *Suppose that $f \pm g \in \mathscr{R}_c$. Define the conformable hyperbolic functions Ch_{fg} and Sh_{fg} as follows*

$$Ch_{fg} = \frac{E_{f+g} + E_{f-g}}{2} \quad and \quad Sh_{fg} = \frac{E_{f+g} - E_{f-g}}{2}.$$

Theorem 1.5.5 *Let $f \pm g \in \mathscr{R}_c$. Then*

$$D^\alpha Ch_{fg} = fCh_{fg} + gSh_{fg}, \quad D^\alpha Sh_{fg} = gCh_{fg} + fSh_{fg},$$

$$Ch_{fg}^2 - Sh_{fg}^2 = E_{(f+g)\oplus_c(f-g)}.$$

Proof 1.5.6 *We have*

$$
\begin{aligned}
D^\alpha Ch_{fg} &= D^\alpha \left(\frac{E_{f+g} + E_{f-g}}{2} \right) \\
&= \frac{D^\alpha E_{f+g} + D^\alpha E_{f-g}}{2} \\
&= \frac{(f+g)E_{f+g} + (f-g)E_{f-g}}{2} \\
&= f\frac{E_{f+g} + E_{f-g}}{2} + g\frac{E_{f+g} - E_{f-g}}{2} \\
&= fCh_{fg} + gSh_{fg},
\end{aligned}
$$

$$
\begin{aligned}
D^\alpha Sh_{fg} &= D^\alpha \left(\frac{E_{f+g} - E_{f-g}}{2} \right) \\
&= \frac{D^\alpha E_{f+g} - D^\alpha E_{f-g}}{2} \\
&= \frac{(f+g)E_{f+g} - (f-g)E_{f-g}}{2} \\
&= f\frac{E_{f+g} - E_{f-g}}{2} + g\frac{E_{f+g} + E_{f-g}}{2} \\
&= fSh_{fg} + gCh_{fg},
\end{aligned}
$$

$$Ch_{fg}^2 - Sh_{fg}^2 = \left(\frac{E_{f+g}+E_{f-g}}{2}\right)^2 - \left(\frac{E_{f+g}-E_{f-g}}{2}\right)^2$$

$$= \frac{1}{4}\left(E_{f+g}^2 + 2E_{f+g}E_{f-g} + E_{f-g}^2\right.$$

$$\left. -E_{f+g}^2 + 2E_{f+g}E_{f-g} - E_{f-g}^2\right)$$

$$= E_{f+g}E_{f-g}$$

$$= E_{(f+g)\oplus_c(f-g)}.$$

This completes the proof. □

Definition 1.5.7 *Let* $\pm if \in \mathcal{R}_c$. *Define the conformable trigonometric functions* Cos_f *and* Sin_f *as follows:*

$$Cos_f = \frac{E_{if}+E_{-if}}{2}, \quad Sin_f = \frac{E_{if}-E_{-if}}{2i}.$$

Theorem 1.5.8 *Let* $\pm if \in \mathcal{R}_c$. *Then*

$$D^\alpha Cos_f = -fSin_f, \quad D^\alpha Sin_f = fCos_f,$$

$$Cos_f^2 + Sin_f^2 = E_{(if)\oplus_c(-if)}.$$

Proof 1.5.9 *We have*

$$D^\alpha Cos_f = D^\alpha\left(\frac{E_{if}+E_{-if}}{2}\right)$$

$$= \frac{D^\alpha E_{if}+D^\alpha E_{-if}}{2}$$

$$= \frac{ifE_{if}-ifE_{-if}}{2}$$

$$= f\frac{E_{-if}-E_{if}}{2i}$$

$$= -fSin_f,$$

$$D^\alpha Sin_f = D^\alpha \left(\frac{E_{if} - E_{-if}}{2i} \right)$$

$$= \frac{D^\alpha E_{if} - D^\alpha E_{-if}}{2i}$$

$$= \frac{if E_{if} + if E_{-if}}{2i}$$

$$= f \frac{E_{if} + E_{-if}}{2}$$

$$= f Cos_f,$$

$$Cos_f^2 + Sin_f^2 = \left(\frac{E_{if} + E_{-if}}{2} \right)^2 + \left(\frac{E_{if} - E_{-if}}{2i} \right)^2$$

$$= \frac{1}{4} \left(E_{if}^2 + 2 E_{if} E_{-if} + E_{-if}^2 \right.$$

$$\left. - E_{if}^2 + 2 E_{if} E_{-if} - E_{-if}^2 \right)$$

$$= E_{if} E_{-if}$$

$$= E_{(if) \oplus_c (-if)}.$$

This completes the proof. □

Definition 1.5.10 *Let $f \pm ig \in \mathcal{R}_c$. Define the conformable trigonometric follows: C_{fg} and S_{fg} as follows:*

$$C_{fg} = \frac{E_{f+ig} + E_{f-ig}}{2}, \quad S_{fg} = \frac{E_{f+ig} - E_{f-ig}}{2i}.$$

Theorem 1.5.11 *Let $f \pm ig \in \mathcal{R}_c$. Then*

$$D^\alpha C_{fg} = f C_{fg} - g S_{fg}, \quad D^\alpha S_{fg} = g C_{fg} + f S_{fg},$$

$$C_{fg}^2 + S_{fg}^2 = E_{(f+ig) \oplus_c (f-ig)}.$$

Proof 1.5.12 *We have*

$$D^\alpha C_{fg} = D^\alpha \left(\frac{E_{f+ig} + E_{f-ig}}{2} \right)$$

$$= \frac{D^{\alpha}E_{f+ig} + D^{\alpha}E_{f-ig}}{2}$$

$$= \frac{(f+ig)E_{f+ig} + (f-ig)E_{f-ig}}{2}$$

$$= f\frac{E_{f+ig} + E_{f-ig}}{2} - g\frac{E_{f+ig} - E_{f-ig}}{2i}$$

$$= fC_{fg} - gS_{fg},$$

$$D^{\alpha}S_{fg} = D^{\alpha}\left(\frac{E_{f+ig} - E_{f-ig}}{2i}\right)$$

$$= \frac{D^{\alpha}E_{f+ig} - D^{\alpha}E_{f-ig}}{2i}$$

$$= \frac{(f+ig)E_{f+ig} - (f-ig)E_{f-ig}}{2i}$$

$$= f\frac{E_{f+ig} - E_{f-ig}}{2i} + g\frac{E_{f+ig} + E_{f-ig}}{2}$$

$$= gC_{fg} + fS_{fg},$$

$$C_{fg}^2 + S_{fg}^2 = \left(\frac{E_{f+ig} + E_{f-ig}}{2}\right)^2 + \left(\frac{E_{f+ig} - E_{f-ig}}{2i}\right)^2$$

$$= \frac{1}{4}\left(E_{f+ig}^2 + 2E_{f+ig}E_{f-ig} + E_{f-ig}^2\right.$$

$$\left. -E_{f+ig}^2 + 2E_{f+ig}E_{f-ig} - E_{f-ig}^2\right)$$

$$= E_{f+ig}E_{f-ig}$$

$$= E_{(f+ig)\oplus_c(f-ig)}.$$

This completes the proof. □

1.6   THE CONFORMABLE LOGARITHM FUNCTION

We now set the foundation for offering a definition of conformable logarithms on time scales. This definition will be of a multi-valued function.

Definition 1.6.1 (Conformable Cylinder Transformation) *Let $\alpha \in (0,1]$, and fix $t \in \mathbb{T}$. For $h > 0$, define the multi-valued conformable cylinder transformation $\zeta_h^c : \mathbb{C}_h^c \to \mathbb{C}$ by*

$$\zeta_h^c(z) = \begin{cases} \dfrac{1}{h} \log\left(1 + h\left(\dfrac{z - k_1(\alpha,t)}{k_0(\alpha,t)}\right)\right) & \text{for } h \neq 0 \\ \dfrac{z - k_1(\alpha,t)}{k_0(\alpha,t)} & \text{for } h = 0, \end{cases} \tag{1.2}$$

where $\mathbb{C}$ is the set of complex numbers, $\mathbb{C}_h^c$ is given by

$$\mathbb{C}_h^c = \left\{ z \in \mathbb{C} : z \neq k_1(\alpha,t) - \frac{k_0(\alpha,t)}{h} \right\}, \tag{1.3}$$

and $\log$ is the multi-valued complex logarithm function.

Lemma 1.6.2 *Fix $\alpha \in (0,1]$. Let $f,g : \mathbb{T} \to \mathbb{C}$ be Δ-differentiable functions with $f,g \neq 0$ on $\mathbb{T}$, and let the multi-valued conformable cylinder transformation ζ^c be given by (1.2). Then, for fixed $\tau \in \mathbb{T}^\kappa$,*

$$\zeta_{\mu(\tau)}^c\left(\left(\frac{D^\alpha f}{f} \oplus_c \frac{D^\alpha g}{g}\right)(\tau)\right) = \zeta_{\mu(\tau)}^c\left(\frac{D^\alpha f(\tau)}{f(\tau)}\right) + \zeta_{\mu(\tau)}^c\left(\frac{D^\alpha g(\tau)}{g(\tau)}\right).$$

Proof 1.6.3 *First, note that the simple useful formula $f^\sigma = f + \mu f^\Delta$ (suppressing the variable) implies*

$$\begin{aligned} \frac{D^\alpha(fg)}{fg} &= \frac{f^\sigma D^\alpha g + g D^\alpha f - k_1(\alpha,\cdot)f^\sigma g}{fg} \\ &= \frac{(f + \mu f^\Delta)(D^\alpha g - k_1(\alpha,\cdot)g)}{fg} + \frac{D^\alpha f}{f} \\ &= \frac{D^\alpha f}{f} + \frac{D^\alpha g}{g} - k_1(\alpha,\cdot) + \mu\frac{f^\Delta}{f}\left(\frac{D^\alpha g}{g} - k_1(\alpha,\cdot)\right) \\ &= \frac{D^\alpha f}{f} \oplus_c \frac{D^\alpha g}{g}. \end{aligned}$$

It follows that for fixed $\tau \in \mathbb{T}^\kappa$,

$$\zeta_{\mu(\tau)}^c\left(\left(\frac{D^\alpha f}{f} \oplus_c \frac{D^\alpha g}{g}\right)(\tau)\right)$$

$$= \zeta_{\mu(\tau)}^c\left(\frac{D^\alpha(fg)(\tau)}{(fg)(\tau)}\right)$$

$$= \begin{cases} \dfrac{1}{\mu(\tau)} \log\left(1 + \mu(\tau)\dfrac{(fg)^\Delta(\tau)}{(fg)(\tau)}\right) & \text{for } \mu(\tau) \neq 0 \\ \dfrac{1}{k_0(\alpha,\tau)}\left(\dfrac{D^\alpha(fg)(\tau)}{(fg)(\tau)} - k_1(\alpha,\tau)\right) & \text{for } \mu(\tau) = 0 \end{cases}$$

$$= \begin{cases} \dfrac{1}{\mu(\tau)} \log\left(\dfrac{(fg)^\sigma(\tau)}{(fg)(\tau)}\right) & \text{for } \mu(\tau) \neq 0 \\ \dfrac{1}{k_0(\alpha,\tau)}\left(\dfrac{D^\alpha f(\tau)}{f(\tau)} + \dfrac{k_0(\alpha,\tau)f^\sigma(\tau)g^\Delta(\tau)}{(fg)(\tau)} - k_1(\alpha,\tau)\right) & \text{for } \mu(\tau) = 0 \end{cases}$$

$$= \begin{cases} \dfrac{1}{\mu(\tau)} \log\left(\dfrac{f^{\sigma}(\tau)}{f(\tau)}\right) + \dfrac{1}{\mu(\tau)} \log\left(\dfrac{g^{\sigma}(\tau)}{g(\tau)}\right) & \text{for } \mu(\tau) \neq 0 \\ \dfrac{1}{k_0(\alpha,\tau)}\left(\dfrac{D^{\alpha} f(\tau)}{f(\tau)} - k_1(\alpha,\tau)\right) + \dfrac{g^{\Delta}(\tau)}{g(\tau)} & \text{for } \mu(\tau) = 0 \end{cases}$$

$$= \begin{cases} \dfrac{1}{\mu(\tau)} \log\left(\dfrac{(f + \mu f^{\Delta})(\tau)}{f(\tau)}\right) + \dfrac{1}{\mu(\tau)} \log\left(\dfrac{(g + \mu g^{\Delta})(\tau)}{g(\tau)}\right) & \text{for } \mu(\tau) \neq 0 \\ \dfrac{1}{k_0(\alpha,\tau)}\left(\dfrac{D^{\alpha} f(\tau)}{f(\tau)} - k_1(\alpha,\tau)\right) + \dfrac{1}{k_0(\alpha,\tau)}\left(\dfrac{D^{\alpha} g(\tau)}{g(\tau)} - k_1(\alpha,\tau)\right) & \text{for } \mu(\tau) = 0 \end{cases}$$

$$= \zeta^c_{\mu(\tau)}\left(\dfrac{D^{\alpha} f(\tau)}{f(\tau)}\right) + \zeta^c_{\mu(\tau)}\left(\dfrac{D^{\alpha} g(\tau)}{g(\tau)}\right).$$

This completes the proof. ☐

Definition 1.6.4 *Fix* $\alpha \in (0,1]$. *For* $\beta \in \mathbb{R}$ *and* $f \in \mathscr{R}(\beta)$, *define the operation* $\odot_c$ *via*

$$\beta \odot_c f := k_1(\alpha,\cdot) + \beta(f - k_1(\alpha,\cdot)) \int_0^1 \left(1 + \mu\left(\frac{f - k_1(\alpha,\cdot)}{k_0(\alpha,\cdot)}\right)h\right)^{\beta-1} dh. \quad (1.4)$$

Lemma 1.6.5 *Let* $\alpha \in (0,1]$, $\beta \in \mathbb{R}$, *and* $p : \mathbb{T} \to \mathbb{C}$ *be such that* $p(t) \neq 0$ *for all* $t \in \mathbb{T}$, *and if* $\beta \notin \mathbb{N}$, *then* $p(t)p^{\sigma}(t) > 0$ *for all* $t \in \mathbb{T}$. *For the multi-valued conformable cylinder transformation* ζ^c *given by* (1.2), *we have*

$$\beta \odot_c \left(\frac{D^{\alpha} p}{p}\right) = \frac{D^{\alpha}\left(p^{\beta}\right)}{p^{\beta}}$$

on $\mathbb{T}^{\kappa}$.

Proof 1.6.6 *Using Pötzsche's chain rule, we have (suppressing the variable)*

$$D^{\alpha}(p^{\beta}) = k_1(\alpha,\cdot)p^{\beta} + k_0(\alpha,\cdot)(p^{\beta})^{\Delta}$$

$$= k_1(\alpha,\cdot)p^{\beta} + \beta k_0(\alpha,\cdot)p^{\Delta} \int_0^1 \left(p + \mu p^{\Delta} h\right)^{\beta-1} dh$$

$$= \begin{cases} k_1(\alpha,\cdot)p^{\beta} + \beta k_0(\alpha,\cdot)p^{\Delta}\left(\dfrac{1}{\beta \mu p^{\Delta}}\right)\left[(p + \mu p^{\Delta})^{\beta} - p^{\beta}\right] & \text{for } \mu(\tau) \neq 0 \\ k_1(\alpha,\cdot)p^{\beta} + \beta k_0(\alpha,\cdot)p^{\Delta} p^{\beta-1} & \text{for } \mu(\tau) = 0 \end{cases}$$

$$= p^{\beta} \begin{cases} k_1(\alpha,\cdot) + \dfrac{k_0(\alpha,\cdot)}{\mu}\left[\left(1 + \mu \dfrac{p^{\Delta}}{p}\right)^{\beta} - 1\right] & \text{for } \mu(\tau) \neq 0 \\ k_1(\alpha,\cdot) + \beta k_0(\alpha,\cdot)\dfrac{p^{\Delta}}{p} & \text{for } \mu(\tau) = 0 \end{cases}$$

$$= p^{\beta} \begin{cases} k_1(\alpha,\cdot) + \dfrac{k_0(\alpha,\cdot)}{\mu}\left[\left(1 + \mu\left(\dfrac{D^{\alpha} p - k_1(\alpha,\cdot)p}{k_0(\alpha,\cdot)p}\right)\right)^{\beta} - 1\right] & \text{for } \mu(\tau) \neq 0 \\ k_1(\alpha,\cdot) + \beta k_0(\alpha,\cdot)\left(\dfrac{D^{\alpha} p - k_1(\alpha,\cdot)p}{k_0(\alpha,\cdot)p}\right) & \text{for } \mu(\tau) = 0. \end{cases}$$

Therefore,

$$
\frac{D^\alpha(p^\beta)}{p^\beta} = \begin{cases} k_1(\alpha,\cdot) + \dfrac{k_0(\alpha,\cdot)}{\mu}\left[\left(1+\mu\left(\dfrac{D^\alpha p - k_1(\alpha,\cdot)p}{k_0(\alpha,\cdot)p}\right)\right)^\beta - 1\right] & \text{for } \mu(\tau) \neq 0 \\[4mm] k_1(\alpha,\cdot) + \beta\left(\dfrac{D^\alpha p - k_1(\alpha,\cdot)p}{p}\right) & \text{for } \mu(\tau) = 0. \end{cases}
$$

By Definition 1.6.4,

$$
\beta \odot_c \left(\frac{D^\alpha p}{p}\right)
$$

$$
= k_1(\alpha,\cdot) + \beta\left(\frac{D^\alpha p}{p} - k_1(\alpha,\cdot)\right)\int_0^1 \left(1+\mu\left(\frac{\frac{D^\alpha p}{p} - k_1(\alpha,\cdot)}{k_0(\alpha,\cdot)}\right)h\right)^{\beta-1} dh
$$

$$
= \begin{cases} k_1(\alpha,\cdot) + \dfrac{k_0(\alpha,\cdot)}{\mu}\left[\left(1+\mu\left(\dfrac{D^\alpha p - k_1(\alpha,\cdot)p}{k_0(\alpha,\cdot)p}\right)\right)^\beta - 1\right] & \text{for } \mu(\tau) \neq 0 \\[4mm] k_1(\alpha,\cdot) + \beta\left(\dfrac{D^\alpha p - k_1(\alpha,\cdot)p}{p}\right) & \text{for } \mu(\tau) = 0. \end{cases}
$$

Hence,

$$
\beta \odot_c \left(\frac{D^\alpha p}{p}\right) = \frac{D^\alpha(p^\beta)}{p^\beta},
$$

and the proof is complete. □

Lemma 1.6.7 *Let* $\alpha \in (0,1]$, $\beta \in \mathbb{R}$, *and* $f \in \mathcal{R}(\beta)$. *For the multi-valued conformable cylinder transformation* ζ^c *given by (1.2) and for fixed* $\tau \in \mathbb{T}$,

$$
\zeta^c_{\mu(\tau)}((\beta \odot_c f)(\tau)) = \beta\zeta^c_{\mu(\tau)}(f(\tau)).
$$

Proof 1.6.8 *Assume* $\alpha \in (0,1]$, $\beta \in \mathbb{R}$, *and* $f \in \mathcal{R}(\beta)$. *Fix* $\tau \in \mathbb{T}^\kappa$. *If* $\mu(\tau) \neq 0$, *then*

$$
k_0(\alpha,\tau) + \mu(\tau)(\beta \odot_c f)(\tau) - \mu(\tau)k_1(\alpha,\tau)
$$

$$
= k_0(\alpha,\tau) + \mu(\tau)\beta(f(\tau) - k_1(\alpha,\tau))\int_0^1\left(1+\mu(\tau)\left(\frac{f(\tau)-k_1(\alpha,\tau)}{k_0(\alpha,\tau)}\right)h\right)^{\beta-1}dh
$$

$$
= k_0(\alpha,\tau)\left(1+\mu(\tau)\frac{f(\tau)-k_1(\alpha,\tau)}{k_0(\alpha,\tau)}\right)^\beta.
$$

Therefore, after dividing by $k_0(\alpha,\tau)$, *we have*

$$
1 + \frac{\mu(\tau)}{k_0(\alpha,\tau)}((\beta \odot_c f)(\tau) - k_1(\alpha,\tau)) = \left(1+\mu(\tau)\frac{f(\tau)-k_1(\alpha,\tau)}{k_0(\alpha,\tau)}\right)^\beta
$$

for any $\tau \in \mathbb{T}^\kappa$.

It follows that for fixed $\tau \in \mathbb{T}^\kappa$,

$$
\zeta^c_{\mu(\tau)}((\beta \odot_c f)(\tau))
$$

$$
= \begin{cases} \dfrac{1}{\mu(\tau)} \log\left(1 + \mu(\tau)\left(\dfrac{\beta \odot_c f - k_1(\alpha,\cdot)}{k_0(\alpha,\cdot)}\right)(\tau)\right) & \text{for } \mu(\tau) \neq 0 \\[3mm] \left(\dfrac{\beta \odot_c f - k_1(\alpha,\cdot)}{k_0(\alpha,\cdot)}\right)(\tau) & \text{for } \mu(\tau) = 0 \end{cases}
$$

$$
= \begin{cases} \dfrac{1}{\mu(\tau)} \log\left(1 + \mu(\tau)\dfrac{f(\tau) - k_1(\alpha,\tau)}{k_0(\alpha,\tau)}\right)^{\beta} & \text{for } \mu(\tau) \neq 0 \\[3mm] \beta\left(\dfrac{f - k_1(\alpha,\cdot)}{k_0(\alpha,\cdot)}\right)(\tau) & \text{for } \mu(\tau) = 0 \end{cases}
$$

$$
= \beta \begin{cases} \dfrac{1}{\mu(\tau)} \log\left(1 + \mu(\tau)\dfrac{f(\tau) - k_1(\alpha,\tau)}{k_0(\alpha,\tau)}\right) & \text{for } \mu(\tau) \neq 0 \\[3mm] \left(\dfrac{f - k_1(\alpha,\cdot)}{k_0(\alpha,\cdot)}\right)(\tau) & \text{for } \mu(\tau) = 0 \end{cases}
$$

$$
= \beta \zeta^c_{\mu(\tau)}(f(\tau)).
$$

This completes the proof. □

Definition 1.6.9 (Logarithm Function) *For a Δ-differentiable function $p : \mathbb{T} \to \mathbb{C}$ with $p \neq 0$ on $\mathbb{T}$, the multi-valued conformable logarithm function on time scales is given by*

$$
\ell^c_p(t,s) = \int_s^t \zeta^c_{\mu(\tau)}\left(\frac{D^\alpha p(\tau)}{p(\tau)}\right) \Delta\tau \quad \text{for} \quad s,t \in \mathbb{T},
$$

where $\zeta^c_h(z)$ is the multi-valued cylinder transformation given in (1.2). Define the principal logarithm on time scales to be

$$
L^c_p(t,s) = \int_s^t \xi^c_{\mu(\tau)}\left(\frac{D^\alpha p(\tau)}{p(\tau)}\right) \Delta\tau \quad \text{for} \quad s,t \in \mathbb{T},
$$

where $\xi^c_h(z)$ is the single-valued cylinder transformation given by

$$
\xi^c_h(z) = \begin{cases} \dfrac{1}{h} \operatorname{Log}\left(1 + h\left(\dfrac{z - k_1(\alpha,t)}{k_0(\alpha,t)}\right)\right) & \text{for } h \neq 0 \\[3mm] \dfrac{z - k_1(\alpha,t)}{k_0(\alpha,t)} & \text{for } h = 0, \end{cases} \tag{1.5}
$$

where Log is the principal logarithm.

Using the definition of the multi-valued conformable logarithm on time scales given above, we establish the following properties.

Theorem 1.6.10 *Let $p : \mathbb{T} \to \mathbb{C}$ be a Δ-differentiable function with $p \neq 0$ on $\mathbb{T}$. Then, for $s,t \in \mathbb{T}$, we have*

$$
\exp\left(L^c_p(t,s)\right) = E_{\frac{D^\alpha p}{p}}(t,s).
$$

Proof 1.6.11 *Let $p : \mathbb{T} \to \mathbb{C}$ be a Δ-differentiable function with $p \neq 0$ on $\mathbb{T}$. Then, for $s,t \in \mathbb{T}$, we have*

$$
L^c_p(t,s) = \int_s^t \xi^c_{\mu(\tau)}\left(\frac{D^\alpha p(\tau)}{p(\tau)}\right) \Delta\tau.
$$

Now exponentiate both sides and use the definition of the exponential function $E_p(t,s)$. □

Corollary 1.6.12 *Let $p \in \mathscr{R}_c$ and $s,t \in \mathbb{T}$. Then*

$$\exp\left(L^c_{E_p}(t,s)\right) = E_p(t,s).$$

Theorem 1.6.13 (Logarithm of Product & Quotient) *Let $f,g : \mathbb{T} \to \mathbb{C}$ be Δ-differentiable functions with $f,g \neq 0$ on $\mathbb{T}$. Then, for $s,t \in \mathbb{T}$, we have*

$$\ell^c_{fg}(t,s) = \ell^c_f(t,s) + \ell^c_g(t,s)$$

and

$$\ell^c_{\frac{f}{g}}(t,s) = \ell^c_f(t,s) - \ell^c_g(t,s).$$

Proof 1.6.14 *Let $f,g : \mathbb{T} \to \mathbb{R}$ be Δ-differentiable functions with $f,g \neq 0$ on $\mathbb{T}$. Then, for $s,t \in \mathbb{T}$, we have via Lemma 1.6.2 and its proof that*

$$
\begin{aligned}
\ell^c_{fg}(t,s) &= \int_s^t \zeta^c_{\mu(\tau)}\left(\frac{D^\alpha(fg)(\tau)}{(fg)(\tau)}\right)\Delta\tau \\
&= \int_s^t \zeta^c_{\mu(\tau)}\left(\left(\frac{D^\alpha f}{f} \oplus_c \frac{D^\alpha g}{g}\right)(\tau)\right)\Delta\tau \\
&= \int_s^t \zeta^c_\mu\left(\frac{D^\alpha f(\tau)}{f(\tau)}\right)\Delta\tau + \int_s^t \zeta^c_\mu\left(\frac{D^\alpha g(\tau)}{g(\tau)}\right)\Delta\tau \\
&= \ell^c_f(t,s) + \ell^c_g(t,s).
\end{aligned}
$$

In a similar manner,

$$
\begin{aligned}
\ell^c_{\frac{f}{g}}(t,s) &= \int_s^t \zeta^c_{\mu(\tau)}\left(\frac{D^\alpha\left(\frac{f}{g}\right)(\tau)}{\left(\frac{f}{g}\right)(\tau)}\right)\Delta\tau \\
&= \int_s^t \zeta^c_{\mu(\tau)}\left(\left(\frac{D^\alpha f}{f} \ominus_c \frac{D^\alpha g}{g}\right)(\tau)\right)\Delta\tau \\
&= \int_s^t \zeta^c_\mu\left(\frac{D^\alpha f(\tau)}{f(\tau)}\right)\Delta\tau - \int_s^t \zeta^c_\mu\left(\frac{D^\alpha g(\tau)}{g(\tau)}\right)\Delta\tau \\
&= \ell^c_f(t,s) - \ell^c_g(t,s).
\end{aligned}
$$

This completes the proof. □

Theorem 1.6.15 *Let $\alpha \in (0,1]$, $\beta \in \mathbb{R}$, and $p : \mathbb{T} \to \mathbb{C}$ be a Δ-differentiable function such that $p(t) \neq 0$ for all $t \in \mathbb{T}$, and if $\beta \notin \mathbb{N}$, then $p(t)p^\sigma(t) > 0$ for all $t \in \mathbb{T}$. Then, for $s,t \in \mathbb{T}$, we have*

$$\ell^c_{p^\beta}(t,s) = \beta\ell^c_p(t,s).$$

Proof 1.6.16 *For the multi-valued cylinder transformation ζ^c given by (1.2) and for fixed $\tau \in \mathbb{T}^\kappa$,*

$$\zeta^c_{\mu(\tau)}\left(\left(\beta \odot_c \frac{D^\alpha p}{p}\right)(\tau)\right) = \beta\zeta^c_{\mu(\tau)}\left(\frac{D^\alpha p(\tau)}{p(\tau)}\right)$$

using Lemma 1.6.7. Moreover, by Lemma 1.6.5, we have

$$\beta \odot_c \left(\frac{D^\alpha p}{p} \right) = \frac{D^\alpha \left(p^\beta \right)}{p^\beta}.$$

Consequently,

$$
\begin{aligned}
\ell_{p^\beta}^c (t,s) &= \int_s^t \zeta_{\mu(\tau)}^c \left(\frac{D^\alpha \left(p^\beta \right) (\tau)}{p^\beta (\tau)} \right) \Delta\tau \\
&= \int_s^t \zeta_{\mu(\tau)}^c \left(\left(\beta \odot_c \frac{D^\alpha p}{p} \right) (\tau) \right) \Delta\tau \\
&= \int_s^t \beta \zeta_{\mu(\tau)}^c \left(\frac{D^\alpha p(\tau)}{p(\tau)} \right) \Delta\tau \\
&= \beta \ell_p^c (t,s).
\end{aligned}
$$

This ends the proof. □

Theorem 1.6.17 *Let $p : \mathbb{T} \to \mathbb{R}$ be a Δ-differentiable function with $p \neq 0$ on $\mathbb{T}$. Then, for $s,t \in \mathbb{T}$, we have*

$$
D^\alpha \left(\ell_p^c \right) (t,s) = k_1(\alpha,t) \ell_p^c (t,s) + k_0(\alpha,t)
\begin{cases}
\dfrac{1}{\mu(t)} \log \left(\dfrac{p^\sigma(t)}{p(t)} \right) & \text{for } \mu(t) \neq 0 \\
\dfrac{p^\Delta(t)}{p(t)} & \text{for } \mu(t) = 0,
\end{cases}
$$

where Δ-differentiation is with respect to t.

Proof 1.6.18 *Using the definition of the conformable logarithm and applying the D^α operator with respect to t,*

$$
\begin{aligned}
D^\alpha \left(\ell_p^c \right) (t,s) &= k_1(\alpha,t) \ell_p^c (t,s) + k_0(\alpha,t) \zeta_{\mu(t)}^c \left(\frac{D^\alpha p(t)}{p(t)} \right) \\
&= k_1(\alpha,t) \ell_p^c (t,s) + k_0(\alpha,t)
\begin{cases}
\dfrac{1}{\mu(t)} \log \left(1 + \mu(t) \dfrac{p^\Delta(t)}{p(t)} \right) & \text{for } \mu(t) \neq 0 \\
\dfrac{p^\Delta(t)}{p(t)} & \text{for } \mu(t) = 0.
\end{cases}
\end{aligned}
$$

Now substitute $\mu p^\Delta = p^\sigma - p$. This ends the proof. □

Example 1.6.19 *Let $t \in \mathbb{T}$ with $t \neq 0$, and set $p(t) = t$. For $s \in \mathbb{T}$, we have*

$$
D^\alpha \left(\ell_p^c \right) (t,s) = k_1(\alpha,t) \ell_p^c (t,s) + k_0(\alpha,t)
\begin{cases}
\dfrac{1}{\mu(t)} \log \left(\dfrac{\sigma(t)}{t} \right) & \text{for } \mu(t) \neq 0 \\
\dfrac{1}{t} & \text{for } \mu(t) = 0,
\end{cases}
$$

where Δ-differentiation is with respect to t. Thus,

$$D^\alpha \left(\ell_p^c \right) (t,s) = \begin{cases} k_1(\alpha,t) \log \left(\dfrac{t}{s} \right) + \dfrac{k_0(\alpha,t)}{t} & \text{for } \mathbb{T} = \mathbb{R} \\[3ex] k_1(\alpha,t) \log \left(\dfrac{t}{s} \right) + \dfrac{k_0(\alpha,t)}{h} \log \left(1 + \dfrac{h}{t} \right) & \text{for } \mathbb{T} = h\mathbb{Z} \\[3ex] k_1(\alpha,t) \log \left(\dfrac{t}{s} \right) + \dfrac{k_0(\alpha,t) \log(q)}{(q-1)t} & \text{for } \mathbb{T} = q^{\mathbb{N}_0}, \end{cases}$$

where $h > 0$ and $q > 1$.

1.7 CONFORMABLE INTEGRATION

Let $a,b \in \mathbb{T}$, $a < b$. Suppose that

$$k_0(\alpha,t) - \mu(t)k_1(\alpha,t) \neq 0, \quad \alpha \in (0,1], \quad t \in \mathbb{T}. \tag{1.6}$$

Definition 1.7.1 *Let $f \in \mathscr{C}_{rd}(\mathbb{T})$ and assume (1.6) holds. Define a conformable antiderivative via*

$$\int D^\alpha f(t) \Delta_\alpha t = f(t) + cE_0(t,t_0), \quad c \in \mathbb{R}, \quad t \in \mathbb{T}.$$

Define the conformable Δ-integral of f over $[a,b]$ as follows

$$\int_a^t f(s) \Delta_{\alpha,t} s = \int_a^t f(s) \frac{E_0(t,\sigma(s))}{k_0(\alpha,s)} \Delta s, \quad t \in [a,b].$$

Theorem 1.7.2 *Let $\alpha \in (0,1]$, $f \in \mathscr{C}_{rd}(\mathbb{T})$ and (1.6) hold. Then*

$$D^\alpha \left(\int_a^t f(s) \Delta_{\alpha,t} s \right) = f(t), \quad t \in [a,b]^\kappa.$$

Proof 1.7.3 *We have*

$$
\begin{aligned}
D^\alpha \left(\int_a^t f(s) \Delta_{\alpha,t} s \right) &= k_1(\alpha,t) \left(\int_a^t f(s) \Delta_{\alpha,t} s \right) \\
&\quad + k_0(\alpha,t) \left(\int_a^t f(s) \frac{E_0(t,\sigma(s))}{k_0(\alpha,s)} \Delta s \right)^\Delta \\
&= k_1(\alpha,t) \left(\int_a^t f(s) \Delta_{\alpha,t} s \right) \\
&\quad + k_0(\alpha,t) f(t) \frac{E_0(\sigma(t),\sigma(t))}{k_0(\alpha,t)} \\
&\quad + k_0(\alpha,t) \int_a^t f(s) \frac{E_0^\Delta(t,\sigma(s))}{k_0(\alpha,s)} \Delta s
\end{aligned}
$$

$$= k_1(\alpha,t)\left(\int_a^t f(s)\Delta_{\alpha,t}s\right) + f(t)$$

$$+ k_0(\alpha,t)\int_a^t f(s)\frac{e^{\Delta_t}_{-\frac{k_1}{k_0}}(t,\sigma(s))}{k_0(\alpha,s)}\Delta s$$

$$= k_1(\alpha,t)\left(\int_a^t f(s)\Delta_{\alpha,t}s\right) + f(t)$$

$$- k_1(\alpha,t)\left(\int_a^t f(s)\frac{E_0(t,\sigma(s))}{k_0(\alpha,s)}\Delta s\right)$$

$$= f(t), \quad t \in [a,b]^\kappa.$$

This completes the proof. □

Example 1.7.4 *Let* $\mathbb{T} = \mathbb{Z}$,

$$k_1(\alpha,t) = (1-\alpha)t^{2\alpha}, \quad k_0(\alpha,t) = \alpha t^{2(1-\alpha)}, \quad \alpha \in (0,1], \quad t \in \mathbb{T},$$

$$f(t) = \frac{t^2+1}{t^2+3}, \quad t \in \mathbb{T}.$$

We will compute

$$\int_0^{10} f(s)\Delta_{\alpha,10}s$$

for

$$\alpha t^{2(1-\alpha)} - (1-\alpha)t^{2\alpha} \neq 0, \quad t \in [0,10].$$

Here

$$\sigma(t) = t+1,$$

$$\mu(t) = 1, \quad t \in \mathbb{T},$$

$$E_0(10,\sigma(s)) = e_{-\frac{k_1}{k_0}}(10,\sigma(s))$$

$$= e^{\int_{\sigma(s)}^{10} \frac{1}{\mu(\tau)}\mathrm{Log}\left(1-\mu(\tau)\frac{k_1(\alpha,\tau)}{k_0(\alpha,\tau)}\right)\Delta\tau}$$

$$= e^{-\int_s^{\sigma(s)} \frac{1}{\mu(\tau)}\mathrm{Log}\left(1-\mu(\tau)\frac{k_1(\alpha,\tau)}{k_0(\alpha,\tau)}\right)\Delta\tau + \int_s^{10} \frac{1}{\mu(\tau)}\mathrm{Log}\left(1-\mu(\tau)\frac{k_1(\alpha,\tau)}{k_0(\alpha,\tau)}\right)\Delta\tau}$$

$$= e^{-\mathrm{Log}\left(1-\frac{k_1(\alpha,s)}{k_0(\alpha,s)}\right)}e^{\sum_{l=s}^9 \mathrm{Log}\left(1-\frac{k_1(\alpha,l)}{k_0(\alpha,l)}\right)}$$

$$= \frac{1}{1 - \frac{k_1(\alpha,s)}{k_0(\alpha,s)}} \prod_{l=s}^{9} \left(1 - \frac{k_1(\alpha,l)}{k_0(\alpha,l)}\right)$$

$$= \frac{k_0(\alpha,s)}{k_0(\alpha,s) - k_1(\alpha,s)} \prod_{l=s}^{9} \frac{k_0(\alpha,l) - k_1(\alpha,l)}{k_0(\alpha,l)}$$

$$= \frac{\alpha s^{2(1-\alpha)}}{-(1-\alpha)s^{2\alpha} + \alpha s^{2(1-\alpha)}} \prod_{l=s}^{9} \frac{\alpha l^{2(1-\alpha)} - (1-\alpha)l^{2\alpha}}{\alpha l^{2(1-\alpha)}}, \quad s \in [0,9].$$

Then

$$\int_0^{10} f(s)\Delta_{\alpha,10}s = \int_0^{10} \frac{s^2+1}{s^2+3} \frac{E_0(10,\sigma(s))}{k_0(\alpha,s)} \Delta s$$

$$= \sum_{s=0}^{9} \frac{s^2+1}{s^2+3} \frac{E_0(10,\sigma(s))}{k_0(\alpha,s)}$$

$$= \sum_{s=0}^{9} \left(\frac{s^2+1}{s^2+3} \frac{1}{\alpha s^{2(1-\alpha)}} \frac{\alpha s^{2(1-\alpha)}}{-(1-\alpha)s^{2\alpha} + \alpha s^{2(1-\alpha)}}\right.$$

$$\times \prod_{l=s}^{9} \frac{\alpha l^{2(1-\alpha)} - (1-\alpha)l^{2\alpha}}{\alpha l^{2(1-\alpha)}}$$

$$= \sum_{s=0}^{9} \left(\frac{s^2+1}{(s^2+3)\left(-(1-\alpha)s^{2\alpha} + \alpha s^{2(1-\alpha)}\right)} \prod_{l=s}^{9} \frac{\alpha l^{2(1-\alpha)} - (1-\alpha)l^{2\alpha}}{\alpha l^{2(1-\alpha)}}\right)$$

for

$$\alpha t^{2(1-\alpha)} - (1-\alpha)t^{2\alpha} \neq 0, \quad t \in [0,10].$$

Example 1.7.5 *Let* $\mathbb{T} = 2^{\mathbb{N}_0}$,

$$k_1(\alpha,t) = (1-\alpha)t^{4\alpha}, \quad k_0(\alpha,t) = \alpha t^{4(1-\alpha)}, \quad \alpha \in (0,1], \quad t \in \mathbb{T},$$

$$f(t) = t^2, \quad t \in \mathbb{T}.$$

We will compute

$$\int_1^{16} f(t)\Delta_{\alpha,16}s$$

for

$$\alpha - (1-\alpha)t^{8\alpha-3} \neq 0, \quad \alpha \in (0,1], \quad t \in \mathbb{T}.$$

We have

$$\sigma(t) = 2t,$$

$$\mu(t) = t,$$

$$E_0(16, \sigma(s)) = e_{-\frac{k_1}{k_0}}(t, \sigma(s))$$

$$= e^{\int_{\sigma(s)}^{16} \frac{1}{\mu(\tau)} \operatorname{Log}\left(1 - \mu(\tau)\frac{k_1(\alpha,\tau)}{k_0(\alpha,\tau)}\right)\Delta\tau}$$

$$= e^{-\int_s^{\sigma(s)} \frac{1}{\mu(\tau)} \operatorname{Log}\left(1 - \mu(\tau)\frac{k_1(\alpha,\tau)}{k_0(\alpha,\tau)}\right)\Delta\tau}$$

$$\times e^{\int_s^{16} \frac{1}{\mu(\tau)} \operatorname{Log}\left(1 - \mu(\tau)\frac{k_1(\alpha,\tau)}{k_0(\alpha,\tau)}\right)\Delta\tau}$$

$$= e^{-\operatorname{Log}\left(1 - \frac{1-\alpha}{\alpha}s^{8\alpha-3}\right)}$$

$$\times e^{\sum_{l=s}^{8} \operatorname{Log}\left(1 - \frac{1-\alpha}{\alpha}l^{8\alpha-3}\right)}$$

$$= \frac{\alpha}{\alpha - (1-\alpha)s^{8\alpha-3}} \prod_{l=s}^{8}\left(1 - \frac{1-\alpha}{\alpha}l^{8\alpha-3}\right),$$

$\alpha \in (0,1]$, $s \in [1,8]$. *Then*

$$\int_1^{16} f(s)\Delta_{\alpha,16}s = \int_1^{16} f(s)\frac{E_0(16, \sigma(s))}{k_0(\alpha,s)}\Delta s$$

$$= \int_1^{16} s^2 \frac{E_0(16, \sigma(s))}{\alpha s^{4-4\alpha}}\Delta s$$

$$= \frac{1}{\alpha}\int_1^{16} s^{4\alpha-2}E_0(16, \sigma(s))\Delta s$$

$$= \frac{1}{\alpha}\sum_{s=1}^{8}\left(s^{4\alpha-1}E_0(16, \sigma(s))\right)$$

$$= \frac{1}{\alpha}\sum_{s=1}^{8}\left(s^{4\alpha-1}\frac{\alpha}{\alpha - (1-\alpha)s^{8\alpha-3}}\prod_{l=s}^{8}\left(1 - \frac{1-\alpha}{\alpha}l^{8\alpha-3}\right)\right)$$

$$= \sum_{s=1}^{8}\left(\frac{s^{4\alpha-1}}{\alpha - (1-\alpha)s^{8\alpha-3}}\prod_{l=s}^{8}\left(1 - \frac{1-\alpha}{\alpha}l^{8\alpha-3}\right)\right),$$

$\alpha \in (0,1]$, *for*

$$\alpha - (1-\alpha)t^{8\alpha-3} \neq 0, \quad \alpha \in (0,1], \quad t \in [1,16].$$

Theorem 1.7.6 *Let* $\alpha \in (0,1]$, $f \in \mathscr{C}_{rd}(\mathbb{T})$ *and* (1.6) *hold. Then*

$$\int_a^t D^\alpha f(s)\Delta_{\alpha,t}s = f(t) - f(a)E_0(t,a), \quad t \in [a,b]^\kappa.$$

Proof 1.7.7 *We have*

$$\int_a^t D^\alpha f(s)\Delta_{\alpha,t}s = \int_a^t \left(k_1(\alpha,s)f(s)+k_0(\alpha,s)f^\Delta(s)\right)\frac{E_0(t,\sigma(s))}{k_0(\alpha,s)}\Delta s$$

$$= \int_a^t k_1(\alpha,s)f(s)\frac{E_0(t,\sigma(s))}{k_0(\alpha,s)}\Delta s + \int_a^t f^\Delta(s)E_0(t,\sigma(s))\Delta s$$

$$= \int_a^t k_1(\alpha,s)f(s)\frac{E_0(t,\sigma(s))}{k_0(\alpha,s)}\Delta s + f(s)E_0(t,s)\Big|_{s=a}^{s=t}$$

$$\quad - \int_a^t f(s)E_0^{\Delta s}(t,s)\Delta s$$

$$= \int_a^t k_1(\alpha,s)f(s)\frac{E_0(t,\sigma(s))}{k_0(\alpha,s)}\Delta s + f(t) - f(a)E_0(t,a)$$

$$\quad - \int_a^t f(s)e_{-\frac{k_1}{k_0}}^{\Delta s}(t,s)\Delta s$$

$$= \int_a^t k_1(\alpha,s)f(s)\frac{E_0(t,\sigma(s))}{k_0(\alpha,s)}\Delta s + f(t) - f(a)E_0(t,a)$$

$$\quad - \int_a^t f(s)\frac{\frac{k_1(\alpha,s)}{k_0(\alpha,s)}}{1-\mu(s)\frac{k_1(\alpha,s)}{k_0(\alpha,s)}}e_{-\frac{k_1}{k_0}}(t,s)\Delta s$$

$$= \int_a^t k_1(\alpha,s)f(s)\frac{E_0(t,\sigma(s))}{k_0(\alpha,s)}\Delta s + f(t) - f(a)E_0(t,a)$$

$$\quad - \int_a^t k_1(\alpha,s)f(s)\frac{E_0(t,\sigma(s))}{k_0(\alpha,s)}\Delta s$$

$$= f(t) - f(a)E_0(t,a), \quad t \in [a,b]^\kappa.$$

This completes the proof. □

Theorem 1.7.8 (Integration by Parts) *Let* $f,g \in \mathscr{C}_{rd}(\mathbb{T})$ *and assume* (1.6) *holds. Then*

$$\int_a^t D^\alpha f(s)g(\sigma(s))\Delta_{\alpha,t}s = f(t)g(t) - f(a)g(a)E_0(t,a)$$

$$\quad - \int_a^t (f(s)D^\alpha g(s) - k_1(\alpha,s)f(s)g(\sigma(s)))\Delta_{\alpha,t}s,$$

$$\int_a^t D^\alpha f(s)g(s)\Delta_{\alpha,t}s = f(t)g(t) - f(a)g(a)E_0(t,a)$$

$$\quad - \int_a^t (f(\sigma(s))D^\alpha g(s) - k_1(\alpha,s)f(\sigma(s))g(s))$$

$$\quad \times \Delta_{\alpha,t}s, \quad t \in [a,b]^\kappa.$$

Proof 1.7.9 *We have*

$$\int_a^t D^\alpha f(s)g(\sigma(s))\Delta_{\alpha,t}s = \int_a^t \left(D^\alpha(fg)(s) - f(s)D^\alpha g(s) + k_1(\alpha,s)f(s)g(\sigma(s))\right)\Delta_{\alpha,t}s$$

$$= \int_a^t D^\alpha(fg)(s)\Delta_{\alpha,t}s$$

$$- \int_a^t \left(f(s)D^\alpha g(s) - k_1(\alpha,s)f(s)g(\sigma(s))\right)\Delta_{\alpha,t}s$$

$$= f(t)g(t) - f(a)g(a)E_0(t,a)$$

$$- \int_a^t \left(f(s)D^\alpha g(s) - k_1(\alpha,s)f(s)g(\sigma(s))\right)\Delta_{\alpha,t}s, \quad t \in [a,b]^\kappa.$$

As above, one can prove the second part of the assertion. This completes the proof. □

Theorem 1.7.10 *Let* $f : \mathbb{T} \times \mathbb{T} \to \mathbb{R}$, $f \in \mathscr{C}_{rd}(\mathbb{T} \times \mathbb{T})$, $f(t,\cdot), D_t^\alpha f(t,\cdot) \in \mathscr{C}_{rd}(\mathbb{T})$, $f(\cdot,t) \in \mathscr{C}_{rd}^1(\mathbb{T})$ *for any* $t \in [a,b]^\kappa$, *and (1.6) hold. Then*

$$D_t^\alpha \left(\int_a^t f(t,s)\Delta_{\alpha,t}s\right) = \int_a^t D_t^\alpha f(t,s)\Delta_{\alpha,t}s - k_1(\alpha,t)\int_a^t f(\sigma(t),s)\Delta_{\alpha,t}s$$

$$+ f(\sigma(t),t), \quad t \in [a,b]^\kappa.$$

Proof 1.7.11 *We have*

$$D^\alpha \left(\int_a^t f(t,s)\Delta_{\alpha,t}s\right) = k_1(\alpha,t)\left(\int_a^t f(t,s)\Delta_{\alpha,t}s\right)$$

$$+ k_0(\alpha,t)\left(\int_a^t f(t,s)\Delta_{\alpha,t}s\right)^{\Delta_t}$$

$$= k_1(\alpha,t)\left(\int_a^t f(t,s)\Delta_{\alpha,t}s\right)$$

$$+ k_0(\alpha,t)\left(\int_a^t f(t,s)\frac{E_0(t,\sigma(s))}{k_0(\alpha,s)}\Delta s\right)^{\Delta_t}$$

$$= k_1(\alpha,t)\left(\int_a^t f(t,s)\Delta_{\alpha,t}s\right)$$

$$+ k_0(\alpha,t)\left(\int_a^t f^{\Delta_t}(t,s)\frac{E_0(t,\sigma(s))}{k_0(\alpha,s)}\Delta s\right)$$

$$+ k_0(\alpha,t)\left(\int_a^t f(\sigma(t),s)\frac{E_0^{\Delta_t}(t,\sigma(s))}{k_0(\alpha,s)}\Delta s\right)$$

$$+k_0(\alpha,t)f(\sigma(t),t)\frac{E_0(\sigma(t),\sigma(t))}{k_0(\alpha,t)}$$

$$=\int_a^t \left(k_1(\alpha,t)f(t,s)+k_0(\alpha,t)f^{\Delta_t}(t,s)\right)\Delta_{\alpha,t}s$$

$$+k_0(\alpha,t)\left(\int_a^t f(\sigma(t),s)\frac{e^{\Delta_t}_{-\frac{k_1}{k_0}}(t,\sigma(s))}{k_0(\alpha,s)}\Delta s\right)$$

$$+f(\sigma(t),t)$$

$$=\int_a^t D_t^\alpha f(t,s)\Delta_{\alpha,t}s$$

$$-k_1(\alpha,t)\int_a^t f(\sigma(t),s)\Delta_{\alpha,t}s$$

$$+f(\sigma(t),t),\quad t\in[a,b]^\kappa.$$

This completes the proof. □

Theorem 1.7.12 *Let* $f\in\mathscr{C}_{rd}(\mathbb{T})$ *and* (1.6) *hold. Then*

$$\int_a^t f(s)\Delta_{\alpha,t}s=-E_0(t,a)\int_t^a f(s)\Delta_{\alpha,a}s,\quad t\in[a,b]^\kappa.$$

Proof 1.7.13 *We have*

$$\int_a^t f(s)\Delta_{\alpha,t}s = \int_a^t f(s)\frac{E_0(t,\sigma(s))}{k_0(\alpha,s)}\Delta s$$

$$= -\int_t^a f(s)\frac{E_0(t,\sigma(s))}{k_0(\alpha,s)}\Delta s$$

$$= -\int_t^a f(s)\frac{E_0(t,a)E_0(a,\sigma(s))}{k_0(\alpha,s)}\Delta s$$

$$= -E_0(t,a)\int_t^a f(s)\frac{E_0(a,\sigma(s))}{k_0(\alpha,s)}\Delta s$$

$$= -E_0(t,a)\int_t^a f(s)\Delta_{\alpha,a}s.$$

This completes the proof. □

Theorem 1.7.14 *Let* $f\in\mathscr{C}_{rd}(\mathbb{T})$ *and* (1.6) *hold. Then*

$$\int_a^t f(s)\Delta_{\alpha,t}s=E_0(t,c)\int_a^c f(s)\Delta_{\alpha,c}s+\int_c^t f(s)\Delta_{\alpha,t}s,\quad c,t\in[a,b]^\kappa.$$

Proof 1.7.15 *We have*

$$
\int_a^t f(s)\Delta_{\alpha,t}s = \int_a^t f(s)\frac{E_0(t,\sigma(s))}{k_0(\alpha,s)}\Delta s
$$

$$
= \int_a^c f(s)\frac{E_0(t,c)E_0(c,\sigma(s))}{k_0(\alpha,s)}\Delta s
$$

$$
+ \int_c^t f(s)\frac{E_0(t,\sigma(s))}{k_0(\alpha,s)}\Delta s
$$

$$
= E_0(t,c)\int_a^c f(s)\Delta_{\alpha,c}s + \int_c^t f(s)\Delta_{\alpha,t}s, \quad c,t \in [a,b]^\kappa.
$$

This completes the proof. □

1.8 TAYLOR'S FORMULA

Definition 1.8.1 *Suppose that* $\alpha \in (0,1]$ *and* (1.6) *hold. Define*

$$
g_0(t,s) = 1,
$$

$$
g_n(t,s) = \int_s^t \frac{g_{n-1}(\sigma(\tau),s)k_0(\alpha,\tau)E_0(\sigma(\tau),t)}{k_0(\alpha,\tau) - \mu(\tau)k_1(\alpha,\tau)}\Delta_{\alpha,t}\tau.
$$

We have

$$
g_n(t,s) = \int_s^t \frac{g_{n-1}(\sigma(\tau),s)k_0(\alpha,\tau)E_0(\sigma(\tau),t)E_0(t,\sigma(\tau))}{(k_0(\alpha,\tau) - \mu(\tau)k_1(\alpha,\tau))k_0(\alpha,\tau)}\Delta\tau
$$

$$
= \int_s^t \frac{g_{n-1}(\sigma(\tau),s)}{k_0(\alpha,\tau) - \mu(\tau)k_1(\alpha,\tau)}\Delta\tau,
$$

whereupon

$$
g_n^{\Delta_t}(t,s) = \frac{g_{n-1}(\sigma(t),s)}{k_0(\alpha,t) - \mu(t)k_1(\alpha,t)},
$$

or

$$
g_{n-1}(\sigma(t),s) = k_0(\alpha,t)g_n^{\Delta_t}(t,s) - \mu(t)g_n^{\Delta_t}(t,s)k_1(\alpha,t)
$$

$$
= k_0(\alpha,t)g_n^{\Delta_t}(t,s)
$$

$$
- (g_n(\sigma(t),s) - g_n(t,s))k_1(\alpha,t)
$$

$$
= k_0(\alpha,t)g_n^{\Delta_t}(t,s) + g_n(t,s)k_1(\alpha,t)
$$

$$
- k_1(\alpha,t)g_n(\sigma(t),s)
$$

$$= D_t^\alpha g_n(t,s) - k_1(\alpha,t) g_n(\sigma(t),s),$$

$$D_t^\alpha g_0(t,s) - k_1(\alpha,t) = k_1(\alpha,t) - k_1(\alpha,t)$$

$$= 0, \quad s,t \in \mathbb{T}.$$

Theorem 1.8.2 *Let $\alpha \in (0,1]$ and (1.6) hold. If f is n-times conformable Δ-differentiable on $\mathbb{T}$, then*

$$D^\alpha \left(\sum_{k=0}^{n-1} (-1)^k \left((D^\alpha)^k f \right) g_k \right) = (-1)^{n-1} \left((D^\alpha)^n f \right) g_{n-1}^\sigma. \tag{1.7}$$

Proof 1.8.3 *We have*

$$D^\alpha \left(\sum_{k=0}^{n-1} (-1)^k \left((D^\alpha)^k f \right) g_k \right) = \sum_{k=0}^{n-1} (-1)^k D^\alpha \left(\left((D^\alpha)^k f \right) g_k \right)$$

$$= \sum_{k=0}^{n-1} (-1)^k \left(\left((D^\alpha)^{k+1} f \right) g_k^\sigma + \left((D^\alpha)^k f \right) D^\alpha g_k - k_1 \left((D^\alpha)^k f \right) g_k^\sigma \right)$$

$$= \sum_{k=0}^{n-1} (-1)^k \left(\left((D^\alpha)^{k+1} f \right) g_k^\sigma + \left((D^\alpha)^k f \right) \left(D^\alpha g_k - k_1 g_k^\sigma \right) \right)$$

$$= \sum_{k=0}^{n-1} (-1)^k \left((D^\alpha)^{k+1} f \right) g_k^\sigma + \sum_{k=1}^{n-1} (-1)^k \left((D^\alpha)^k f \right) g_{k-1}^\sigma + f \left(D^\alpha g_0 - k_1 \right)$$

$$= (D^\alpha f) g_0^\sigma - \left((D^\alpha)^2 f \right) g_1^\sigma + \cdots + (-1)^{n-1} \left((D^\alpha)^n f \right) g_{n-1}^\sigma$$

$$- (D^\alpha f) g_0^\sigma + \left((D^\alpha)^2 f \right) g_1^\sigma + \cdots + (-1)^{n-1} \left((D^\alpha)^{n-1} f \right) g_{n-2}^\sigma$$

$$= (-1)^{n-1} \left((D^\alpha)^n f \right) g_{n-1}^\sigma.$$

This completes the proof. □

Theorem 1.8.4 (Taylor's Formula) *Let $\alpha \in (0,1]$ and (1.6) hold. If f is n-times conformable Δ-differentiable on $\mathbb{T}$, then*

$$f(t) = \sum_{k=0}^{n-1} (-1)^k \left((D^\alpha)^k f \right)(s) g_k(s,t) E_0(t,s)$$

$$+ (-1)^{n-1} \int_s^t \left((D^\alpha)^n f \right)(\tau) g_{n-1}(\sigma(\tau),t) \Delta_{\alpha,t} \tau, \quad s,t \in \mathbb{T}. \tag{1.8}$$

Proof 1.8.5 *We integrate (1.7) from s to t and we get*

$$\int_s^t (-1)^{n-1} \left((D^\alpha)^n f \right)(\tau) g_{n-1}(\sigma(\tau),t) \Delta_{\alpha,t} \tau$$

$$= \int_s^t D^\alpha \left(\sum_{k=0}^{n-1} (-1)^k \left((D^\alpha)^k f \right)(\tau) g_k(\tau,t) \right) \Delta_{\alpha,t} \tau$$

$$= \sum_{k=0}^{n-1} (-1)^k \left((D^\alpha)^k f \right)(t) g_k(t,t)$$

$$- \sum_{k=0}^{n-1} (-1)^k \left((D^\alpha)^k f \right)(s) g_k(s,t) E_0(t,s)$$

$$= f(t) - \sum_{k=0}^{n-1} (-1)^k \left((D^\alpha)^k f \right)(s) g_k(s,t) E_0(t,s), \quad s,t \in \mathbb{T},$$

whereupon we get (1.8). This completes the proof. □

Now we define the polynomials

$$h_0(t,s) = E_0(t,s),$$

$$h_n(t,s) = \int_s^t h_{n-1}(\tau,s) \Delta_{\alpha,t} \tau, \quad \alpha \in (0,1], \quad s,t \in \mathbb{T}.$$

Note that

$$h_0(s,s) = 1,$$

$$D^\alpha h_0(t,s) = 0,$$

$$D^\alpha h_n(t,s) = h_{n-1}(t,s), \quad n \in \mathbb{N}, \quad s,t \in \mathbb{T}.$$

Theorem 1.8.6 *Suppose that $\alpha \in (0,1]$ and assume (1.6) holds. Then*

$$h_n(t,s) = (-1)^n g_n(s,t) E_0(t,s), \quad s,t \in \mathbb{T}, \quad n \in \mathbb{N}.$$

Proof 1.8.7 *Note that*

$$(D_t^\alpha)^k h_n(t,s) = h_{n-k}(t,s), \quad s,t \in \mathbb{T}, \quad k \in \mathbb{T}, \quad 0 \le k \le n, \quad \alpha \in (0,1].$$

Hence,

$$(D_t^\alpha)^k h_n(s,s) = 0, \quad s \in \mathbb{T}, \quad k \in \mathbb{N}, \quad 0 \le k \le n-1, \quad \alpha \in (0,1],$$

$$(D_t^\alpha)^n h_n(s,s) = 1,$$

$$(D_t^\alpha)^{n+1} h_n(t,s) = D_t^\alpha h_0(t,s)$$

$$= 0, \quad t,s \in \mathbb{T}, \quad \alpha \in (0,1].$$

From here and from Taylor's formula (1.8), we get

$$
\begin{aligned}
h_n(t,s) &= \sum_{k=0}^{n} (-1)^k \left((D^\alpha)^k h_n \right)(s,s) g_k(s,t) E_0(t,s) \\
&\quad + (-1)^n \int_s^t \left((D^\alpha)^{n+1} h_n \right)(\tau,s) g_n(\sigma(\tau),t) \Delta_{\alpha,t} \tau \\
&= \sum_{k=0}^{n-1} (-1)^k \left((D^\alpha)^k h_n \right)(s,s) g_k(s,t) E_0(t,s) \\
&\quad + (-1)^n g_n(s,t) E_0(t,s) \\
&= (-1)^n g_n(s,t) E_0(t,s), \quad s,t \in \mathbb{T}, \quad \alpha \in (0,1], \quad n \in \mathbb{N}.
\end{aligned}
$$

This completes the proof. $\qquad\qquad\qquad\qquad\qquad\qquad\qquad\qquad\qquad\qquad\qquad$ □

Corollary 1.8.8 (Taylor's Formula) *Let $\alpha \in (0,1]$ and assume (1.6) holds. If f is n-times conformable Δ-differentiable on $\mathbb{T}$, then*

$$
\begin{aligned}
f(t) &= \sum_{k=0}^{n-1} \left((D^\alpha)^k f \right)(s) h_k(t,s) \\
&\quad + \int_s^t \left((D^\alpha)^n f \right)(\tau) h_{n-1}(t,\sigma(\tau)) \left(E_0(t,\sigma(\tau)) \right)^{-1} \Delta_{\alpha,t} \tau, \quad s,t \in \mathbb{T}.
\end{aligned}
$$

1.9 CALCULUS FOR THE NABLA CONFORMABLE DERIVATIVE

Suppose that k_0 and k_1 satisfy $(A1)$. In this section we consider the nabla version of the conformable derivative. In particular, the nabla conformable derivative on time scales is given by

$$\widehat{D}_\alpha f(t) = k_1(\alpha,t) f(t) + k_0(\alpha,t) f^\nabla(t), \qquad t \in \mathbb{T}_\kappa \qquad (1.9)$$

provided the right-hand side exists at t, where f^∇ is the time scale nabla derivative.

We continue with the next important definition, which establishes a type of exponential function for derivative (1.9). We assume throughout that the reader is familiar with time scale notation, and the basics of nabla derivatives on time scales found in [1]. In particular, recall the graininess function $v(t) = t - \rho(t)$, where ρ is the backward jump operator $\rho(t) := \sup\{s \in \mathbb{T} : s < t\}$, and the nabla derivative [4] of f at t, denoted $f^\nabla(t)$, to be the number (provided it exists) with the property that given any $\varepsilon > 0$, there is a neighborhood U of t such that

$$|f(\rho(t)) - f(s) - f^{\nabla}(t)[\rho(t) - s]| \leq \varepsilon|\rho(t) - s|$$

for all $s \in U$.

Definition 1.9.1 (ν-Regressive) *A function $f : \mathbb{T} \to \mathbb{R}$ is conformable ν-regressive if and only if*

$$k_0(\alpha, t) - \nu(t)(f(t) - k_1(\alpha, t)) \neq 0$$

for any $\alpha \in (0, 1]$ and for any $t \in \mathbb{T}$. The set of all conformable ν-regressive functions on $\mathbb{T}$ will be denoted by $\widehat{\mathscr{R}_c}$.

Remark 1.9.2 *We will often need to divide by the expression $k_0 + \nu k_1$. Since all three functions are non-negative, and for most of the development to follow, we assume $\alpha \in (0, 1]$, we then have*

$$k_0(\alpha, t) + \nu(t)k_1(\alpha, t) \neq 0$$

for all $t \in \mathbb{T}_k$ for all $\alpha \in (0, 1]$.

Definition 1.9.3 *For $f, g \in \widehat{\mathscr{R}_c}$, the nabla conformable circle plus is given by*

$$(f \widehat{\oplus}_c g)(t) = f(t) + g(t) - k_1(\alpha, t) - \nu(t)\frac{(f(t) - k_1(\alpha, t))(g(t) - k_1(\alpha, t))}{k_0(\alpha, t)},$$

and the nabla conformable circle minus is given by

$$\widehat{\ominus}_c f(t) = k_1(\alpha, t) - \frac{k_0(\alpha, t)(f(t) - k_1(\alpha, t))}{k_0(\alpha, t) - \nu(t)(f(t) - k_1(\alpha, t))}.$$

The following theorem is straightforward.

Theorem 1.9.4 *The set with binary operation $\left(\widehat{\mathscr{R}_c}, \widehat{\oplus}_c\right)$ is an Abelian group.*

Definition 1.9.5 (Nabla Conformable Exponential Function) *Let $\alpha \in (0, 1]$, the points $s, t \in \mathbb{T}$, and let the function $p \in \widehat{\mathscr{R}_c}$. Then the nabla conformable exponential function with respect to $\widehat{D}_\alpha$ in (1.9) is defined to be*

$$\widehat{E}_p(t, s) := \widehat{e}_{\frac{p - k_1}{k_0}}(t, s), \tag{1.10}$$

where $\widehat{e}$ is the nabla time scale exponential [1, Section 3]. Note that $\widehat{E}_p$ satisfies

$$\widehat{D}_\alpha \widehat{E}_p(t, s) = p(t)\widehat{E}_p(t, s), \qquad t \in \mathbb{T}_k, \qquad s \in \mathbb{T}, \tag{1.11}$$

using (1.9).

The following properties of the nabla exponential function are useful direct results of (1.10).

Lemma 1.9.6 (Nabla Exponential Function Properties) *Let* $\widehat{D}_\alpha$ *satisfy* (1.9), *let* $p, q \in$ $\widehat{\mathscr{R}}_c$, *and let* $t, s, r \in \mathbb{T}$. *For the nabla conformable exponential function given in* (1.10), *the following properties hold.*

(i) $\widehat{E}_p(t, t) \equiv 1.$

(ii) $\widehat{E}_p(t, s)\widehat{E}_p(s, r) = \widehat{E}_p(t, r).$

(iii) $\dfrac{1}{\widehat{E}_p(t, s)} = \widehat{E}_p(s, t) = \widehat{E}_{\widehat{\ominus}_c p}(t, s).$

(iv) $\widehat{E}_{k_1}(t, s) \equiv 1 \equiv \widehat{E}_{\widehat{\ominus}_c k_1}(t, s).$

(v) $\widehat{E}_p(\rho(t), s) = \left(1 - \nu(t)\dfrac{p(t) - k_1(\alpha, t)}{k_0(\alpha, t)}\right)\widehat{E}_p(t, s).$

(vi) $\widehat{E}_p(t, s)\widehat{E}_q(t, s) = \widehat{E}_{p \widehat{\oplus}_c q}(t, s).$

(vii) $\dfrac{\widehat{E}_p(t, s)}{\widehat{E}_q(t, s)} = \widehat{E}_{p \widehat{\ominus}_c q}(t, s).$

(viii) *If* p *is nabla differentiable and* $\dfrac{\widehat{D}_\alpha p}{p} \in \widehat{\mathscr{R}}_c$, *then*

$$\widehat{E}_{\left(\frac{\widehat{D}_\alpha p}{p}\right)}(t, s) = \frac{p(t)}{p(s)}.$$

Theorem 1.9.7 (Fundamental Theorem of Integral Calculus) *Let* $\alpha \in (0, 1]$. *Suppose* $f : [a, b]_\mathbb{T} \to \mathbb{R}$ *is differentiable on* $[a, b]_\mathbb{T}$ *and* $\widehat{D}_\alpha$ *is integrable on* $[a, b]_\mathbb{T}$. *Then*

$$\int_a^t \widehat{D}_\alpha[f(s)]\frac{\widehat{E}_0(t, \rho(s))}{k_0(\alpha, s)}\nabla s = f(t) - f(a)\widehat{E}_0(t, a),$$

and

$$\widehat{D}_\alpha\left[\int_a^t f(s)\frac{\widehat{E}_0(t, \rho(s))}{k_0(\alpha, s)}\nabla s\right] = f(t).$$

Proof 1.9.8 *Since* $k_0 + \nu k_1 \neq 0$, *we have* $0 \in \widehat{\mathscr{R}}_c$ *and* $\widehat{E}_0$ *is well defined. The equalities then follow.* $\qquad\square$

Lemma 1.9.9 (Basic Derivatives) *Let* $\widehat{D}_\alpha$ *satisfy* (1.9) *for* $\alpha \in [0, 1]$. *Let the functions* $f, g : \mathbb{T} \to \mathbb{R}$ *be nabla differentiable as needed. Moreover, define*

$$f^\rho(t) := f(\rho(t)).$$

Then

1. $\widehat{D}_\alpha c = ck_1(\alpha, t)$ *for all constants* $c \in \mathbb{R}$;

2. $\widehat{D}_\alpha[af + bg] = a\widehat{D}_\alpha[f] + b\widehat{D}_\alpha[g]$ for all $a, b \in \mathbb{R}$;

3. $\widehat{D}_\alpha[fg] = f\widehat{D}_\alpha[g] + g^\rho\widehat{D}_\alpha[f] - fg^\rho k_1$;

4. $\widehat{D}_\alpha[fg] = g\widehat{D}_\alpha[f] + f^\rho\widehat{D}_\alpha[g] - f^\rho gk_1$;

5. $\widehat{D}_\alpha[f/g] = \dfrac{g\widehat{D}_\alpha[f] - f\widehat{D}_\alpha[g]}{g^\rho g} + \dfrac{f}{g}k_1.$

Theorem 1.9.10 (Equivalence of Exponential Functions) *Let* $\alpha \in (0,1]$. *If* f *is continuous and* $\kappa_0^\rho + \nu(f^\rho - \kappa_1^\rho) \neq 0$, *then*

$$E_f(t,s) = \widehat{E}_{\frac{\kappa_0(f-\kappa_1)^\rho}{\kappa_0^\rho + \nu(f-\kappa_1)^\rho}+\kappa_1}(t,s).$$

If g *is continuous and* $\kappa_0^\sigma - \mu(f^\sigma - \kappa_1^\sigma) \neq 0$, *then*

$$\widehat{E}_g(t,s) = E_{\frac{\kappa_0(g-\kappa_1)^\sigma}{\kappa_0^\sigma - \mu(g-\kappa_1)^\sigma}+\kappa_1}(t,s).$$

Proof 1.9.11 *It is known on time scales* $(\alpha = 1)$ *that if a function* p *is continuous and regressive, then*

$$e_p(t,s) = \widehat{e}_{\frac{p\rho}{1+\nu p^\rho}}(t,s).$$

Hence, suppressing the arguments, and letting $F := \dfrac{f - \kappa_1}{\kappa_0}$, *we see that*

$$E_f = e_F = \widehat{e}_{\frac{F\rho}{1+\nu F^\rho}} = \widehat{E}_{\frac{\kappa_0 F\rho}{1+\nu F^\rho}+\kappa_1},$$

which yields the first equation. We must also check that the resulting base of $\widehat{E}$ *is in* $\widehat{\mathscr{R}_c}$. *We have*

$$\kappa_0 - \nu\left[\left(\frac{\kappa_0(f-\kappa_1)^\rho}{\kappa_0^\rho + \nu(f-\kappa_1)^\rho}+\kappa_1\right)-\kappa_1\right] = \frac{\kappa_0\kappa_0^\rho}{\kappa_0^\rho + \nu(f-\kappa_1)^\rho} \neq 0,$$

and the result holds.

Next, if a function q *is continuous and* ν-*regressive, then*

$$\widehat{e}_q(t,s) = e_{\frac{q^\sigma}{1-\mu q^\sigma}}(t,s).$$

Similar calculations then yield the second equation and the appropriate regressivity. This completes the proof. □

1.10 CONFORMABLE PARTIAL DERIVATIVES

Definition 1.10.1 *Let* $u: \mathbb{T} \times \mathbb{T} \to \mathbb{C}$ *be delta differentiable with respect to the first variable or with respect to the second variable at some point* $(t,s) \in \mathbb{T} \times \mathbb{T}$. *Define the conformable* Δ-*partial derivative with respect to the first variable or the conformable* Δ-*partial derivative with respect to the second variable at* (t,s) *as follows:*

$$D_t^\alpha u(t,s) = k_1(\alpha,t)u(t,s) + k_0(\alpha,t)u_t^\Delta(t,s),$$

$$D_s^\alpha u(t,s) = k_1(\alpha,s)u(t,s) + k_0(\alpha,s)u_s^\Delta(t,s),$$

respectively.

Example 1.10.2 *Let* $\mathbb{T} = \mathbb{Z}$,

$$k_1(\alpha,t) = (1-\alpha)t^\alpha, \quad k_0(\alpha,t) = \alpha t^{1-\alpha}, \quad \alpha \in (0,1], \quad t \in \mathbb{T},$$

$$u(t,s) = s^2 + st, \quad (t,s) \in \mathbb{T} \times \mathbb{T}.$$

We will find

$$D_t^{\frac{1}{4}}u(t,s), \quad D_s^{\frac{1}{4}}u(t,s), \quad (t,s) \in \mathbb{T} \times \mathbb{T}.$$

We have

$$\sigma(t) = t+1, \quad t \in \mathbb{T},$$

$$k_1\left(\frac{1}{4},t\right) = \frac{3}{4}t^{\frac{1}{4}},$$

$$k_0\left(\frac{1}{4},t\right) = \frac{1}{4}t^{\frac{3}{4}}, \quad t \in \mathbb{T},$$

$$u_t^\Delta(t,s) = s,$$

$$u_s^\Delta(t,s) = \sigma(s) + s + t$$

$$= s+1+s+t$$

$$= 2s+t+1, \quad (t,s) \in \mathbb{T} \times \mathbb{T}.$$

Then

$$D_t^{\frac{1}{4}}u(t,s) = k_1\left(\frac{1}{4},t\right)u(t,s) + k_0\left(\frac{1}{4},t\right)u_t^\Delta(t,s)$$

$$= \frac{3}{4}t^{\frac{1}{4}}\left(s^2+st\right) + \frac{1}{4}t^{\frac{3}{4}}s$$

$$= \frac{t^{\frac{1}{4}}s}{4}\left(3s+3t+t^{\frac{1}{2}}\right),$$

$$D_s^{\frac{1}{4}}u(t,s) = k_1\left(\frac{1}{4},s\right)u(t,s) + k_0\left(\frac{1}{4},s\right)u_s^\Delta(t,s)$$

$$= \frac{3}{4}s^{\frac{1}{4}}\left(s^2 + st\right) + \frac{1}{4}s^{\frac{3}{4}}(2s + t + 1)$$

$$= \frac{s^{\frac{1}{4}}}{4}\left(3s^2 + 3st + 2s^{\frac{3}{2}} + ts^{\frac{1}{2}} + s^{\frac{1}{2}}\right), \quad (t,s) \in \mathbb{T} \times \mathbb{T}.$$

Exercise 1.10.3 *Let* $\mathbb{T} = 2^{\mathbb{N}_0}$,

$$k_1(\alpha,t) = (1-\alpha)t^{2\alpha}, \quad k_0(\alpha,t) = \alpha t^{2(1-\alpha)}, \quad \alpha \in (0,1], \quad t \in \mathbb{T},$$

$$u(t,s) = s^3 + 3s^2t + st^2 + t^3, \quad (t,s) \in \mathbb{T}| \times \mathbb{T}.$$

Find

$$D_t^{\frac{1}{2}}u(t,s), \quad D_t^{\frac{1}{3}}u(t,s), \quad D_s^{\frac{1}{4}}u(t,s), \quad (t,s) \in \mathbb{T} \times \mathbb{T}.$$

Theorem 1.10.4 *Let* $u,v: \mathbb{T} \times \mathbb{T} \to \mathbb{C}$ *be delta differentiable with respect to the first variable at some point* $(t,s) \in \mathbb{T} \times \mathbb{T}$. *Then*

1. $D_t^{\alpha}(u+v)(t,s) = D_t^{\alpha}u(t,s) + D_t^{\alpha}v(t,s)$,

2. $D_t^{\alpha}(au)(t,s) = aD_t^{\alpha}u(t,s)$.

Proof 1.10.5 *1. We have*

$$D_t^{\alpha}(u+v)(t,s) = k_1(\alpha,t)(u+v)(t,s) + k_0(\alpha,t)(u+v)_t^{\Delta}(t,s)$$

$$= k_1(\alpha,t)u(t,s) + k_1(\alpha,t)v(t,s)$$

$$+ k_0(\alpha,t)u_t^{\Delta}(t,s) + k_1(\alpha,t)v_t^{\Delta}(t,s)$$

$$= D_t^{\alpha}u(t,s) + D_t^{\alpha}v(t,s).$$

2. We have

$$D_t^{\alpha}(au)(t,s) = k_1(\alpha,t)(au)(t,s) + k_0(\alpha,t)(au)_t^{\Delta}(t,s)$$

$$= ak_1(\alpha,t)u(t,s) + ak_0(\alpha,t)u_t^{\Delta}(t,s)$$

$$= aD_t^{\alpha}u(t,s).$$

This completes the proof. $\qquad\qquad\qquad\qquad\qquad\qquad\qquad\qquad \square$

Theorem 1.10.6 *Let $u, v : \mathbb{T} \times \mathbb{T} \to \mathbb{C}$ be delta differentiable with respect to the first variable at some point $(t, s) \in \mathbb{T} \times \mathbb{T}$. Then*

$$
\begin{aligned}
D_t^\alpha (uv)(t,s) &= (D_t^\alpha u(t,s)) v(t,s) + u(\sigma(t),s) D_t^\alpha v(t,s) \\[2ex]
&\quad - k_1(\alpha,t) u(\sigma(t),s) v(t,s) \\[2ex]
&= (D_t^\alpha u(t,s)) v(\sigma(t),s) + u(t,s) D_t^\alpha v(t,s) \\[2ex]
&\quad - k_1(\alpha,t) u(t,s) v(\sigma(t),s).
\end{aligned}
$$

Proof 1.10.7 *We have*

$$
\begin{aligned}
D_t^\alpha (uv)(t,s) &= k_1(\alpha,t) u(t,s) v(t,s) + k_0(\alpha,t)(uv)_t^\Delta(t,s) \\[2ex]
&= k_1(\alpha,t) u(t,s) v(t,s) \\[2ex]
&\quad + k_0(\alpha,t) u_t^\Delta(t,s) v(t,s) \\[2ex]
&\quad + k_0(\alpha,t) u(\sigma(t),s) v_t^\Delta(t,s) \\[2ex]
&= v(t,s) D_t^\alpha u(t,s) \\[2ex]
&\quad + k_0(\alpha,t) u(\sigma(t),s) v_t^\Delta(t,s) \\[2ex]
&\quad + k_1(\alpha,t) u(\sigma(t),s) v(t,s) \\[2ex]
&\quad - k_1(\alpha,t) u(\sigma(t),s) v(t,s) \\[2ex]
&= v(t,s) D_t^\alpha u(t,s) + u(\sigma(t),s) D_t^\alpha v(t,s) \\[2ex]
&\quad - k_1(\alpha,t) u(\sigma(t),s) v(t,s) \\[2ex]
&= k_1(\alpha,t) u(t,s) v(t,s) \\[2ex]
&\quad + k_0(\alpha,t) u_t^\Delta(t,s) v(\sigma(t),s) \\[2ex]
&\quad + k_0(\alpha,t) u(t,s) v_t^\Delta(t,s)
\end{aligned}
$$

$$= u(t,s)D_t^\alpha v(t,s)$$

$$+k_0(\alpha,t)u_t^\Delta(t,s)v(\sigma(t),s)$$

$$+k_1(\alpha,t)u(t,s)v(\sigma(t),s)$$

$$-k_1(\alpha,t)u(t,s)v(\sigma(t),s)$$

$$= u(t,s)D_t^\alpha v(t,s) + v(\sigma(t),s)D_t^\alpha u(t,s)$$

$$-k_1(\alpha,t)u(t,s)v(\sigma(t),s).$$

This completes the proof. □

Theorem 1.10.8 *Suppose that* $u,v : \mathbb{T} \times \mathbb{T} \to \mathbb{C}$ *are delta differentiable with respect to the first variable at some point* $(t,s) \in \mathbb{T} \times \mathbb{T}$. *Then*

$$D_t^\alpha\left(\frac{u}{v}\right)(t,s) = \frac{v(t,s)D_t^\alpha u(t,s) - u(t,s)D_t^\alpha v(t,s)}{v(t,s)v(\sigma(t),s)}$$

$$+k_1(\alpha,t)\frac{u(t,s)}{v(t,s)}$$

provided that $v(t,s)v(\sigma(t),s) \neq 0$.

Proof 1.10.9 *We have*

$$D_t^\alpha\left(\frac{u}{v}\right)(t,s) = k_1(\alpha,t)\frac{u(t,s)}{v(t,s)} + k_0(\alpha,t)\left(\frac{u}{v}\right)_t^\Delta(t,s)$$

$$= k_1(\alpha,t)\frac{u(t,s)}{v(t,s)}$$

$$+k_0(\alpha,t)\frac{u_t^\Delta(t,s)v(t,s) - u(t,s)v_t^\Delta(t,s)}{v(t,s)v(\sigma(t),s)}$$

$$= \frac{1}{v(t,s)v(\sigma(t),s)}\bigg(k_1(\alpha,t)u(t,s)v(\sigma(t),s)$$

$$+k_0(\alpha,t)u_t^\Delta(t,s)v(t,s)$$

$$-k_0(\alpha,t)u(t,s)v_t^\Delta(t,s)\bigg)$$

$$= \frac{1}{v(t,s)v(\sigma(t),s)}\bigg(k_1(\alpha,t)u(t,s)v(\sigma(t),s)$$

$$+k_0(\alpha,t)u_t^\Delta(t,s)v(t,s)$$

$$+k_1(\alpha,t)u(t,s)v(t,s)$$

$$-k_1(\alpha,t)u(t,s)v(t,s)$$

$$-k_0(\alpha,t)u(t,s)v_t^\Delta(t,s)\Bigg)$$

$$= \frac{1}{v(t,s)v(\sigma(t),s)}\Bigg(k_1(\alpha,t)u(t,s)v(\sigma(t),s)$$

$$+v(t,s)D_t^\alpha u(t,s)-u(t,s)D_t^\alpha v(t,s)\Bigg)$$

$$= \frac{v(t,s)D_t^\alpha u(t,s)-u(t,s)D_t^\alpha v(t,s)}{v(t,s)v(\sigma(t),s)}$$

$$+k_1(\alpha,t)\frac{u(t,s)}{v(t,s)}.$$

This completes the proof. □

Example 1.10.10 *Let* $\mathbb{T}=2^{\mathbb{N}_0}$,

$$k_1(\alpha,t) = (1-\alpha)t^{2\alpha}, \quad k_0(\alpha,t)=\alpha t^{2(1-\alpha)}, \quad \alpha\in(0,1], \quad t\in\mathbb{T},$$

$$u(t,s) = 1+ts,$$

$$v(t,s) = t^2+s^2, \quad (t,s)\in\mathbb{T}\times\mathbb{T}.$$

We will find

$$D_t^{\frac{1}{2}}\left(\frac{u}{v}\right)(t,s), \quad (t,s)\in\mathbb{T}\times\mathbb{T}.$$

We have

$$\sigma(t) = 2t,$$

$$u_t^\Delta(t,s) = s,$$

$$v_t^\Delta(t,s) = \sigma(t)+t=2t+t=3t,$$

$$k_1\left(\frac{1}{2},t\right) = \frac{1}{2}t,$$

$$k_0\left(\frac{1}{2},t\right) = \frac{1}{2}t,$$

$$D_t^{\frac{1}{2}}u(t,s) = k_1\left(\frac{1}{2},t\right)u(t,s) + k_0\left(\frac{1}{2},t\right)u_t^{\Delta}(t,s)$$

$$= \frac{1}{2}t(1+ts) + \frac{1}{2}t(3t)$$

$$= \frac{1}{2}t(1+3t+st),$$

$$D_t^{\frac{1}{2}}v(t,s) = k_1\left(\frac{1}{2},t\right)v(t,s) + k_0\left(\frac{1}{2},t\right)v_t^{\Delta}(t,s)$$

$$= \frac{1}{2}t(t^2+s^2) + \frac{1}{2}t(3t)$$

$$= \frac{1}{2}t(t^2+3t+s^2),$$

$$v(\sigma(t),s) = (\sigma(t))^2 + s^2$$

$$= (2t)^2 + s^2$$

$$= 4t^2 + s^2, \quad (t,s) \in \mathbb{T} \times \mathbb{T}.$$

Then

$$D_t^{\frac{1}{2}}\left(\frac{u}{v}\right)(t,s) = \frac{1}{(t^2+s^2)(4t^2+s^2)}\left((t^2+s^2)\left(\frac{1}{2}t(1+st+3t)\right)\right.$$

$$\left. -(1+ts)\left(\frac{1}{2}t(t^2+3t+s^2)\right)\right) + \frac{1}{2}t\frac{1+ts}{t^2+s^2}$$

$$= \frac{1}{2(t^2+s^2)(4t^2+s^2)}\left(t^2+st^3+3t^3+s^2+s^3t+3s^2t\right.$$

$$-t^2 - 3t - s^2 - t^3s - 3t^2s - ts^2$$

$$\left. +(1+ts)(4t^2+s^2)\right)$$

$$= \frac{t}{2(t^2+s^2)(4t^2+s^2)}\left(3t^3 + 3s^2t - 3t^2s - 3t + 4t^2 + s^2 + 4t^3s + ts^3\right),$$

$(t,s) \in \mathbb{T} \times \mathbb{T}.$

Exercise 1.10.11 *Let* $\mathbb{T} = 2\mathbb{Z}$,

$$k_1(\alpha,t) = (1-\alpha)t^{3\alpha}, \quad k_0(\alpha,t) = \alpha t^{3(1-\alpha)}, \quad \alpha \in (0,1], \quad t \in \mathbb{T},$$

$$u(t,s) = t^3 + 3ts^2 + s^3,$$

$$v(t,s) = t^4 + s^4, \quad (t,s) \in \mathbb{T} \times \mathbb{T}.$$

Find

$$D_t^{\frac{1}{5}}\left(\frac{u}{v}\right)(t,s), \quad (t,s) \in \mathbb{T} \times \mathbb{T}.$$

1.11 ADVANCED PRACTICAL PROBLEMS

Problem 1.11.1 *Let* $\mathbb{T} = 2^{\mathbb{N}_0}$,

$$k_1(\alpha,t) = \left(\sin\left(\frac{\pi}{2}(1-\alpha)t^\alpha\right)\right)^2, \quad k_0(\alpha,t) = t^{1-\alpha}\left(\cos\left(\frac{\pi}{2}(1-\alpha)t^\alpha\right)\right)^2,$$

$\alpha \in [0,1]$, $t \in \mathbb{T}$,

$$f(t) = t^2, \quad t \in \mathbb{T}.$$

Find

$$D^{\frac{1}{3}}f(t), \quad t \in \mathbb{T}.$$

Problem 1.11.2 *Let* $\mathbb{T} = 2^{\mathbb{N}_0}$,

$$k_1(\alpha,t) = (1-\alpha)t^\alpha, \quad k_0(\alpha,t) = \alpha t^{1-\alpha}, \quad \alpha \in [0,1], \quad t \in \mathbb{T},$$

$$f(t) = t^3 + 3t, \quad g(t) = t^2 + t, \quad t \in \mathbb{T}.$$

Find

1. $D^{\frac{1}{2}}(f+g)(t)$,

2. $D^{\frac{1}{2}}(fg - g^2)(t)$,

3. $D^{\frac{1}{4}}\left(\dfrac{f-g}{f+g}\right)(t)$,

4. $D^{\frac{1}{6}}\left(\dfrac{f}{g}\right)(t)$.

Problem 1.11.3 *Let* $\mathbb{T} = 2^{\mathbb{N}_0}$,

$$k_1(\alpha,t) = (1-\alpha)(1+t)^\alpha, \quad k_0(\alpha,t) = \alpha(1+t)^{1-\alpha}, \quad \alpha \in [0,1], \quad t \in \mathbb{T},$$

$$f(t) = t^2, \quad t \in \mathbb{T}.$$

Find

   1. $D^{\frac{1}{6}} f(t),$

   2. $D^{\frac{1}{6}} \left(D^{\frac{1}{6}} f \right) (t),$

   3. $D^{\frac{1}{6}} \left(D^{\frac{1}{6}} \left(D^{\frac{1}{6}} f \right) \right) (t).$

Problem 1.11.4 *Let* $\mathbb{T} = 4\mathbb{Z}$,

$$k_1(\alpha,t) = (1-\alpha)\left(1+2t^2\right)^{\alpha}, \quad k_0(\alpha,t) = \alpha\left(1+2t^2\right)^{1-\alpha}, \quad \alpha \in (0,1], \quad t \in \mathbb{T},$$

$$f(t) = t^2 + t, \quad t \in \mathbb{T}.$$

Compute

$$\int_0^{40} f(s)\Delta_{\alpha,40}s \qquad \text{-}$$

for

$$\alpha\left(1+2t^2\right)^{1-\alpha} - 4(1-\alpha)\left(1+2t^2\right)^{\alpha} \neq 0, \quad \alpha \in (0,1], \quad t \in [0,40].$$

Problem 1.11.5 *Let* $\mathbb{T} = q^{\mathbb{N}_0}, q > 1$. *Find*

$$g_1(t,s), \quad g_2(t,s), \quad h_1(t,s), \quad h_2(t,s), \quad t,s \in \mathbb{T}.$$

Problem 1.11.6 *Let* $\mathbb{T} = 3^{\mathbb{N}_0}$,

$$k_1(\alpha,t) = (1-\alpha)t^{\alpha}, \quad k_0(\alpha,t) = \alpha t^{1-\alpha}, \quad \alpha \in (0,1], \quad t \in \mathbb{T},$$

$$u(t,s) = (s+t)^2 + t^4, \quad (t,s) \in \mathbb{T} \times \mathbb{T}.$$

Find

$$D_t^{\frac{1}{2}} u(t,s), \quad D_s^{\frac{1}{3}} u(t,s), \quad D_s^{\frac{1}{4}} u(t,s).$$

Problem 1.11.7 *Let* $\mathbb{T} = 3^{\mathbb{N}_0}$,

$$k_1(\alpha,t) = (1-\alpha)t^{2\alpha}, \quad k_0(\alpha,t) = \alpha t^{4(1-\alpha)}, \quad \alpha \in (0,1], \quad t \in \mathbb{T},$$

$$u(t,s) = 2t^2 + 3ts^2 + 4s^2,$$

$$v(t,s) = 3t^2 + 5s^2, \quad (t,s) \in \mathbb{T} \times \mathbb{T}.$$

Find

$$D_t^{\frac{1}{3}} \left(\frac{u}{v} \right)(t,s), \quad (t,s) \in \mathbb{T} \times \mathbb{T}.$$

1.12 NOTES AND REFERENCES

This chapter introduces the concepts of conformable delta (Hilger) and nabla derivatives on time scales, and some of their properties. Results in this chapter include the basic conformable delta derivative, the conformable exponential function, the conformable logarithm function, conformable trigonometric and hyperbolic functions, the conformable delta integral and integral rules and Taylor's formula.

CHAPTER 2

First-Order Linear Dynamic Equations

Suppose that $\mathbb{T}$ is a time scale with forward jump operator and delta differentiation operator σ and Δ, respectively. Let $\alpha \in (0,1]$, k_0 and k_1 satisfy $(A1)$, and assume (1.6) holds.

2.1 LINEAR FIRST-ORDER DYNAMIC EQUATIONS

Let $p, q \in \mathscr{C}_{rd}(\mathbb{T})$. Consider the IVP

$$D^{\alpha}y = (p(t) + k_1(\alpha,t))y + q(t), \quad t \in \mathbb{T}^{\kappa}, \tag{2.1}$$

$$y(t_0) = y_0, \tag{2.2}$$

where $t_0 \in \mathbb{T}$, $y_0 \in \mathbb{R}$.

Theorem 2.1.1 *Suppose that*

$$k_0(\alpha,t) + \mu(t)p(t) \neq 0, \quad \alpha \in (0,1], \quad t \in \mathbb{T}.$$

Then the problem (2.1), (2.2) *has a unique solution represented in the form*

$$y(t) = y_0 E_{p+k_1}(t,t_0) + \int_{t_0}^{t} q(s)E_g(\sigma(s),t)\Delta_{\alpha,t}s, \quad s,t \in \mathbb{T}^{\kappa}, \tag{2.3}$$

where

$$g = \frac{p(\mu k_1 - k_0)}{k_0 + \mu p}.$$

Proof 2.1.2 *Let $t \in \mathbb{T}^{\kappa}$. We multiply both sides of equation* (2.1) *by $E_g(\sigma(t),t_0)$ to get*

$$(D^{\alpha}y(t))E_g(\sigma(t),t_0) = (p(t)+k_1(\alpha,t))y(t)E_g(\sigma(t),t_0) + q(t)E_g(\sigma(t),t_0). \tag{2.4}$$

Note that

$$p(t)E_g(\sigma(t),t_0) = p(t)\left(1 + \mu(t)\frac{g(t)-k_1(\alpha,t)}{k_0(\alpha,t)}\right)E_g(t,t_0)$$

$$= p(t)\left(1+\mu(t)\frac{\frac{p(t)(\mu(t)k_1(\alpha,t)-k_0(\alpha,t))}{k_0(\alpha,t)+\mu(t)p(t)}-k_1(\alpha,t)}{k_0(\alpha,t)}\right)E_g(t,t_0)$$

$$= p(t)\left(1+\mu(t)\frac{\mu(t)p(t)k_1(\alpha,t)-p(t)k_0(\alpha,t)-k_0(\alpha,t)k_1(\alpha,t)-\mu(t)p(t)k_1(\alpha,t)}{k_0(\alpha,t)\left(k_0(\alpha,t)+\mu(t)p(t)\right)}\right)$$

$$\times E_g(t,t_0)$$

$$= p(t)\frac{(k_0(\alpha,t))^2+\mu(t)k_0(\alpha,t)p(t)-\mu(t)k_0(\alpha,t)p(t)-\mu(t)k_0(\alpha,t)k_1(\alpha,t)}{k_0(\alpha,t)\left(k_0(\alpha,t)+\mu(t)p(t)\right)}E_g(t,t_0)$$

$$= p(t)\frac{k_0(\alpha,t)-\mu(t)k_1(\alpha,t)}{k_0(\alpha,t)+\mu(t)p(t)}E_g(t,t_0)$$

$$= -g(t)E_g(t,t_0)$$

$$= -D^\alpha E_g(t,t_0),\quad t\in\mathbb{T}^\kappa.$$

Then (2.4) takes the form

$$(D^\alpha y(t))E_g(\sigma(t),t_0)+(D^\alpha E_g(t,t_0))y(t)-k_1(\alpha,t)E_g(\sigma(t),t_0)y(t) = q(t)E_g(\sigma(t),t_0),$$

$t\in\mathbb{T}^\kappa$, *or*

$$D^\alpha\left(yE_g(\cdot,t_0)\right)(t)=q(t)E_g(\sigma(t),t_0),\quad t\in\mathbb{T}^\kappa.$$

We integrate the last equality from t_0 to t and we get

$$\int_{t_0}^t D^\alpha\left(yE_g(\cdot,t_0)\right)(s)\Delta_{\alpha,t}s=\int_{t_0}^t q(s)E_g(\sigma(s),t_0)\Delta_{\alpha,t}s,\quad t\in\mathbb{T}^\kappa,$$

or

$$y(t)E_g(t,t_0)-y(t_0)E_g(t_0,t_0)E_0(t,t_0)=\int_{t_0}^t q(s)E_g(\sigma(s),t_0)\Delta_{\alpha,t}s,\quad t\in\mathbb{T}^\kappa,$$

or

$$y(t)E_g(t,t_0)=y_0E_0(t,t_0)+\int_{t_0}^t q(s)E_g(\sigma(s),t_0)\Delta_{\alpha,t}s,\quad t\in\mathbb{T}^\kappa,$$

or

$$y(t) = y_0\frac{E_0(t,t_0)}{E_g(t,t_0)}+\int_{t_0}^t q(s)E_g(\sigma(s),t_0)E_g(t_0,t)\Delta_{\alpha,t}s$$

$$= y_0E_{0\ominus_c g}(t,t_0)+\int_{t_0}^t q(s)E_g(\sigma(s),t)\Delta_{\alpha,t}s,\quad t\in\mathbb{T}^\kappa.$$

We have

$$g-k_1 = \frac{p(\mu k_1-k_0)}{k_0+\mu p}-k_1$$

$$= \frac{p(\mu k_1 - k_0) - k_0 k_1 - \mu p k_1}{k_0 + \mu p}$$

$$= \frac{\mu p k_1 - p k_0 - k_0 k_1 - \mu p k_1}{k_0 + \mu p}$$

$$= \frac{-p k_0 - k_0 k_1}{k_0 + \mu p},$$

$$\ominus_c g = -\frac{k_0 (g - k_1)}{k_0 + \mu (g - k_1)} + k_1,$$

$$\ominus_c g - k_1 = -\frac{k_0 (g - k_1)}{k_0 + \mu (g - k_1)}$$

$$= \frac{\frac{k_0^2 (p + k_1)}{k_0 + \mu p}}{k_0 - \frac{\mu k_0 (p + k_1)}{k_0 + \mu p}}$$

$$= \frac{k_0 (p + k_1)}{k_0 + \mu p - \mu p - \mu k_1}$$

$$= \frac{k_0 (p + k_1)}{k_0 - \mu k_1},$$

$$0 \ominus_c g = 0 \oplus_c (\ominus_c g)$$

$$= 0 + (\ominus_c g) - k_1 + \mu \frac{(0 - k_1)(\ominus_c g - k_1)}{k_0}$$

$$= (\ominus_c g) - k_1 - \frac{\mu k_1}{k_0}(\ominus_c g - k_1)$$

$$= (\ominus_c g - k_1)\frac{k_0 - \mu k_1}{k_0}$$

$$= \frac{k_0 (p + k_1)}{k_0 - \mu k_1}\frac{k_0 - \mu k_1}{k_0}$$

$$= p + k_1.$$

Consequently,

$$y(t) = y_0 E_{p+k_1}(t, t_0) + \int_{t_0}^{t} q(s) E_g(\sigma(s), t) \Delta_{\alpha, t} s, \quad t \in \mathbb{T}^\kappa.$$

We differentiate with respect to t the last equality and we get

$$D^\alpha y(t) = y_0 D^\alpha \left(E_{p+k_1}(\cdot, t_0)\right)(t)$$

$$+ D^\alpha \left(\frac{1}{E_g(t,t_0)} \int_{t_0}^t q(s) E_g(\sigma(s),t_0) \Delta_{\alpha,t} s \right)$$

$$= (p(t) + k_1(\alpha,t)) y_0 E_{p+k_1}(t,t_0)$$

$$+ D^\alpha \left(\frac{1}{E_g(t,t_0)} \right) \int_{t_0}^t q(s) E_g(\sigma(s),t_0) \Delta_{\alpha,t} s$$

$$+ \frac{1}{E_g(\sigma(t),t_0)} q(t) E_g(\sigma(t),t_0)$$

$$- k_1(\alpha,t) \frac{1}{E_g(\sigma(t),t_0)} \int_{t_0}^t q(s) E_g(\sigma(s),t_0) \Delta_{\alpha,t} s$$

$$= (p(t) + k_1(\alpha,t)) y_0 E_{p+k_1}(t,t_0) + q(t)$$

$$+ \frac{E_g(t,t_0) k_1(\alpha,t) - g(t) E_g(t,t_0)}{E_g(t,t_0) E_g(\sigma(t),t_0)} \int_{t_0}^t q(s) E_g(\sigma(s),t_0) \Delta_{\alpha,t} s$$

$$+ k_1(\alpha,t) \frac{1}{E_g(t,t_0)} \int_{t_0}^t q(s) E_g(\sigma(s),t_0) \Delta_{\alpha,t} s$$

$$- \frac{k_1(\alpha,t)}{E_g(\sigma(t),t_0)} \int_{t_0}^t q(s) E_g(\sigma(s),t_0) \Delta_{\alpha,t} s$$

$$= (p(t) + k_1(\alpha,t)) y_0 E_{p+k_1}(t,t_0) + q(t)$$

$$- \frac{g(t)}{E_g(\sigma(t),t_0)} \int_{t_0}^t q(s) E_g(\sigma(s),t_0) \Delta_{\alpha,t} s$$

$$+ k_1(\alpha,t) \frac{1}{E_g(t,t_0)} \int_{t_0}^t q(s) E_g(\sigma(s),t_0) \Delta_{\alpha,t} s$$

$$= (p(t) + k_1(\alpha,t)) y_0 E_{p+k_1}(t,t_0) + q(t)$$

$$+ k_1(\alpha,t) \frac{1}{E_g(t,t_0)} \int_{t_0}^t q(s) E_g(\sigma(s),t_0) \Delta_{\alpha,t} s$$

$$- \frac{g(t)}{\left(1 + \mu(t) \frac{g(t) - k_1(\alpha,t)}{k_0(\alpha,t)}\right) E_g(t,t_0)} \int_{t_0}^t q(s) E_g(\sigma(s),t_0) \Delta_{\alpha,t} s$$

$$= (p(t) + k_1(\alpha,t)) y_0 E_{p+k_1}(t,t_0) + q(t)$$

$$+ k_1(\alpha,t) \frac{1}{E_g(t,t_0)} \int_{t_0}^t q(s) E_g(\sigma(s),t_0) \Delta_{\alpha,t} s$$

$$-\frac{g(t)}{1+\mu(t)\frac{g(t)-k_1(\alpha,t)}{k_0(\alpha,t)}}\int_{t_0}^{t}q(s)E_g(\sigma(s),t)\Delta_{\alpha,t}s$$

$$= (p(t)+k_1(\alpha,t))y_0E_{p+k_1}(t,t_0)+q(t)$$

$$+k_1(\alpha,t)\frac{1}{E_g(t,t_0)}\int_{t_0}^{t}q(s)E_g(\sigma(s),t_0)\Delta_{\alpha,t}s$$

$$-\frac{g(t)k_0(\alpha,t)}{k_0(\alpha,t)+\mu(t)(g(t)-k_1(\alpha,t))}\int_{t_0}^{t}q(s)E_g(\sigma(s),t)\Delta_{\alpha,t}s, \quad t\in\mathbb{T}^\kappa.$$

Note that

$$-\frac{gk_0}{k_0+\mu(g-k_1)} = -\frac{\frac{p(\mu k_1-k_0)}{k_0+\mu p}k_0}{k_0-\mu\frac{k_0(p+k_1)}{k_0+\mu p}}$$

$$= -\frac{\frac{p(\mu k_1-k_0)}{k_0+\mu p}}{1-\frac{\mu(p+k_1)}{k_0+\mu p}}$$

$$= -\frac{p(\mu k_1-k_0)}{k_0-\mu k_1}$$

$$= p.$$

Therefore

$$D^\alpha y(t) = (p(t)+k_1(\alpha,t))y_0E_{p+k_1}(t,t_0)+q(t)$$

$$+p(t)\int_{t_0}^{t}q(s)E_g(\sigma(s),t)\Delta_{\alpha,t}s$$

$$+k_1(\alpha,t)\frac{1}{E_g(t,t_0)}\int_{t_0}^{t}q(s)E_g(\sigma(s),t_0)\Delta_{\alpha,t}s$$

$$= (p(t)+k_1(\alpha,t))y(t)+q(t),, \quad t\in\mathbb{T}^\kappa.$$

Assume that the IVP (2.1), (2.2) has two solutions y_1 and y_2. We set

$$v=y_1-y_2.$$

Then v is a solution of the IVP

$$D^\alpha v = (p(t)+k_1(\alpha,t))v,$$

$$v(t_0) = 0.$$

Hence, and considering (2.3), we get that

$$v = 0 \quad on \quad \mathbb{T},$$

i.e.,

$$y_1 = y_2 \quad on \quad \mathbb{T}.$$

This completes the proof. □

Remark 2.1.3 *Let $p \in \mathscr{C}_{rd}(\mathbb{T}) \bigcap \mathscr{R}_c$ and $q \in \mathscr{C}_{rd}(\mathbb{T})$. Consider the IVP*

$$
\begin{aligned}
D^\alpha y &= p(t)y + q(t), \quad t \in \mathbb{T}^\kappa, \\
y(t_0) &= y_0,
\end{aligned}
\tag{2.5}
$$

where $t_0 \in \mathbb{T}$, $y_0 \in \mathbb{R}$. We can rewrite the equation in the form

$$D^\alpha y = (p(t) - k_1(\alpha,t) + k_1(\alpha,t))y + q(t), \quad t \in \mathbb{T}^\kappa.$$

Because $p \in \mathscr{R}_c$, we have

$$1 + \mu(t)(p(t) - k_1(\alpha,t)) \neq 0, \quad \alpha \in (0,1], \quad t \in \mathbb{T}^\kappa.$$

Therefore, the solution y of the IVP (2.5) can be represented in the form

$$y(t) = y_0 E_p(t,t_0) + \int_{t_0}^t q(s) E_g(\sigma(s),t) \Delta_{\alpha,t} s, \quad t \in \mathbb{T}^\kappa,$$

where

$$g = \frac{(p - k_1)(\mu k_1 - k_0)}{k_0 + \mu(p - k_1)}.$$

Example 2.1.4 *Let $\mathbb{T} = 2^{\mathbb{N}_0}$,*

$$k_1(\alpha,t) = (1-\alpha)t^{\frac{4}{3}\alpha}, \quad k_0(\alpha,t) = \alpha t^{\frac{8}{3}\alpha}, \quad \alpha \in (0,1], \quad t \in \mathbb{T}.$$

Consider the IVP

$$
\begin{aligned}
D^{\frac{3}{4}} y &= \frac{1}{2} t y + t^2 + 1, \quad t \in \mathbb{T}, \\
y(1) &= 2.
\end{aligned}
$$

Here

$$
\begin{aligned}
\sigma(t) &= 2t, \\
\mu(t) &= t, \\
p(t) &= \frac{1}{2}t,
\end{aligned}
$$

$$q(t) \quad = \quad t^2 + 1, \quad t \in \mathbb{T}.$$

Note that

$$
\begin{aligned}
g - k_1 &= \frac{(p - k_1)(\mu k_1 - k_0)}{k_0 + \mu(p - k_1)} - k_1 \\
&= \frac{\mu k_1(p - k_1) - k_0(p - k_1) - k_0 k_1 - \mu k_1(p - k_1)}{k_0 + \mu(p - k_1)} \\
&= -\frac{p k_0}{k_0 + \mu(p - k_1)}, \\
\frac{g - k_1}{k_0} &= -\frac{p}{k_0 + \mu(p - k_1)}, \\
1 + \mu\frac{g - k_1}{k_0} &= 1 - \frac{\mu p}{k_0 + \mu(p - k_1)} \\
&= \frac{k_0 - \mu k_1}{k_0 + \mu(p - k_1)}.
\end{aligned}
$$

Then

$$
\begin{aligned}
1 + \mu(t)\frac{g(t) - k_1\left(\frac{3}{4}, t\right)}{k_0\left(\frac{3}{4}, t\right)} &= \frac{k_0\left(\frac{3}{4}, t\right) - \mu(t)k_1\left(\frac{3}{4}, t\right)}{k_0\left(\frac{3}{4}, t\right) + \mu(t)\left(p(t) - k_1\left(\frac{3}{4}, t\right)\right)} \\
&= \frac{\frac{3}{4}t^2 - t\left(\frac{1}{4}t\right)}{\frac{3}{4}t^2 + t\left(\frac{1}{2}t - \frac{1}{4}t\right)} \\
&= \frac{\frac{3}{4}t^2 - \frac{1}{4}t^2}{\frac{3}{4}t^2 - \frac{1}{4}t^2} \\
&= 1, \quad t \in \mathbb{T}.
\end{aligned}
$$

Hence,

$$
\begin{aligned}
E_g(\sigma(s), t) &= e_{\frac{g - k_1}{k_0}}(\sigma(s), t) \\
&= e^{\int_t^{\sigma(s)} \frac{1}{\mu(\tau)} \log\left(1 + \mu(\tau)\frac{g(\tau) - k_1\left(\frac{3}{4}, \tau\right)}{k_0\left(\frac{3}{4}, \tau\right)}\right)\Delta\tau} \\
&= e^{\int_t^{\sigma(s)} \frac{1}{\tau} \log 1 \Delta\tau} \\
&= 1, \quad s, t \in \mathbb{T},
\end{aligned}
$$

$$-\frac{k_1\left(\frac{3}{4},t\right)}{k_0\left(\frac{3}{4},t\right)} = -\frac{\frac{1}{4}t}{\frac{3}{4}t^2} = -\frac{1}{3t}, \quad t \in \mathbb{T},$$

and

$$E_0(t,\sigma(s)) = e_{-\frac{k_1}{k_0}}(t,\sigma(s))$$

$$= e^{\int_{\sigma(s)}^{t} \frac{1}{\mu(\tau)} \log\left(1-\mu(\tau)\frac{k_1\left(\frac{3}{4},\tau\right)}{k_0\left(\frac{3}{4},\tau\right)}\right)\Delta\tau}$$

$$= e^{\int_{\sigma(s)}^{t} \frac{1}{\tau} \log\left(1-\frac{1}{3}\right)\Delta\tau}$$

$$= e^{\left(\log\frac{2}{3}\right)\left(-\int_{s}^{\sigma(s)} \frac{1}{\tau}\Delta\tau + \int_{s}^{t} \frac{1}{\tau}\Delta\tau\right)}$$

$$= e^{\left(\log\frac{2}{3}\right)\left(-1+\Sigma_{l=s}^{\frac{t}{2}}1\right)}, \quad t,s \in \mathbb{T}, \quad s \leq \frac{t}{2},$$

and

$$\frac{p(t)-k_1\left(\frac{3}{4},t\right)}{k_0\left(\frac{3}{4},t\right)} = \frac{\frac{1}{2}t-\frac{1}{4}t}{\frac{3}{4}t^2} = \frac{\frac{1}{4}t}{\frac{3}{4}t^2} = \frac{1}{3t},$$

$$E_p(t,t_0) = e_{\frac{p-k_1}{k_0}}(t,t_0)$$

$$= e^{\int_{t_0}^{t} \frac{1}{\mu(\tau)} \log\left(1+\mu(\tau)\frac{p(\tau)-k_1\left(\frac{3}{4},\tau\right)}{k_0\left(\frac{3}{4},\tau\right)}\right)\Delta\tau}$$

$$= e^{\int_{t_0}^{t} \frac{1}{\tau} \log\left(1+\frac{1}{3}\right)\Delta\tau}$$

$$= e^{\left(\log\frac{4}{3}\right)\int_{t_0}^{t} \frac{1}{\tau}\Delta\tau}$$

$$= e^{\left(\log\frac{4}{3}\right)\Sigma_{l=t_0}^{\frac{t}{2}}1}, \quad t_0 \leq \frac{t}{2}, \quad t_0,t \in \mathbb{T}.$$

Therefore

$$y(t) = y(1)E_p(t,1) + \int_1^t q(s)E_g(\sigma(s),t)\Delta_{\alpha,t}s$$

$$= 2E_p(t,1) + \int_1^t q(s)E_g(\sigma(s),t)\frac{E_0(t,\sigma(s))}{k_0\left(\frac{3}{4},s\right)}\Delta s$$

$$= 2E_p(t,1) + \frac{4}{3}\int_1^t \frac{s^2+1}{s^2}E_g(\sigma(s),t)E_0(t,\sigma(s))\Delta s$$

$$= 2E_p(t,1) + \frac{4}{3}\sum_{s=1}^{\frac{t}{2}}\frac{s^2+1}{s}E_g(\sigma(s),t)E_0(t,\sigma(s))$$

$$= 2e^{\left(\log\frac{4}{3}\right)\Sigma_{l=1}^{\frac{t}{2}}1}$$

$$+\frac{4}{3}\sum_{s=1}^{\frac{t}{2}}\frac{s^2+1}{s}e^{\left(\log\frac{2}{3}\right)\left(-1+\Sigma_{l=s}^{\frac{t}{2}}1\right)}, \quad t\geq 2.$$

If $t = 2^k$, $k \in \mathbb{N}$, then

$$y\left(2^k\right) = 2e^{\left(\log\frac{4}{3}\right)\Sigma_{l=2^0}^{2^{k-1}}1}$$

$$+\frac{4}{3}\sum_{2^m=2^0}^{2^{k-1}}\frac{4^m+1}{2^m}e^{\left(\log\frac{2}{3}\right)\left(-1+\Sigma_{l=2^m}^{2^{k-1}}1\right)}$$

$$= 2e^{k\left(\log\frac{4}{3}\right)}+\frac{4}{3}\sum_{2^m=2^0}^{2^{k-1}}\frac{4^m+1}{2^m}e^{\left(\log\frac{2}{3}\right)(k-m-1)}$$

$$= 2\left(\frac{4}{3}\right)^k+\frac{4}{3}\sum_{2^m=2^0}^{2^{k-1}}\frac{4^m+1}{2^m}\left(\frac{2}{3}\right)^{k-m-1}.$$

This ends the example.

Now we consider the conformable dynamic equation

$$D^\alpha y = (-p(t)+k_1(\alpha,t))y^\sigma + q(t), \quad t\in\mathbb{T}^\kappa, \tag{2.6}$$

subject to the initial condition (2.2), where $p,q\in\mathscr{C}_{rd}(\mathbb{T})$, $p\in\mathscr{R}_c$, $t_0\in\mathbb{T}$, $y_0\in\mathbb{R}$.

Theorem 2.1.5 *The problem* (2.6), (2.2) *has a unique solution represented in the form*

$$y(t) = y_0 E_g(t,t_0) + \int_{t_0}^t q(s)E_p(s,t)\Delta_{\alpha,t}s, \quad t\in\mathbb{T}^\kappa, \tag{2.7}$$

where

$$g(t) = -\frac{(p(t)-k_1(\alpha,t))(k_0(\alpha,t)-\mu(t)k_1(\alpha,t))}{k_0(\alpha,t)+\mu(t)(p(t)-k_1(\alpha,t))}, \quad t\in\mathbb{T}.$$

Proof 2.1.6 *We multiply both sides of equation* (2.6) *by $E_p(t,t_0)$ to get*

$$(D^\alpha y(t))E_p(t,t_0)+p(t)E_p(t,t_0)y^\sigma(t)-k_1(\alpha,t)E_p(t,t_0)y^\sigma(t)=q(t)E_p(t,t_0), \quad t\in\mathbb{T}^\kappa,$$

or

$$D^\alpha\left(yE_p(\cdot,t_0)\right)(t)=q(t)E_p(t,t_0), \quad t\in\mathbb{T}^\kappa.$$

The last equality we integrate from t_0 to t and we obtain

$$y(t)E_p(t,t_0) - y(t_0)E_p(t_0,t_0)E_0(t,t_0) = \int_{t_0}^t q(s)E_p(s,t_0)\Delta_{\alpha,t}s, \quad t \in \mathbb{T}^\kappa,$$

or

$$y(t)E_p(t,t_0) = y_0 E_0(t,t_0) + \int_{t_0}^t q(s)E_p(s,t_0)\Delta_{\alpha,t}s, \quad t \in \mathbb{T}^\kappa,$$

or

$$y(t) = y_0\frac{E_0(t,t_0)}{E_p(t,t_0)} + \int_{t_0}^t q(s)E_p(s,t_0)E_p(t_0,t)\Delta_{\alpha,t}s, \quad t \in \mathbb{T}^\kappa,$$

or

$$y(t) = y_0 E_{0 \ominus_c p}(t,t_0) + \int_{t_0}^t q(s)E_p(s,t)\Delta_{\alpha,t}s, \quad t \in \mathbb{T}^\kappa.$$

Note that

$$\ominus_c p = -\frac{k_0(p-k_1)}{k_0 + \mu(p-k_1)} + k_1,$$

$$\ominus_c p - k_1 = -\frac{k_0(p-k_1)}{k_0 + \mu(p-k_1)},$$

$$0 \oplus_c (\ominus_c p) = 0 + (\ominus_c p) - k_1 + \mu\frac{(0-k_1)(\ominus_c p - k_1)}{k_0}$$

$$= (\ominus_c p - k_1) - \mu k_1\frac{(\ominus_c p - k_1)}{k_0}$$

$$= (\ominus_c p - k_1)\frac{k_0 - \mu k_1}{k_0}$$

$$= -\frac{(p-k_1)(k_0 - \mu k_1)}{k_0 + \mu(p-k_1)}$$

$$= g.$$

Therefore

$$y(t) = y_0 E_g(t,t_0) + \int_{t_0}^t q(s)E_p(s,t)\Delta_{\alpha,t}s, \quad t \in \mathbb{T}^\kappa.$$

Now we differentiate with respect to t both sides of the last equality and we obtain

$$D^\alpha y(t) = y_0 D^\alpha E_g(t,t_0) + D^\alpha\left(\frac{1}{E_p(t,t_0)}\int_{t_0}^t q(s)E_p(s,t_0)\Delta_{\alpha,t}s\right)$$

$$= y_0 g(t)E_g(t,t_0)$$

$$+ D^\alpha\left(\frac{1}{E_p(t,t_0)}\right)\left(\int_{t_0}^{\sigma(t)} q(s)E_p(s,t_0)\Delta_{\alpha,t}s\right)$$

$$+\frac{1}{E_p(t,t_0)}D^\alpha\left(\int_{t_0}^t q(s)E_p(s,t_0)\Delta_{\alpha,t}s\right)$$

$$-\frac{k_1(\alpha,t)}{E_p(t,t_0)}\int_{t_0}^{\sigma(t)} q(s)E_p(s,t_0)\Delta_{\alpha,t}s$$

$$= y_0g(t)E_g(t,t_0)$$

$$+\left(\frac{k_1(\alpha,t)E_p(t,t_0)-p(t)E_p(t,t_0)}{E_p(t,t_0)E_p(\sigma(t),t_0)}+\frac{k_1(\alpha,t)}{E_p(t,t_0)}\right)\left(\int_{t_0}^{\sigma(t)} q(s)E_p(s,t_0)\Delta_{\alpha,t}s\right)$$

$$+\frac{q(t)E_p(t,t_0)}{E_p(t,t_0)}-\frac{k_1(\alpha,t)}{E_p(t,t_0)}\int_{t_0}^{\sigma(t)} q(s)E_p(s,t_0)\Delta_{\alpha,t}s$$

$$= y_0g(t)E_g(t,t_0)+q(t)$$

$$+\frac{k_1(\alpha,t)-p(t)}{E_p(\sigma(t),t_0)}\int_{t_0}^{\sigma(t)} q(s)E_p(s,t_0)\Delta_{\alpha,t}s$$

$$= y_0g(t)\frac{k_0(\alpha,t)}{k_0(\alpha,t)+\mu(t)(g(t)-k_1(\alpha,t))}E_g(\sigma(t),t_0)+q(t)$$

$$+\frac{k_1(\alpha,t)-p(t)}{E_p(\sigma(t),t_0)}\int_{t_0}^{\sigma(t)} q(s)E_p(s,t_0)\Delta_{\alpha,t}s$$

$$= y_0g(t)\frac{k_0(\alpha,t)}{k_0(\alpha,t)+\mu(t)(g(t)-k_1(\alpha,t))}E_g(\sigma(t),t_0)+q(t)$$

$$+(k_1(\alpha,t)-p(t))\int_{t_0}^{\sigma(t)} q(s)E_p(s,\sigma(t))\Delta_{\alpha,t}s, \quad t\in\mathbb{T}^\kappa.$$

Note that

$$g-k_1 = -\frac{(p-k_1)(k_0-\mu k_1)}{k_0+\mu(p-k_1)}-k_1$$

$$= -\frac{k_0(p-k_1)-\mu k_1(p-k_1)+k_1k_0+\mu k_1(p-k_1)}{k_0+\mu(p-k_1)}$$

$$= -\frac{k_0p}{k_0+\mu(p-k_1)},$$

$$k_0+\mu(g-k_1) = k_0-\frac{\mu k_0 p}{k_0+\mu(p-k_1)}$$

$$= \frac{k_0(k_0+\mu p-\mu k_1-\mu p)}{k_0+\mu(p-k_1)}$$

$$= \frac{k_0(k_0-\mu k_1)}{k_0+\mu(p-k_1)},$$

$$\frac{gk_0}{k_0 + \mu(g - k_1)} = -\frac{\frac{k_0(p-k_1)(k_0-\mu k_1)}{k_0+\mu(p-k_1)}}{\frac{k_0(k_0-\mu k_1)}{k_0+\mu(p-k_1)}}$$

$$= k_1 - p.$$

Therefore

$$D^\alpha y(t) = y_0(k_1(\alpha,t) - p(t))E_g(\sigma(t), t_0) + q(t)$$

$$+ (k_1(\alpha,t) - p(t)) \int_{t_0}^{\sigma(t)} q(s)E_p(s, \sigma(t))\Delta_{\alpha,t}s$$

$$= (k_1(\alpha,t) - p(t))y^\sigma(t) + q(t), \quad t \in \mathbb{T}^\kappa.$$

Assume that the IVP (2.6), (2.2) has two solutions y_1 and y_2. Let

$$v = y_1 - y_2.$$

Then v solves the IVP

$$D^\alpha v = (-p(t) + k_1(\alpha,t))v^\sigma, \quad t \in \mathbb{T}^\kappa,$$

$$v(t_0) = 0.$$

Hence, and considering (2.7), we get

$$v = 0 \quad on \quad \mathbb{T}^\kappa,$$

i.e.,

$$y_1 = y_2 \quad on \quad \mathbb{T}^\kappa.$$

This completes the proof. □

Remark 2.1.7 *Consider the IVP*

$$D^\alpha y = p(t)y^\sigma + q(t), \quad t \in \mathbb{T}^\kappa,$$

$$(2.8)$$

$$y(t_0) = y_0.$$

Suppose that $t_0 \in \mathbb{T}$, $y_0 \in \mathbb{R}$, $p, q \in \mathscr{C}_{rd}(\mathbb{T})$ and

$$k_0(\alpha,t) - \mu(t)p(t) \neq 0, \quad t \in \mathbb{T}, \quad \alpha \in (0,1].$$

The IVP (2.8) can be rewritten in the form

$$D^\alpha y = -((-p(t) + k_1(\alpha,t)) - k_1(\alpha,t))y^\sigma + q(t), \quad t \in \mathbb{T}^\kappa,$$

$$y(t_0) = y_0.$$

Note that

$$g = -\frac{(-p+k_1-k_1)(k_0-\mu k_1)}{k_0+\mu(-p+k_1-k_1)} = \frac{p(k_0-\mu k_1)}{k_0-\mu p}.$$

Therefore the solution of the IVP (2.8) can be represented in the form

$$y(t) = y_0 E_g(t,t_0) + \int_{t_0}^{t} q(s)E_{-p+k_1}(s,t)\Delta_{\alpha,t}s, \quad t \in \mathbb{T}^\kappa.$$

Example 2.1.8 *Let* $\mathbb{T} = \left(\frac{1}{2}\mathbb{N}\right)\bigcup\{0\}$ *and*

$$k_1(\alpha,t) = (1-\alpha)t^{4\alpha}, \quad k_0(\alpha,t) = \alpha t^{4(1-\alpha)}, \quad t \in \mathbb{T}, \quad \alpha \in (0,1].$$

Consider the IVP

$$D^{\frac{1}{2}}y = \frac{1}{8}t^2 y^\sigma + t^3, \quad t \in \mathbb{T},$$

$$y(0) = 1.$$

Here

$$\sigma(t) = t+\frac{1}{2},$$

$$\mu(t) = \frac{1}{2},$$

$$k_1\left(\frac{1}{2},t\right) = \frac{1}{2}t^2,$$

$$k_0\left(\frac{1}{2},t\right) = \frac{1}{2}t^2,$$

$$p(t) = \frac{1}{8}t^2,$$

$$q(t) = t^3, \quad t \in \mathbb{T}.$$

Then

$$g(t) = \frac{p(t)\left(k_0\left(\frac{1}{2},t\right)-\mu(t)k_1\left(\frac{1}{2},t\right)\right)}{k_0\left(\frac{1}{2},t\right)-\mu(t)p(t)}$$

$$= \frac{\frac{1}{8}t^2\left(\frac{1}{2}t^2-\frac{1}{2}\left(\frac{1}{2}t^2\right)\right)}{\frac{1}{2}t^2-\frac{1}{2}\left(\frac{1}{8}t^2\right)}$$

$$= \frac{\frac{1}{8}t^2 \left(\frac{1}{2}t^2 - \frac{1}{4}t^2\right)}{\frac{1}{2}t^2 - \frac{1}{16}t^2}$$

$$= \frac{\frac{1}{8}t^2 \left(\frac{1}{4}t^2\right)}{\frac{7}{16}t^2}$$

$$= \frac{1}{14}t^2,$$

$$\frac{g(t) - k_1\left(\frac{1}{2},t\right)}{k_0\left(\frac{1}{2},t\right)} = \frac{\frac{1}{14}t^2 - \frac{1}{2}t^2}{\frac{1}{2}t^2} = \frac{-\frac{6}{14}t^2}{\frac{1}{2}t^2} = -\frac{6}{7},$$

$$1 + \mu(t)\frac{g(t) - k_1\left(\frac{1}{2},t\right)}{k_0\left(\frac{1}{2},t\right)} = 1 - \frac{1}{2}\left(\frac{6}{7}\right) = 1 - \frac{3}{7} = \frac{4}{7},$$

$$\frac{-p(t) + k_1\left(\frac{1}{2},t\right) - k_1\left(\frac{1}{2},t\right)}{k_0\left(\frac{1}{2},t\right)} = -\frac{p(t)}{k_0\left(\frac{1}{2},t\right)} = -\frac{\frac{1}{8}t^2}{\frac{1}{2}t^2} = -\frac{1}{4},$$

$$1 + \mu(t)\frac{-p(t) + k_1\left(\frac{1}{2},t\right) - k_1\left(\frac{1}{2},t\right)}{k_0\left(\frac{1}{2},t\right)} = 1 - \frac{1}{2}\left(\frac{1}{4}\right) = 1 - \frac{1}{8} = \frac{7}{8},$$

$$-\frac{k_1\left(\frac{1}{2},t\right)}{k_0\left(\frac{1}{2},t\right)} = -\frac{\frac{1}{2}t^2}{\frac{1}{2}t^2} = -1,$$

$$1 - \mu(t)\frac{k_1\left(\frac{1}{2},t\right)}{k_0\left(\frac{1}{2},t\right)} = 1 - \frac{1}{2} = \frac{1}{2}, \quad t \in \mathbb{T}.$$

Hence,

$$E_g(t,0) = e_{\frac{g-k_1}{k_0}}(t,0)$$

$$= e^{\int_0^t \frac{1}{\mu(s)} \log\left(1 + \mu(s)\frac{g(s) - k_1\left(\frac{1}{2},s\right)}{k_0\left(\frac{1}{2},s\right)}\right)\Delta s}$$

$$= e^{\int_0^t 2\log\frac{4}{7}\Delta s}$$

$$= e^{\sum_{s=0}^{t-\frac{1}{2}} \log\frac{4}{7}}$$

$$= e^{2t\log\frac{4}{7}}$$

$$= \left(\frac{4}{7}\right)^{2t},$$

$$E_{-p+k_1}(s,t) = e_{-\frac{p}{k_0}}(s,t)$$

$$= e^{\int_t^s \frac{1}{\mu(y)} \log\left(1 - \mu(y)\frac{p(y)}{k_0\left(\frac{1}{2},y\right)}\right)\Delta y}$$

$$= e^{\int_t^s 2\log\left(1-\frac{1}{8}\right)\Delta y}$$

$$= e^{-\int_s^t 2\log\frac{7}{8}\Delta y}$$

$$= e^{-\sum_{l=s}^{t-\frac{1}{2}}\left(\log\frac{7}{8}\right)}$$

$$= e^{-2(t-s)\left(\log\frac{7}{8}\right)}$$

$$= \left(\frac{8}{7}\right)^{2(t-s)},$$

$$E_0(t,\sigma(s)) = e_{-\frac{k_1}{k_0}}(t,\sigma(s))$$

$$= e^{\int_{\sigma(s)}^t \frac{1}{\mu(y)} \log\left(1 - \mu(y)\frac{k_1\left(\frac{1}{2},y\right)}{k_0\left(\frac{1}{2},y\right)}\right)\Delta y}$$

$$= e^{\int_{\sigma(s)}^t 2\log\left(1-\frac{1}{2}\right)\Delta y}$$

$$= e^{\int_{\sigma(s)}^t \log\frac{1}{4}\Delta y}$$

$$= e^{-\int_s^{\sigma(s)} \log\frac{1}{4}\Delta y + \int_s^t \log\frac{1}{4}\Delta y}$$

$$= e^{-\log\frac{1}{2} + \sum_{l=s}^{t-\frac{1}{2}}\log\frac{1}{2}}$$

$$= e^{-\log\frac{1}{2} + 2(t-s)\log\frac{1}{2}}$$

$$= e^{\log 2 - \log 4^{t-s}}$$

$$= e^{\log\frac{2}{4^{t-s}}}$$

$$= \frac{2}{2^{2t-2s}}$$

$$= 2^{1+2s-2t}, \quad s,.t \in \mathbb{T}, \quad s \leq t.$$

Then the solution of the considered IVP

$$
\begin{aligned}
y(t) &= E_g(t,0) + \int_0^t q(s)E_{-p+k_1}(s,t)\Delta_{\alpha,t}s \\
&= E_g(t,0) + \int_0^t q(s)E_{-p+k_1}(s,t)\frac{E_0(t,\sigma(s))}{k_0\left(\frac{1}{2},s\right)}\Delta s \\
&= \left(\frac{4}{7}\right)^{2t} + \int_0^t s^3 \left(\frac{8}{7}\right)^{2(t-s)}\frac{1}{2^{2t-2s-1}}\frac{1}{\frac{1}{2}s^2}\Delta s \\
&= \left(\frac{4}{7}\right)^{2t} + 2\int_0^t s\frac{2^{6t-6s}}{7^{2(t-s)}}\frac{1}{2^{2t-2s-1}}\Delta s \\
&= \left(\frac{4}{7}\right)^{2t} + 2\int_0^t s\frac{2^{4t-4s+1}}{7^{2(t-s)}}\Delta s \\
&= \left(\frac{4}{7}\right)^{2t} + \sum_{s=0}^{t-\frac{1}{2}} s\frac{2^{4t-4s+1}}{7^{2(t-s)}} \\
&= \begin{cases}
\left(\dfrac{4}{7}\right)^{2t} & if \quad t = \dfrac{1}{2} \\[2em]
\left(\dfrac{4}{7}\right)^{2t} + \displaystyle\sum_{s=\frac{1}{2}}^{t-\frac{1}{2}} s\dfrac{2^{4t-4s+1}}{7^{2(t-s)}}, & t \geq 1, \quad t \in \mathbb{T}.
\end{cases}
\end{aligned}
$$

This ends the example.

2.2 CONFORMABLE BERNOULLI EQUATIONS

Consider the following IVP

$$
D^\alpha y = -p(t)y^\sigma + k_1(\alpha,t)y + q(t)yy^\sigma, \quad t \in \mathbb{T}^\kappa, \tag{2.9}
$$

$$
y(t_0) = y_0, \tag{2.10}
$$

where $p, q \in \mathscr{C}_{rd}(\mathbb{T})$,

$$
k_0(\alpha,t) + \mu(t)p(t) \neq 0, \quad \alpha \in (0,1], \quad t \in \mathbb{T},
$$

$t_0 \in \mathbb{T}$, $y_0 \in \mathbb{R}$.

Definition 2.2.1 *The equation* (2.9) *will be called the conformable Bernoulli equation.*

Suppose that $y_0 \neq 0$ and $y(t) \neq 0$, $y^\sigma(t) \neq 0$, for some $t \in \mathbb{T}^\kappa$, and y is a solution of the problem (2.9), (2.10). Then

$$
\frac{D^\alpha y}{yy^\sigma} = -p(t)\frac{1}{y} + k_1(\alpha,t)\frac{1}{y^\sigma} + q(t), \quad t \in \mathbb{T}^\kappa. \tag{2.11}
$$

Let
$$v = \frac{1}{y}.$$

Then
$$D^\alpha v = D^\alpha \left(\frac{1}{y} \right)$$
$$= \frac{k_1 y - D^\alpha y}{y y^\sigma} + \frac{k_1}{y}$$

and
$$\frac{D^\alpha y}{y y^\sigma} = \frac{k_1}{y^\sigma} + \frac{k_1}{y} - D^\alpha v \quad \text{on} \quad \mathbb{T}^\kappa.$$

From the equation (2.11), we get

$$\frac{k_1(\alpha,t)}{y^\sigma(t)} + \frac{k_1(\alpha,t)}{y(t)} - D^\alpha v(t) = -p(t)\frac{1}{y(t)} + \frac{k_1(\alpha,t)}{y^\sigma(t)} + q(t), \quad t \in \mathbb{T}^\kappa,$$

or
$$-D^\alpha v(t) = -(p(t) + k_1(\alpha,t))\frac{1}{y(t)} + q(t), \quad t \in \mathbb{T}^\kappa,$$

or
$$D^\alpha v(t) = (p(t) + k_1(\alpha,t)) v(t) - q(t), \quad t \in \mathbb{T}^\kappa.$$

Also,
$$v(t_0) = \frac{1}{y_0}.$$

By Theorem 2.1.1, we get that

$$v(t) = \frac{1}{y_0} E_{p+k_1}(t,t_0) - \int_{t_0}^t q(s) E_g(\sigma(s),t) \Delta_{\alpha,t} s,$$

where
$$g(t) = \frac{p(t)(\mu(t)k_1(\alpha,t) - k_0(\alpha,t))}{k_0(\alpha,t) + \mu(t)p(t)}, \quad t \in \mathbb{T}^\kappa.$$

Consequently,

$$y(t) = \frac{y_0}{E_{p+k_1}(t,t_0) - y_0 \int_{t_0}^t q(s) E_g(\sigma(s),t)\Delta_{\alpha,t}s}, \quad t \in \mathbb{T}^\kappa.$$

By the above computations, we conclude that when we search for a nontrivial solution of equation (2.9), we can reduce equation (2.9) to a first-order linear conformable dynamic equation.

Example 2.2.2 *Let* $\mathbb{T} = 3\mathbb{Z}$,

$$k_1(\alpha,t) = (1-\alpha)(1+t^2)^{3\alpha}, \quad k_0(\alpha,t) = \alpha(1+t^2)^{3(1-\alpha)}, \quad \alpha \in (0,1], \quad t \in \mathbb{T}.$$

Consider the equation

$$D^{\frac{1}{3}}y = -\frac{1}{3}ty^\sigma + \frac{2}{3}(1+t^2)y + tyy^\sigma, \quad t \in \mathbb{T}^\kappa.$$

Here

$$\sigma(t) = t+3,$$

$$\mu(t) = 3,$$

$$p(t) = \frac{1}{3}t,$$

$$q(t) = t,$$

$$k_1\left(\frac{1}{3},t\right) = \frac{2}{3}(1+t^2),$$

$$k_0\left(\frac{1}{3},t\right) = \frac{1}{3}(1+t^2)^2, \quad t \in \mathbb{T}.$$

Let y be a solution of the considered equation and $y(t) \neq 0$, $y^\sigma(t) \neq 0$. *Then, we get*

$$\frac{D^{\frac{1}{3}}y}{yy^\sigma} = -\frac{1}{3}t\frac{1}{y} + \frac{2}{3}(1+t^2)\frac{1}{y^\sigma} + t, \quad t \in \mathbb{T}^\kappa.$$

We set

$$v = \frac{1}{y}.$$

Hence,

$$\begin{aligned}
D^{\frac{1}{3}}v(t) &= D^{\frac{1}{3}}\left(\frac{1}{y}\right)(t) \\[2mm]
&= \frac{k_1\left(\frac{1}{3},t\right)y(t) - D^{\frac{1}{3}}y(t)}{y(t)y^\sigma(t)} + \frac{k_1\left(\frac{1}{3},t\right)}{y(t)} \\[2mm]
&= \frac{\frac{2}{3}(1+t^2)y(t) - D^{\frac{1}{3}}y(t)}{y(t)y^\sigma(t)} + \frac{\frac{2}{3}(1+t^2)}{y(t)} \\[2mm]
&= \frac{2}{3}(1+t^2)\frac{1}{y^\sigma(t)} - \frac{D^{\frac{1}{3}}y(t)}{y(t)y^\sigma(t)} \\[2mm]
&\quad + \frac{2}{3}(1+t^2)\frac{1}{y(t)}, \quad t \in \mathbb{T}^\kappa,
\end{aligned}$$

and

$$\frac{D^{\frac{1}{3}}y(t)}{y(t)y^\sigma(t)} = \frac{2}{3}(1+t^2)\frac{1}{y^\sigma(t)} - D^{\frac{1}{3}}v(t) + \frac{2}{3}(1+t^2)\frac{1}{y(t)}, \quad t \in \mathbb{T}^\kappa.$$

Therefore

$$\frac{2}{3}(1+t^2)\frac{1}{y^\sigma(t)} - D^{\frac{1}{3}}v(t) + \frac{2}{3}(1+t^2)\frac{1}{y(t)} = -\frac{1}{3}t\frac{1}{y(t)} + \frac{2}{3}(1+t^2)\frac{1}{y^\sigma(t)}, \quad t \in \mathbb{T}^\kappa,$$

or

$$-D^{\frac{1}{3}}v(t) = -\frac{1}{3}\left(2t^2+t+2\right)v(t)+t, \quad t \in \mathbb{T}^\kappa,$$

or

$$D^{\frac{1}{3}}v(t) = \frac{1}{3}\left(2t^2++2\right)v(t)-t, \quad t \in \mathbb{T}^\kappa.$$

Exercise 2.2.3 *Let* $\mathbb{T} = 2^{\mathbb{N}_0}$,

$$k_1(\alpha,t) = (1-\alpha)(1+t^4)^{4\alpha}, \quad k_0(\alpha,t) = \alpha(1+t^4)^{4(1-\alpha)}, \quad \alpha \in (0,1], \quad t \in \mathbb{T}.$$

Reduce the following conformable Bernoulli equations to first-order linear conformable dynamic equations.

1.
$$D^{\frac{1}{4}}y = 2y^\sigma + \frac{3}{4}(1+t^4)y + t^2yy^\sigma, \quad t \in \mathbb{T}^\kappa,$$

2.
$$D^{\frac{1}{2}}y = -3y^\sigma + \frac{1}{2}(1+t^4)^2y + (1+t+t^2)yy^\sigma, \quad t \in \mathbb{T}^\kappa,$$

3.
$$D^{\frac{1}{6}}y = (1+t^2)y^\sigma + \frac{1}{6}(1+t^4)^{\frac{2}{3}}y + \frac{t^2+6}{t^2+t+1}yy^\sigma, \quad t \in \mathbb{T}^\kappa.$$

This ends the exercise.

Now we consider the equation

$$D^\alpha y = p(t)y + k_1(\alpha,t)y^\sigma + q(t)yy^\sigma, \quad t \in \mathbb{T}^\kappa, \tag{2.12}$$

where p and q satisfy the conditions as above, subject to the initial condition (2.10). Suppose that $y_0 \neq 0$, $y(t) \neq 0$, $y^\sigma(t) \neq 0$ for some $t \in \mathbb{T}^\kappa$, and y is a solution of the equation (2.12). Then

$$\frac{D^\alpha y(t)}{y(t)y^\sigma(t)} = p(t)\frac{1}{y^\sigma(t)} + k_1(\alpha,t)\frac{1}{y(t)} + q(t),$$

or

$$-\frac{D^\alpha y(t)}{y(t)y^\sigma(t)} = -p(t)\frac{1}{y^\sigma(t)} - k_1(\alpha,t)\frac{1}{y(t)} - q(t),$$

or

$$\frac{k_1(\alpha,t)}{y^\sigma(t)} - \frac{D^\alpha y(t)}{y(t)y^\sigma(t)} + \frac{k_1(\alpha,t)}{y(t)} = (-p(t)+k_1(\alpha,t))\frac{1}{y^\sigma(t)} - q(t).$$

Let

$$v(t) = \frac{1}{y(t)}.$$

Thus we get

$$D^{\alpha}v(t) = (-p(t) + k_1(\alpha,t))v^{\sigma}(t) - q(t).$$

Hence, and considering Theorem 2.1.5, we obtain

$$v(t) = \frac{1}{y_0}E_g(t,t_0) - \int_{t_0}^t q(s)E_p(s,t)\Delta_{\alpha,t}s,$$

where

$$g(t) = -\frac{(p(t) - k_1(\alpha,t))(k_0(\alpha,t) - \mu(t)k_1(\alpha,t))}{k_0(\alpha,t) + \mu(t)(p(t) - k_1(\alpha,t))}.$$

Therefore

$$y(t) = \frac{y_0}{E_g(t,t_0) - y_0\int_{t_0}^t q(s)E_p(s,t)\Delta_{\alpha,t}s}.$$

Example 2.2.4 *Let* $\mathbb{T} = 2^{\mathbb{N}_0}$,

$$k_1(\alpha,t) = (1-\alpha)t^{2\alpha}, \quad k_0(\alpha,t) = \alpha t^{2(1-\alpha)}, \quad \alpha \in (0,1], \quad t \in \mathbb{T}.$$

Consider the equation

$$D^{\frac{1}{4}}y = (t^2+t)y + \frac{3}{4}t^{\frac{1}{2}}y^{\sigma} + (t^2+1)yy^{\sigma}, \quad t \in \mathbb{T}^{\kappa}.$$

Here

$$\sigma(t) = 2t,$$

$$\mu(t) = t,$$

$$p(t) = t^2+t,$$

$$q(t) = t^2+1, \quad t \in \mathbb{T},$$

$$\alpha = \frac{1}{4}.$$

Let y *be a solution of the considered equation,* $y(t) \neq 0$, $y^{\sigma}(t) \neq 0$ *for some* $t \in \mathbb{T}^{\kappa}$. *Then*

$$\frac{D^{\frac{1}{4}}y(t)}{y(t)y^{\sigma}(t)} = (t^2+t)\frac{1}{y^{\sigma}(t)} + \frac{3}{4}t^{\frac{1}{2}}\frac{1}{y(t)} + t^2+1.$$

We set

$$v(t) = \frac{1}{y(t)}.$$

Then

$$D^{\frac{1}{4}}v(t) = D^{\frac{1}{4}}\left(\frac{1}{y(t)}\right)$$

$$= \frac{k_1\left(\frac{1}{4},t\right)y(t) - D^{\frac{1}{4}}y(t)}{y(t)y^\sigma(t)} + \frac{k_1\left(\frac{1}{4},t\right)}{y(t)}$$

$$= \frac{\frac{3}{4}t^{\frac{1}{2}}y(t) - D^{\frac{1}{4}}y(t)}{y(t)y^\sigma(t)} + \frac{\frac{3}{4}t^{\frac{1}{2}}}{y(t)}$$

$$= \frac{3}{4}t^{\frac{1}{2}}\frac{1}{y^\sigma(t)} - \frac{D^{\frac{1}{4}}y(t)}{y(t)y^\sigma(t)} + \frac{3}{4}t^{\frac{1}{2}}v(t)$$

$$= \frac{3}{4}t^{\frac{1}{2}}v^\sigma(t) + \frac{3}{4}t^{\frac{1}{2}}v(t) - \frac{D^{\frac{1}{4}}y(t)}{y(t)y^\sigma(t)}.$$

Therefore

$$\frac{D^{\frac{1}{4}}y(t)}{y(t)y^\sigma(t)} = \frac{3}{4}t^{\frac{1}{2}}v^\sigma(t) + \frac{3}{4}t^{\frac{1}{2}}v(t) - D^{\frac{1}{4}}v(t)$$

and

$$\frac{3}{4}t^{\frac{1}{2}}v^\sigma(t) + \frac{3}{4}t^{\frac{1}{2}}v(t) - D^{\frac{1}{4}}v(t) = (t^2 + t)v^\sigma(t) + \frac{3}{4}t^{\frac{1}{2}}v(t) + t^2 + 1,$$

or

$$-D^{\frac{1}{4}}v(t) = \left(t^2 + t - \frac{3}{4}t^{\frac{1}{2}}\right)v^\sigma(t) + t^2 + 1,$$

or

$$D^{\frac{1}{4}}v(t) = \left(-t^2 - t + \frac{3}{4}t^{\frac{1}{2}}\right)v^\sigma(t) - t^2 - 1.$$

Exercise 2.2.5 *Let* $\mathbb{T} = \mathbb{Z}$,

$$k_1(\alpha,t) = (1-\alpha)(1+t^6)^\alpha, \quad k_0(\alpha,t) = \alpha(1+t^6)^{1-\alpha}, \quad \alpha \in (0,1], \quad t \in \mathbb{T}.$$

Reduce the following conformable Bernoulli's equations to first-order linear conformable dynamic equations.

1.

$$D^{\frac{1}{2}}y = ty + \frac{1}{2}(1+t^6)^{\frac{1}{2}}y^\sigma + yy^\sigma, \quad t \in \mathbb{T}^\kappa,$$

2.

$$D^{\frac{1}{3}}y = \left(1+t^2+t^4\right)y + \frac{2}{3}(1+t^6)^{\frac{1}{3}}y^\sigma + (t+1)yy^\sigma, \quad t \in \mathbb{T}^\kappa,$$

3.

$$D^{\frac{1}{4}}y = (1+t)y + \frac{3}{4}(1+t^6)^{\frac{1}{4}}y^\sigma + (t^2+1)yy^\sigma, \quad t \in \mathbb{T}^\kappa.$$

2.3 CONFORMABLE RICCATI EQUATIONS

Consider the equation

$$D^{\alpha}y = -p(t)y^{\sigma} + k_1(\alpha,t)y + q(t)yy^{\sigma} + f(t), \quad t \in \mathbb{T}^{\kappa}, \tag{2.13}$$

where $p, q \in \mathscr{C}_{rd}(\mathbb{T})$ and

$$k_0(\alpha,t) + \mu(t)p(t) \neq 0, \quad \alpha \in (0,1], \quad t \in \mathbb{T},$$

$f \in \mathscr{C}_{rd}(\mathbb{T})$.

Definition 2.3.1 *The equation (2.13) is called the conformable Riccati equation.*

Suppose that y_p is a particular solution of the considered equation. Set

$$y = z + y_p.$$

We get

$$
\begin{aligned}
D^{\alpha}z + D^{\alpha}y_p &= -p(t)z^{\sigma} - p(t)y_p^{\sigma} \\[2mm]
&\quad + k_1(\alpha,t)z + k_1(\alpha,t)y_p \\[2mm]
&\quad + q(t)(z + y_p)(z^{\sigma} + y_p^{\sigma}) + f(t) \\[2mm]
&= -p(t)z^{\sigma} - p(t)y_p^{\sigma} \\[2mm]
&\quad + k_1(\alpha,t)z + k_1(\alpha,t)y_p + q(t)zz^{\sigma} + q(t)y_p z^{\sigma} \\[2mm]
&\quad + q(t)zy_p^{\sigma} + q(t)y_p y_p^{\sigma} + f(t) \\[2mm]
&= -\left(p(t) - q(t)y_p\right)z^{\sigma} + \left(k_1(\alpha,t) + q(t)y_p^{\sigma}\right)z + q(t)zz^{\sigma} \\[2mm]
&\quad - p(t)y_p^{\sigma} + k_1(\alpha,t)y_p + q(t)y_p y_p^{\sigma} + f(t), \quad t \in \mathbb{T}^{\kappa}.
\end{aligned}
$$

Hence,

$$D^{\alpha}z = -(p(t) - q(t)y_p)z^{\sigma} + (k_1(\alpha,t) + q(t)y_p^{\sigma})z + q(t)zz^{\sigma},$$

i.e., if we know a particular solution of equation (2.13), then we can reduce it to a Bernoulli equation.

Example 2.3.2 *Let* $\mathbb{T} = 2^{\mathbb{N}_0}$,

$$k_1(\alpha,t) = (1-\alpha)t^{2\alpha}, \quad k_0(\alpha,t) = \alpha t^{2(1-\alpha)}, \quad \alpha \in (0,1], \quad t \in \mathbb{T}.$$

Consider the equation

$$D^{\frac{1}{2}}y = -(t^2+t)y^\sigma + \frac{1}{2}ty + tyy^\sigma - 4t^5 + 4t^4 + 4t^3 + \frac{3}{2}t^2, \quad t \in \mathbb{T}^\kappa.$$

Here

$$\sigma(t) = 2t,$$

$$\mu(t) = t,$$

$$k_1\left(\frac{1}{2},t\right) = \frac{1}{2}t,$$

$$k_0\left(\frac{1}{2},t\right) = \frac{1}{2}t,$$

$$p(t) = t^2+t,$$

$$q(t) = t,$$

$$f(t) = -4t^5 + 4t^4 + 4t^3 + \frac{3}{2}t^2, \quad t \in \mathbb{T}.$$

We will prove that

$$y_p(t) = t^2, \quad t \in \mathbb{T}^\kappa,$$

is a particular solution of the considered equation. We have

$$y_p^\Delta(t) = \sigma(t)+t$$

$$= 2t+t$$

$$= 3t, \quad t \in \mathbb{T}^\kappa,$$

$$D^{\frac{1}{2}}y_p(t) = k_1\left(\frac{1}{2},t\right)y_p(t) + k_0\left(\frac{1}{2},t\right)y_p^\Delta(t)$$

$$= \frac{1}{2}t(t^2) + \frac{1}{2}t(3t)$$

$$= \frac{1}{2}t^3 + \frac{3}{2}t^2, \quad t \in \mathbb{T}^\kappa,$$

$$y_p^\sigma(t) = (\sigma(t))^2$$

$$= (2t)^2$$

$$= 4t^2, \quad t \in \mathbb{T}^\kappa,$$

and

$$-(t^2+t)y_p^\sigma(t) + \frac{1}{2}ty_p(t) + ty_p(t)y_p^\sigma(t) - 4t^5 + 4t^4 + 4t^3 + \frac{3}{2}t^2$$

$$= -(t^2+t)(4t^2) + \frac{1}{2}t(t^2) + t(t^2)(4t^2) - 4t^5 + 4t^4 + 4t^3 + \frac{3}{2}t^2$$

$$= -4t^4 - 4t^3 + \frac{1}{2}t^3 + 4t^5 - 4t^5 + 4t^4 + 4t^3 + \frac{3}{2}t^2$$

$$= \frac{1}{2}t^3 + \frac{3}{2}t^2, \quad t \in \mathbb{T}^\kappa.$$

Therefore y_p is a particular solution of the considered equation. Note that

$$p(t) - q(t)y_p(t) = t^2 + t - t(t^2)$$

$$= t^2 + t - t^3, \quad t \in \mathbb{T}^\kappa,$$

$$k_1\left(\frac{1}{2},t\right) + q(t)y_p^\sigma(t) = \frac{1}{2}t + t(4t^2)$$

$$= \frac{1}{2}t + 4t^3, \quad t \in \mathbb{T}^\kappa.$$

Hence, the considered equation can be reduced to the following Bernoulli equation

$$D^{\frac{1}{2}}z = -(t^2 + t - t^3)z^\sigma + \left(\frac{1}{2}t + 4t^3\right)z + tzz^\sigma, \quad t \in \mathbb{T}^\kappa.$$

Exercise 2.3.3 *Let $\mathbb{T} = 2^{\mathbb{N}_0}$,*

$$k_1(\alpha,t) = (1-\alpha)(1+t^2)^{2\alpha}, \quad k_0(\alpha,t) = \alpha(1+t^2)^{2(1-\alpha)}, \quad \alpha \in (0,1], \quad t \in \mathbb{T}.$$

Prove that

$$y_p(t) = 1, \quad t \in \mathbb{T},$$

is a particular solution to the following Riccati equation

$$D^{\frac{1}{2}}y = -y^\sigma + \frac{1}{2}(1+t^2)y + t^2yy^\sigma + 1 - t^2, \quad t \in \mathbb{T}^\kappa.$$

Reduce this equation to Bernoulli's equation. This ends the exercise.

Now we consider the equation

$$D^\alpha y = p(t)y + k_1(\alpha,t)y^\sigma + q(t)yy^\sigma + f(t), \quad t \in \mathbb{T}^\kappa, \tag{2.14}$$

where p, q, and f satisfy the conditions as above.

Definition 2.3.4 *The equation* (2.14) *is called the alternative conformable Riccati equation.*

Let y_p be a particular solution of the equation (2.14). Set

$$y(t) = z(t) + y_p(t), \quad t \in \mathbb{T}.$$

Then we get

$$
\begin{aligned}
D^\alpha z(t) + D^\alpha y_p(t) &= p(t)z(t) + p(t)y_p(t) \\
&\quad + k_1(\alpha,t)z^\sigma(t) + k_1(\alpha,t)y_p^\sigma(t) \\
&\quad + q(t)\left(z(t) + y_p(t)\right)\left(z^\sigma(t) + y_p^\sigma(t)\right) + f(t) \\
&= p(t)z(t) + p(t)y_p(t) \\
&\quad + k_1(\alpha,t)z^\sigma(t) + k_1(\alpha,t)y_p^\sigma(t) \\
&\quad + q(t)z(t)z^\sigma(t) + q(t)z(t)y_p^\sigma(t) \\
&\quad + q(t)y_p(t)z^\sigma(t) + q(t)y_p(t)y_p^\sigma(t) \\
&\quad + f(t) \\
&= \left(p(t) + q(t)y_p^\sigma(t)\right)z(t) \\
&\quad + (q(t)y_p(t) + k_1(\alpha,t))z^\sigma(t) \\
&\quad + q(t)z(t)z^\sigma(t) \\
&\quad + p(t)y_p(t) + k_1(\alpha,t)y_p^\sigma(t) \\
&\quad + q(t)y_p(t)y_p^\sigma(t) + f(t), \quad t \in \mathbb{T}^\kappa.
\end{aligned}
$$

Hence,

$$D^\alpha z(t) = \left(p(t)+q(t)y_p^\sigma(t)\right)z(t)+\left(q(t)y_p(t)+k_1(\alpha,t)\right)z^\sigma(t)$$

$$+q(t)z(t)z^\sigma(t), \quad t \in \mathbb{T}^\kappa,$$

i.e., if we know a particular solution of equation (2.14), then we can reduce it to a conformable Bernoulli equation.

Example 2.3.5 *Let* $\mathbb{T} = 2\mathbb{Z}$,

$$k_1(\alpha,t)=(1-\alpha)t^{4\alpha}, \quad k_0(\alpha,t)=\alpha t^{4(1-\alpha)}, \quad \alpha \in (0,1], \quad t \in \mathbb{T}.$$

Consider the equation

$$D^{\frac{1}{4}}y = \frac{1}{4}t^2 y + \frac{3}{4}t y^\sigma + y y^\sigma - t^2 - \frac{7}{2}t, \quad t \in \mathbb{T}^\kappa.$$

We will prove that

$$y_p(t) = t, \quad t \in \mathbb{T},$$

is its particular solution. We have

$$\sigma(t) = t+2,$$

$$y_p^\sigma(t) = \sigma(t)=t+2,$$

$$k_1\left(\frac{1}{4},t\right) = \frac{3}{4}t,$$

$$k_0\left(\frac{1}{4},t\right) = \frac{1}{4}t^3,$$

$$y_p^\Delta(t) = 1,$$

$$p(t) = \frac{1}{4}t^2,$$

$$q(t) = 1,$$

$$f(t) = -t^2 - \frac{7}{2}t, \quad t \in \mathbb{T},$$

$$D^{\frac{1}{4}}y_p(t) = k_1\left(\frac{1}{4},t\right)y_p(t)+k_0\left(\frac{1}{4},t\right)y_p^\Delta(t)$$

$$= \frac{3}{4}t^2 + \frac{1}{4}t^3, \quad t \in \mathbb{T}^\kappa.$$

Then

$$\frac{1}{4}t^2 y_p(t) + \frac{3}{4}t y_p(t) + y_p(t) y_p^\sigma(t) - t^2 - \frac{7}{2}t$$

$$= \frac{1}{4}t^3 + \frac{3}{4}t(t+2) + t(t+2) - t^2 - \frac{7}{2}t$$

$$= \frac{1}{4}t^3 + \frac{3}{4}t^2 + \frac{3}{2}t + t^2 + 2t - t^2 - \frac{7}{2}t$$

$$= \frac{1}{4}t^3 + \frac{3}{4}t^2, \quad t \in \mathbb{T}.$$

Therefore $y_p(t) = t$, $t \in \mathbb{T}$, is a particular solution of the considered equation. Note that

$$p(t) + q(t)y_p^\sigma(t) = \frac{1}{4}t^2 + t + 2,$$

$$q(t)y_p(t) + k_1\left(\frac{1}{4}, t\right) = t + \frac{3}{4}t$$

$$= \frac{7}{4}t, \quad t \in \mathbb{T}.$$

The considered equation is reduced to the following Bernoulli equation

$$D^{\frac{1}{4}}z = \left(\frac{1}{4}t^2 + t + 2\right)z + \frac{7}{4}tz^\sigma + zz^\sigma, \quad t \in \mathbb{T}^\kappa.$$

This ends the example.

Exercise 2.3.6 *Let $\mathbb{T} = 2^{\mathbb{N}_0}$,*

$$k_1(\alpha, t) = (1-\alpha)t^{6\alpha}, \quad k_0(\alpha, t) = \alpha t^{6(1-\alpha)}, \quad \alpha \in (0,1], \quad t \in \mathbb{T}.$$

Prove that

$$y_p(t) = t^2, \quad t \in \mathbb{T},$$

is a particular solution of the equation

$$D^{\frac{1}{6}}y = y + \frac{5}{6}ty^\sigma + yy^\sigma + \frac{1}{2}t^6 - 4t^4 - \frac{5}{2}t^3 - t^2, \quad t \in \mathbb{T}^\kappa.$$

Reduce this equation to Bernoulli's equation.

2.4   CONFORMABLE LOGISTIC EQUATIONS

Let *u* solve the linear equation

$$D^\alpha u = pu + q,$$

and let $y = 1/u$. Then

$$
\begin{aligned}
D^\alpha y &= D^\alpha \left(\frac{1}{u} \right) \\
&= \frac{uk_1 - D^\alpha u}{uu^\sigma} + k_1 \left(\frac{1}{u} \right) \\
&= k_1(y^\sigma + y) - yy^\sigma \left(\frac{p}{y} + q \right) \\
&= k_1 y - y^\sigma (p + qy - k_1) \\
&= k_1 y - \left(\frac{(k_0 - \mu k_1)y + \mu D^\alpha y}{k_0} \right) (p + qy - k_1).
\end{aligned}
$$

Collecting the $D^\alpha y$ terms on the left, we get

$$
(k_0 + \mu(p + qy - k_1)) D^\alpha y = k_0 k_1 y - (k_0 - \mu k_1)(p + qy - k_1)y.
$$

Now, assuming that $p + qy \in \mathscr{R}_c$, we see that

$$
D^\alpha y = \left(\frac{k_0 k_1 - (k_0 - \mu k_1)(p + qy - k_1)}{k_0 + \mu(p + qy - k_1)} \right) y,
$$

which can be written using the circle minus notation as the conformable logistic equation on time scales

$$
D^\alpha y = [\ominus_c(p + qy)]y, \qquad p + qy \in \mathscr{R}_c. \tag{2.15}
$$

In like manner, let v solve the linear equation

$$
D^\alpha v = (k_1 - p)v^\sigma + q,
$$

and let $x = 1/v$. Then

$$
\begin{aligned}
D^\alpha x &= D^\alpha \left(\frac{1}{v} \right) \\
&= k_1(x^\sigma + x) - xx^\sigma \left(-(p - k_1)\frac{1}{x^\sigma} + q \right) \\
&= px - x^\sigma (qx - k_1) \\
&= px - \left(\frac{(k_0 - \mu k_1)x + \mu D^\alpha x}{k_0} \right) (qx - k_1).
\end{aligned}
$$

Rearranging terms, we arrive at

$$
(k_0 + \mu(qx - k_1)) D^\alpha x = k_0 px - (k_0 - \mu k_1)(qx - k_1)x.
$$

Assuming that $qx \in \mathscr{R}_c$, we have

$$
D^\alpha x = \left(\frac{k_0 p - (k_0 - \mu k_1)(qx - k_1)}{k_0 + \mu(qx - k_1)} \right) x,
$$

which can also be rewritten as

$$D^\alpha x = [p \ominus_c (qx)]x, \qquad qx \in \mathscr{R}_c. \tag{2.16}$$

This equation will also be called a conformable logistic equation on time scales.

Throughout this section we will assume $p \in \mathscr{R}_c$ and q is right-dense continuous. If $u \neq 0$ and $y = 1/u$, then y is a solution to (2.15). To check the regressivity condition for $(p + qy)$, note that

$$k_0 + \mu(p + qy - k_1) = \frac{k_0 u + \mu(pu + q) - k_1 \mu y}{u}.$$

Recalling that $D^\alpha u = pu + q$, we have

$$k_0 + \mu(p + qy - k_1) = \frac{k_0 u + \mu D^\alpha u - k_1 \mu y}{u} = \frac{k_0 u + k_0 u^\sigma - k_0 u}{u} = \frac{k_0 u^\sigma}{u} \neq 0,$$

putting $(p + qy) \in \mathscr{R}_c$. Likewise, if $v \neq 0$ and $x = 1/v$, then x is a solution to (2.16), and

$$k_0 + \mu(qx - k_1) = \frac{(k_0 - \mu k_1)v + \mu q}{v}.$$

Recalling that $D^\alpha v = (k_1 - p)v^\sigma + q$, we have

$$k_0 + \mu(qx - k_1) = \frac{k_0 v^\sigma - \mu D^\alpha v + \mu q}{v} = \frac{v^\sigma}{v}[k_0 + \mu(p - k_1)] \neq 0$$

by the regressivity of p, so that $(qx) \in \mathscr{R}_c$. This leads to the following result.

Theorem 2.4.1 *Suppose $p \in \mathscr{R}_c$ and q is right-dense continuous.*

1. Let $y_0 \neq 0$. If

$$u(t) = u_0 E_p(t, t_0) + \int_{t_0}^{t} q(s) E_f(\sigma(s), t) \Delta_{\alpha, t} s \neq 0$$

for all $t \in \mathbb{T}$, then $y(t) = 1/u(t)$ solves (2.15) with $y(t_0) = y_0$, where

$$f = \frac{(p - k_1)(\mu k_1 - k_0)}{k_0 + \mu(p - k_1)}.$$

2. Let $x_0 \neq 0$. If

$$v(t) = v_0 E_f(t, t_0) + \int_{t_0}^{t} q(s) E_p(s, t) \Delta_{\alpha, t} s \neq 0$$

for all $t \in \mathbb{T}$, then $x(t) = 1/v(t)$ solves (2.16) with $x(t_0) = x_0$.

Proof 2.4.2 *Using Remark 2.1 to solve $D^\alpha u = pu + q$ for u, and using Theorem 2.2 to solve $D^\alpha v = (k_1 - p)v^\sigma + q$ for v, we can obtain the above solutions of the conformable logistic equations (2.15) and (2.16), as stated in the theorem. This completes the proof.* □

2.5 ADVANCED PRACTICAL PROBLEMS

Problem 2.5.1 *Let* $\mathbb{T} = 3^{\mathbb{N}_0}$,

$$k_1(\alpha,t) = (1-\alpha)t^{2\alpha}, \quad k_0(\alpha,t) = \alpha t^{2(1-\alpha)}, \quad t \in \mathbb{T}, \quad \alpha \in (0,1].$$

Find the solution of the IVP

$$D^{\frac{1}{4}}y(t) = \left(t^3 + \frac{3}{4}t^{\frac{1}{2}}\right)y(t) + t^2 + t, \quad t \in \mathbb{T},$$

$$y(1) = 3.$$

Problem 2.5.2 *Let* $\mathbb{T} = 2^{\mathbb{N}_0}$,

$$k_1(\alpha,t) = (1-\alpha)(1+t^2)^{\alpha}, \quad k_0(\alpha,t) = \alpha(1+t^2)^{1-\alpha}, \quad \alpha \in (0,1], \quad t \in \mathbb{T}.$$

Find the solution of the IVP

$$D^{\frac{1}{4}}y = (t+2)y^{\sigma} + t^2, \quad t \in \mathbb{T}, \quad t > 128,$$

$$y(128) = 1.$$

Problem 2.5.3 *Let* $\mathbb{T} = 3^{\mathbb{N}_0}$,

$$k_1(\alpha,t) = (1-\alpha)(1+t)^{\alpha}, \quad k_0(\alpha,t) = \alpha(1+t)^{1-\alpha}, \quad \alpha \in (0,1], \quad t \in \mathbb{T}.$$

Reduce the following conformable Bernoulli equations to first-order linear conformable dynamic equations.

1.
$$D^{\frac{1}{2}}y = (2+t)y^{\sigma} + \frac{1}{2}(1+t)^{\frac{1}{2}}y + t^2 yy^{\sigma}, \quad t \in \mathbb{T},$$

2.
$$D^{\frac{2}{3}}y = (3-t^2)y^{\sigma} + \frac{1}{3}(1+t)^{\frac{2}{3}}y + tyy^{\sigma}, \quad t \in \mathbb{T},$$

3.
$$D^{\frac{3}{4}}y = (t^2+t+1)y^{\sigma} + \frac{1}{4}(1+t)^{\frac{3}{4}}y + (1+t)yy^{\sigma}, \quad t \in \mathbb{T}.$$

Problem 2.5.4 *Let* $\mathbb{T} = 3^{\mathbb{N}_0}$,

$$k_1(\alpha,t) = (1-\alpha)t^{3\alpha}, \quad k_0(\alpha,t) = \alpha t^{3(1-\alpha)}, \quad \alpha \in (0,1], \quad t \in \mathbb{T}.$$

Reduce the following conformable Bernoulli equations to first-order linear conformable dynamic equations.

1.
$$D^{\frac{1}{2}}y = (t^2+2t)y + \frac{1}{2}t^{\frac{3}{2}}y^{\sigma} + tyy^{\sigma} \quad t \in \mathbb{T},$$

2.

$$D^{\frac{1}{4}}y = (1+t)y + \frac{3}{4}t^{\frac{3}{4}}y^{\sigma} + t^2yy^{\sigma}, \qquad t \in \mathbb{T},$$

3.

$$D^{\frac{1}{8}}y = (1+t^2)y + \frac{7}{8}t^{\frac{3}{8}}y^{\sigma} + (t+1)yy^{\sigma}, \quad t \in \mathbb{T}.$$

Problem 2.5.5 *Let* $\mathbb{T} = 3^{\mathbb{N}_0}$,

$$k_1(\alpha,t) = (1-\alpha)t^{4\alpha}, \quad k_0(\alpha,t) = \alpha t^{4(1-\alpha)}, \quad \alpha \in (0,1], \quad t \in \mathbb{T}.$$

Prove that

$$y_p(t) = t, \quad t \in \mathbb{T},$$

is a particular solution of the equation

$$D^{\frac{1}{4}}y = -2ty^{\sigma} + \frac{3}{4}ty + tyy^{\sigma} - \frac{11}{4}t^3 + 6t^2, \quad t \in \mathbb{T}.$$

Reduce this equation to Bernoulli's equation.

Problem 2.5.6 *Let* $\mathbb{T} = 3^{\mathbb{N}_0}$,

$$k_1(\alpha,t) = (1-\alpha)t^{2\alpha}, \quad k_0(\alpha,t) = \alpha t^{2(1-\alpha)}, \quad \alpha \in (0,1], \quad t \in \mathbb{T}.$$

Prove that

$$y_p(t) = t$$

is a particular solution to the equation

$$D^{\frac{1}{4}}y = ty + \frac{3}{4}ty^{\sigma} + t^2yy^{\sigma} - 3t^4 - \frac{13}{4}t^2 + t^{\frac{3}{2}}, \quad t \in \mathbb{T}.$$

Reduce this equation to Bernoulli's equation.

Conformable Dynamic Systems on Time Scales

Suppose that $\mathbb{T}$ is a time scale with forward jump operator and delta differentiation operator σ and Δ, respectively. Let $\alpha \in (0,1]$, k_0 and k_1 satisfy $(A1)$, and assume (1.6) holds.

3.1 STRUCTURE OF CONFORMABLE DYNAMIC SYSTEMS ON TIME SCALES

Suppose that A is an $m \times n$ matrix, $A = (a_{ij})_{1 \le i \le m, 1 \le j \le n}$, shortly $A + (a_{ij})$, $a_{ij} : \mathbb{T} \to \mathbb{R}$, $1 \le i \le m$, $1 \le j \le n$.

Definition 3.1.1 *We say that A is conformable Δ-differentiable at $t \in \mathbb{T}^\kappa$, if each entry is conformable Δ-differentiable at t and*

$$D^\alpha A = (D^\alpha a_{ij}).$$

Example 3.1.2 *Let $\mathbb{T} = 2^{\mathbb{N}_0}$,*

$$k_1(\alpha,t) = (1-\alpha)t^{4\alpha}, \quad k_0(\alpha,t) = \alpha t^{4(1-\alpha)}, \quad \alpha \in (0,1], \quad t \in \mathbb{T},$$

and

$$A(t) = \begin{pmatrix} t^2+t & t \\ t-1 & t^3 \end{pmatrix}, \quad t \in \mathbb{T}.$$

We will find

$$D^{\frac{1}{4}}A(t), \quad t \in \mathbb{T}.$$

Here

$$\sigma(t) = 2t,$$

$$k_1\left(\frac{1}{4},t\right) = \frac{3}{4}t,$$

$$k_0\left(\frac{1}{4},t\right) = \frac{1}{4}t^3,$$

$$a_{11}(t) = t^2+t,$$

$$a_{12}(t) = t,$$

$$a_{21}(t) = t-1,$$

$$a_{22}(t) = t^3, \quad t \in \mathbb{T}.$$

Then

$$a_{11}^{\Delta}(t) = \sigma(t)+t+1$$

$$= 2t+t+1$$

$$= 3t+1,$$

$$a_{12}^{\Delta}(t) = 1,$$

$$a_{21}^{\Delta}(t) = 1,$$

$$a_{22}^{\Delta}(t) = (\sigma(t))^2+t\sigma(t)+t^2$$

$$= (2t)^2+2t^2+t^2$$

$$= 4t^2+3t^2$$

$$= 7t^2, \quad t \in \mathbb{T}.$$

Hence,

$$D^{\frac{1}{4}}a_{11}(t) = k_1\left(\frac{P1}{4},t\right)a_{11}(t)+k_0\left(\frac{1}{4},t\right)a_{11}^{\Delta}(t)$$

$$= \frac{3}{4}t(t^2+t)+\frac{1}{4}t^3(3t+1)$$

$$= \frac{3}{4}t^3+\frac{3}{4}t^2+\frac{3}{4}t^4+\frac{1}{4}t^3$$

$$= \frac{3}{4}t^4 + t^3 + \frac{3}{4}t^2,$$

$$F^{\frac{1}{4}}a_{12}(t) = k_1\left(\frac{1}{4},t\right)a_{12}(t) + k_0\left(\frac{1}{4},t\right)a_{12}^\Delta(t)$$

$$= \frac{3}{4}t^2 + \frac{1}{4}t^3,$$

$$D^{\frac{1}{4}}a_{21}(t) = k_1\left(\frac{1}{4},t\right)a_{21}(t) + k_0\left(\frac{1}{4},t\right)a_{21}^\Delta(t)$$

$$= \frac{3}{4}t(t-1) + \frac{1}{4}t^3$$

$$= \frac{3}{4}t^2 - \frac{3}{4}t + \frac{1}{4}t^3,$$

$$D^{\frac{1}{4}}a_{22}(t) = k_1\left(\frac{1}{4},t\right)a_{22}(t) + k_0\left(\frac{1}{4},t\right)a_{22}^\Delta(t)$$

$$= \frac{3}{4}t^4 + \frac{1}{4}t^3(7t^2)$$

$$= \frac{7}{4}t^5 + \frac{3}{4}t^4, \quad t \in \mathbb{T}.$$

Therefore

$$D^{\frac{1}{4}}A(t) = \begin{pmatrix} \frac{3}{4}t^4 + t^3 + \frac{3}{4}t^2 & \frac{1}{4}t^3 + \frac{3}{4}t^2 \\ \frac{1}{4}t^3 + \frac{3}{4}t^2 - \frac{3}{4}t & \frac{7}{4}t^5 + \frac{3}{4}t^4 \end{pmatrix}, \quad t \in \mathbb{T}.$$

Exercise 3.1.3 *Let* $\mathbb{T} = \mathbb{Z}$,

$$k_1(\alpha,t) = (1-\alpha)t^{2\alpha}, \quad k_0(\alpha,t) = \alpha t^{2(1-\alpha)}, \quad \alpha \in (0,1], \quad t \in \mathbb{T},$$

and

$$A(t) = \begin{pmatrix} t^2 + 2t + 4 & t^3 - t \\ t^2 - 3t & t^2 \end{pmatrix}, \quad t \in \mathbb{T}.$$

Find

$$D^{\frac{1}{3}}A(t), \quad t \in \mathbb{T}.$$

Definition 3.1.4 *If A is conformable Δ-differentiable at* $t \in \mathbb{T}^\kappa$, *then we define*

$$A^\sigma(t) = (a_{ij}^\sigma(t)).$$

Theorem 3.1.5 *If A is conformable Δ-differentiable at* $t \in \mathbb{T}^\kappa$, *then*

$$A^\sigma(t) = \left(1 - \frac{\mu(t)}{k_0(\alpha,t)}k_1(\alpha,t)\right)A(t) + \frac{\mu(t)}{k_0(\alpha,t)}D^\alpha A(t), \quad \alpha \in (0,1].$$

Proof 3.1.6 *We have*

$$A^{\sigma}(t) = (a_{ij}^{\sigma}(t))$$

$$= \left(\left(1 - \frac{\mu(t)}{k_0(\alpha,t)} k_1(\alpha,t) \right) a_{ij}(t) + \frac{\mu(t)}{a_{ij}(t)} D^{\alpha} a_{ij}(t) \right)$$

$$= \left(1 - \frac{\mu(t)}{k_0(\alpha,t)} k_1(\alpha,t) \right) (a_{ij}(t)) + \frac{\mu(t)}{a_{ij}(t)} (D^{\alpha} a_{ij}(t))$$

$$= \left(1 - \frac{\mu(t)}{k_0(\alpha,t)} k_1(\alpha,t) \right) A(t) + \frac{\mu(t)}{a_{ij}(t)} D^{\alpha} A(t).$$

This completes the proof. □

Below we suppose that $B = (b_{ij})_{1 \le i \le m, 1 \le j \le n}$, where $b_{ij} : \mathbb{T} \to \mathbb{R}$, $1 \le i \le m$, $1 \le j \le n$.

Theorem 3.1.7 *Let A and B be conformable Δ-differentiable at $t \in \mathbb{T}^{\kappa}$. Then*

$$D^{\alpha}(A+B)(t) = D^{\alpha}A(t) + D^{\alpha}B(t), \quad \alpha \in (0,1].$$

Proof 3.1.8 *We have*

$$D^{\alpha}(A+B)(t) = (D^{\alpha}(a_{ij} + b_{ij})(t))$$

$$= (D^{\alpha}a_{ij}(t) + D^{\alpha}b_{ij}(t))$$

$$= (D^{\alpha}a_{ij}(t)) + (D^{\alpha}b_{ij}(t))$$

$$= D^{\alpha}A(t) + D^{\alpha}B(t).$$

This completes the proof. □

Theorem 3.1.9 *Let $a \in \mathbb{R}$ and A be conformable Δ-differentiable at $t \in \mathbb{T}^{\kappa}$. Then*

$$D^{\alpha}(aA)(t) = aD^{\alpha}A(t).$$

Proof 3.1.10 *We have*

$$D^{\alpha}(aA)(t) = D^{\alpha}(aa_{ij})(t)$$

$$= (D^{\alpha}(aa_{ij}))(t)$$

$$= (aD^{\alpha}a_{ij})(t)$$

$$= a\left(D^{\alpha}a_{ij}\right)(t)$$

$$= aD^{\alpha}A(t).$$

This completes the proof. □

Theorem 3.1.11 *Let $m = n$ and A, B be conformable Δ-differentiable at $t \in \mathbb{T}^{\kappa}$. Then*

$$D^{\alpha}(AB)(t) = D^{\alpha}A(t)B(t) + A^{\sigma}(t)D^{\alpha}B(t) - k_1(\alpha,t)A^{\sigma}(t)B(t)$$

$$= D^{\alpha}A(t)B^{\sigma}(t) + A(t)D^{\alpha}B(t) - k_1(\alpha,t)A(t)B^{\sigma}(t).$$

Proof 3.1.12 *We have*

$$D^{\alpha}(AB)(t) = D^{\alpha}\left(\sum_{l=1}^{n} a_{il}b_{lj}\right)(t)$$

$$= \left(D^{\alpha}\left(\sum_{l=1}^{n} a_{il}b_{lj}\right)\right)(t)$$

$$= \left(\sum_{l=1}^{n} D^{\alpha}(a_{il}b_{lj})\right)(t)$$

$$= \left(\sum_{l=1}^{n} (D^{\alpha}a_{il})b_{lj} + \sum_{l=1}^{n} a_{il}^{\sigma}D^{\alpha}b_{lj} - \sum_{l=1}^{n} k_1(\alpha,\cdot)a_{il}^{\sigma}b_{lj}\right)(t)$$

$$= \left(\sum_{l=1}^{n} (D^{\alpha}a_{il})b_{lj}\right)(t) + \left(\sum_{l=1}^{n} a_{il}^{\sigma}D^{\alpha}b_{lj}\right)(t)$$

$$\quad - k_1(\alpha,t)\left(\sum_{l=1}^{n} a_{il}^{\sigma}b_{lj}\right)(t)$$

$$= D^{\alpha}A(t)B(t) + A^{\sigma}(t)B(t) - k_1(\alpha,t)A^{\sigma}(t)B(t)$$

$$= \left(\sum_{l=1}^{n} (D^{\alpha}a_{il})b_{lj}^{\sigma} + \sum_{l=1}^{n} a_{il}D^{\alpha}b_{lj} - \sum_{l=1}^{n} k_1(\alpha,\cdot)a_{il}b_{lj}^{\sigma}\right)(t)$$

$$= \left(\sum_{l=1}^{n} (D^{\alpha}a_{ij})b_{lj}^{\sigma}\right)(t) + \left(\sum_{l=1}^{n} a_{il}D^{\alpha}b_{lj}\right)(t)$$

$$\quad - k_1(\alpha,t)\left(\sum_{l=1}^{n} a_{il}b_{lj}^{\sigma}\right)(t)$$

$$= D^{\alpha}A(t)B^{\sigma}(t) + A(t)D^{\alpha}B(t) - k_1(\alpha,t)A(t)B^{\sigma}(t).$$

This completes the proof. □

Example 3.1.13 *Let* $\mathbb{T} = 2^{\mathbb{N}_0}$,

$$k_1(\alpha,t) = (1-\alpha)t^{4\alpha}, \quad k_0(\alpha,t) = \alpha t^{4(1-\alpha)}, \quad \alpha \in (0,1], \quad t \in \mathbb{T},$$

and

$$A(t) = \begin{pmatrix} t & 2t \\ t+1 & 1 \end{pmatrix}, \quad B(t) = \begin{pmatrix} t & 1 \\ 2 & 3 \end{pmatrix}, \quad t \in \mathbb{T}.$$

We will find

$$D^{\frac{1}{2}}(AB)(t), \quad t \in \mathbb{T}.$$

We have

$$\sigma(t) = 2t,$$

$$k_1\left(\frac{1}{2},t\right) = \frac{1}{2}t^2,$$

$$k_0\left(\frac{1}{2},t\right) = \frac{1}{2}t^2, \quad t \in \mathbb{T}.$$

Let

$$C(t) = A(t)B(t), \quad t \in \mathbb{T}.$$

Then

$$C(t) = \begin{pmatrix} t & 2t \\ t+1 & 1 \end{pmatrix}\begin{pmatrix} t & 1 \\ 2 & 3 \end{pmatrix} = \begin{pmatrix} t^2+4t & 7t \\ t^2+t+2 & t+4 \end{pmatrix}, \quad t \in \mathbb{T}.$$

We have

$$c_{11}(t) = t^2 + 4t,$$

$$c_{12}(t) = 7t,$$

$$c_{21}(t) = t^2 + t + 2,$$

$$c_{22}(t) = t + 4,$$

$$c_{11}^{\Delta}(t) = \sigma(t) + t + 4$$

$$= 2t + t + 4$$

$$= 3t + 4,$$

$$c_{12}^{\Delta}(t) = 7,$$

$$c_{21}^{\Delta}(t) \;=\; \sigma(t) + t + 1$$

$$=\; 2t + t + 1$$

$$=\; 3t + 1,$$

$$c_{22}^{\Delta}(t) \;=\; 1,$$

$$D^{\frac{1}{2}}c_{11}(t) \;=\; k_1\left(\frac{1}{2},t\right)c_{11}(t) + k_0\left(\frac{1}{2},t\right)c_{11}^{\Delta}(t)$$

$$=\; \frac{1}{2}t^2\left(t^2 + 4t\right) + \frac{1}{2}t^2(3t + 4)$$

$$=\; \frac{1}{2}t^4 + 2t^3 + \frac{3}{2}t^3 + 2t^2$$

$$=\; \frac{1}{2}t^4 + \frac{7}{2}t^3 + 2t^2,$$

$$D^{\frac{1}{2}}c_{12}(t) \;=\; k_1\left(\frac{1}{2},t\right)c_{12}(t) + k_0\left(\frac{1}{2},t\right)c_{12}^{\Delta}(t)$$

$$=\; \frac{7}{2}t^3 + \frac{7}{2}t^2,$$

$$D^{\frac{1}{2}}c_{21}(t) \;=\; k_1\left(\frac{1}{2},t\right)c_{21}(t) + k_0\left(\frac{1}{2},t\right)c_{21}^{\Delta}(t)$$

$$=\; \frac{1}{2}t^2\left(t^2 + t + 2\right) + \frac{1}{2}t^2(3t + 1)$$

$$=\; \frac{1}{2}t^4 + \frac{1}{2}t^3 + t^2 + \frac{3}{2}t^3 + \frac{1}{2}t^2$$

$$=\; \frac{1}{2}t^4 + 2t^3 + \frac{3}{2}t^2,$$

$$D^{\frac{1}{2}}cx_{22}(t) \;=\; k_1\left(\frac{1}{2},t\right)c_{22}(t) + k_0\left(\frac{1}{2},t\right)c_{22}^{\Delta}(t)$$

$$=\; \frac{1}{2}t^2(t + 4) + \frac{1}{2}t^3$$

$$=\; \frac{1}{2}t^3 + 2t^2 + \frac{1}{2}t^2$$

$$= \frac{1}{2}t^3 + \frac{5}{2}t^2, \quad t \in \mathbb{T}.$$

Therefore

$$D^{\frac{1}{2}}(AB)(t) = D^{\frac{1}{2}}C(t)$$

$$= \begin{pmatrix} \frac{1}{2}t^4 + \frac{7}{2}t^3 + 2t^2 & \frac{7}{2}t^3 + \frac{7}{2}t^2 \\ \frac{1}{2}t^4 + 2t^3 + \frac{3}{2}t^2 & \frac{1}{2}t^3 + \frac{5}{2}t^2 \end{pmatrix}, \quad t \in \mathbb{T}.$$

Next,

$$a_{11}(t) = t,$$

$$a_{12}(t) = 2t,$$

$$a_{21}(t) = t+1,$$

$$a_{22}(t) = 1,$$

$$b_{11}(t) = t,$$

$$b_{12}(t) = 1,$$

$$b_{21}(t) = 2,$$

$$b_{22}(t) = 3,$$

$$a_{11}^{\Delta}(t) = 1,$$

$$a_{12}^{\Delta}(t) = 2,$$

$$a_{21}^{\Delta}(t) = 1,$$

$$a_{22}^{\Delta}(t) = 0,$$

$$b_{11}^{\Delta}(t) = 1,$$

$$b_{12}^{\Delta}(t) = 0,$$

$$b_{21}^{\Delta}(t) \;=\; 0,$$

$$b_{22}^{\Delta}(t) \;=\; 0,$$

$$D^{\frac{1}{2}}a_{11}(t) \;=\; k_1\left(\frac{1}{2},t\right)a_{11}(t)+k_0\left(\frac{1}{2},t\right)a_{11}^{\Delta}(t)$$

$$\;=\; \frac{1}{2}t^3+\frac{1}{2}t^2,$$

$$D^{\frac{1}{2}}a_{12}(t) \;=\; k_1\left(\frac{1}{2},t\right)a_{12}(t)+k_0\left(\frac{1}{2},t\right)a_{12}^{\Delta}(t)$$

$$\;=\; t^3+t^2,$$

$$D^{\frac{1}{2}}a_{21}(t) \;=\; k_1\left(\frac{1}{2},t\right)a_{21}(t)+k_0\left(\frac{1}{2},t\right)a_{21}^{\Delta}(t)$$

$$\;=\; \frac{1}{2}t^2(t+1)+\frac{1}{2}t^2$$

$$\;=\; \frac{1}{2}t^3+t^2,$$

$$D^{\frac{1}{2}}a_{22}(t) \;=\; k_1\left(\frac{1}{2},t\right)a_{22})(t)+k_0\left(\frac{1}{2},t\right)a_{22}^{\Delta}(t)$$

$$\;=\; \frac{1}{2}t^2,$$

$$D^{\frac{1}{2}}b_{11}(t) \;=\; k_1\left(\frac{1}{2},t\right)b_{11}(t)+k_0\left(\frac{1}{2},t\right)b_{11}^{\Delta}(t)$$

$$\;=\; \frac{1}{2}t^3+\frac{1}{2}t^2,$$

$$D^{\frac{1}{2}}b_{12}(t) \;=\; k_1\left(\frac{1}{2},t\right)b_{12}(t)+k_0\left(\frac{1}{2},t\right)b_{12}^{\Delta}(t)$$

$$\;=\; \frac{1}{2}t^2,$$

$$D^{\frac{1}{2}}b_{21}(t) \;=\; k_1\left(\frac{1}{2},t\right)b_{21}(t)+k_0\left(\frac{1}{2},t\right)b_{21}^{\Delta}(t)$$

$$\;=\; t^2,$$

$$D^{\frac{1}{2}}b_{22}(t) = k_1\left(\frac{1}{2},t\right)b_{22}(t) + k_0\left(\frac{1}{2},t\right)b_{22}^{\Delta}(t)$$

$$= \frac{3}{2}t^2,$$

$$b_{11}^{\sigma}(t) = \sigma(t)$$

$$= 2t,$$

$$b_{12}^{\sigma}(t) = 1,$$

$$b_{21}^{\sigma}(t) = 2,$$

$$b_{22}^{\sigma}(t) = 3.$$

Hence,

$$D^{\frac{1}{2}}A(t) = \begin{pmatrix} \frac{1}{2}t^3 + \frac{1}{2}t^2 & t^3 + t^2 \\ \frac{1}{2}t^3 + t^2 & \frac{1}{2}t^2 \end{pmatrix},$$

$$D^{\frac{1}{2}}B(t) = \begin{pmatrix} \frac{1}{2}t^3 + \frac{1}{2}t^2 & \frac{1}{2}t^2 \\ t^2 & \frac{3}{2}t^2 \end{pmatrix},$$

$$B^{\sigma}(t) = \begin{pmatrix} 2t & 1 \\ 2 & 3 \end{pmatrix},$$

$$D^{\frac{1}{2}}A(t)B^{\sigma}(t) = \begin{pmatrix} \frac{1}{2}t^3 + \frac{1}{2}t^2 & t^3 + t^2 \\ \frac{1}{2}t^3 + t^2 & \frac{1}{2}t^2 \end{pmatrix}\begin{pmatrix} 2t & 1 \\ 2 & 3 \end{pmatrix}$$

$$= \begin{pmatrix} t^4 + 3t^3 + 2t^2 & \frac{7}{2}t^3 + \frac{7}{2}t^2 \\ t^4 + 2t^3 + t^2 & \frac{1}{2}t^3 + \frac{5}{2}t^2 \end{pmatrix},$$

$$A(t)D^{\frac{1}{2}}B(t) = \begin{pmatrix} t & 2t \\ t+1 & 1 \end{pmatrix}\begin{pmatrix} \frac{1}{2}t^3 + \frac{1}{2}t^2 & \frac{1}{2}t^2 \\ t^2 & \frac{3}{2}t^2 \end{pmatrix}$$

$$= \begin{pmatrix} \frac{1}{2}t^4 + \frac{5}{2}t^3 & \frac{7}{2}t^3 \\ \frac{1}{2}t^4 + t^3 + \frac{3}{2}t^2 & \frac{1}{2}t^3 + 2t^2 \end{pmatrix},$$

$$D^{\frac{1}{2}}A(t)B^{\sigma}(t)+A(t)D^{\frac{1}{2}}B(t) = \begin{pmatrix} t^4+3t^3+2t^2 & \frac{7}{2}t^3+\frac{7}{2}t^2 \\ t^4+2t^3+t^2 & \frac{1}{2}t^3+\frac{5}{2}t^2 \end{pmatrix}$$

$$+ \begin{pmatrix} \frac{1}{2}t^4+\frac{5}{2}t^3 & \frac{7}{2}t^3 \\ \frac{1}{2}t^4+t^3+\frac{3}{2}t^2 & \frac{1}{2}t^3+2t^2 \end{pmatrix}$$

$$= \begin{pmatrix} \frac{3}{2}t^4+\frac{11}{2}t^3+2t^2 & 7t^3+\frac{7}{2}t^2 \\ \frac{3}{2}t^4+3t^3+\frac{5}{2}t^2 & t^3+\frac{9}{2}t^2 \end{pmatrix},$$

$$A(t)B^{\sigma}(t) = \begin{pmatrix} t & 2t \\ t+1 & 1 \end{pmatrix}\begin{pmatrix} 2t & 1 \\ 2 & 3 \end{pmatrix}$$

$$= \begin{pmatrix} 2t^2+4t & 7t \\ 2t^2+2t+2 & t+4 \end{pmatrix},$$

$$k_1\left(\frac{1}{2},t\right)A(t)B^{\sigma}(t) = \frac{1}{2}t^2\begin{pmatrix} 2t^2+4t & 7t \\ 2t^2+2t+2 & t+4 \end{pmatrix}$$

$$= \begin{pmatrix} t^4+2t^3 & \frac{7}{2}t^3 \\ t^4+t^3+t^2 & \frac{1}{2}t^3+2t^2 \end{pmatrix},$$

and

$$D^{\frac{1}{2}}A(t)B^{\sigma}(t)+A(t)D^{\frac{1}{2}}B(t)-k_1\left(\frac{1}{2},t\right)A(t)B^{\sigma}(t)$$

$$= \begin{pmatrix} \frac{3}{2}t^4+\frac{11}{2}t^3+2t^2 & 7t^3+\frac{7}{2}t^2 \\ \frac{3}{2}t^4+3t^3+\frac{5}{2}t^2 & t^3+\frac{9}{2}t^2 \end{pmatrix} - \begin{pmatrix} t^4+2t^3 & \frac{7}{2}t^3 \\ t^4+t^3+t^2 & \frac{1}{2}t^3+2t^2 \end{pmatrix}$$

$$= \begin{pmatrix} \frac{1}{2}t^4+\frac{7}{2}t^3+2t^2 & \frac{7}{2}t^3+\frac{7}{2}t^2 \\ \frac{1}{2}t^4+2t^3+\frac{3}{2}t^2 & \frac{1}{2}t^3+\frac{5}{2}t^2 \end{pmatrix}$$

$$= D^{\frac{1}{2}}(AB)(t), \quad t\in\mathbb{T}.$$

This ends the example.

Exercise 3.1.14 *Let* $\mathbb{T}=3^{\mathbb{N}_0}$,

$$k_1(\alpha,t)=(1-\alpha)(1+t)^{3\alpha}, \quad k_0(\alpha,t)=\alpha(1+t)^{3(1-\alpha)}, \quad \alpha\in(0,1], \quad t\in\mathbb{T},$$

and

$$A(t)=\begin{pmatrix} t & 2t+3 \\ t+4 & t \end{pmatrix}, \quad B(t)=\begin{pmatrix} t^2 & 1 \\ -1 & t \end{pmatrix}, \quad t\in\mathbb{T}.$$

Find

$$D^{\frac{1}{3}}(AB)(t), \quad t \in \mathbb{T}.$$

Theorem 3.1.15 *Let $m = n$, and assume A^{-1} exists on $\mathbb{T}$. Then*

$$(A^\sigma)^{-1} = (A^{-1})^\sigma \quad on \quad \mathbb{T}.$$

Proof 3.1.16 *For any $t \in \mathbb{T}$ we have*

$$A(t)A^{-1}(t) = I, \quad t \in \mathbb{T}.$$

Then

$$A^\sigma(t)\left(A^{-1}\right)^\sigma(t) = I,$$

whereupon

$$(A^\sigma)^{-1}(t) = (A^{-1})^\sigma(t), \quad t \in \mathbb{T}.$$

This completes the proof. □

Example 3.1.17 *Let $\mathbb{T} = l\mathbb{N}_0, l > 0$,*

$$A(t) = \begin{pmatrix} t+1 & t+2 \\ 1 & t+3 \end{pmatrix}, \quad t \in \mathbb{T}.$$

Then

$$
\begin{aligned}
A^\sigma(t) &= \begin{pmatrix} \sigma(t)+1 & \sigma(t)+2 \\ 1 & \sigma(t)+3 \end{pmatrix} \\
&= \begin{pmatrix} t+l+1 & t+l+2 \\ 1 & t+l+3 \end{pmatrix}, \\
(A^\sigma)^{-1}(t) &= \frac{1}{(t+l)(t+l+3)+1} \begin{pmatrix} t+l+3 & -t-l-2 \\ -1 & t+l+1 \end{pmatrix}, \quad t \in \mathbb{T}.
\end{aligned}
$$

Next,

$$A^{-1}(t) = \frac{1}{t(t+3)+1} \begin{pmatrix} t+3 & -t-2 \\ -1 & t+1 \end{pmatrix}, \quad t \in \mathbb{T},$$

whereupon

$$
\begin{aligned}
(A^{-1})^\sigma(t) &= \frac{1}{\sigma(t)(\sigma(t)+3)+1} \begin{pmatrix} \sigma(t)+3 & -\sigma(t)-2 \\ -1 & \sigma(t)+1 \end{pmatrix} \\
&= \frac{1}{(t+l)(t+l+3)+1} \begin{pmatrix} t+l+3 & -t-l-2 \\ -1 & t+l+1 \end{pmatrix}, \quad t \in \mathbb{T}.
\end{aligned}
$$

Consequently,

$$(A^\sigma)^{-1}(t) = (A^{-1})^\sigma(t), \quad t \in \mathbb{T}.$$

This ends the example.

Exercise 3.1.18 *Let* $\mathbb{T} = 2^{\mathbb{N}_0}$ *and*

$$A(t) = \begin{pmatrix} t+2 & \dfrac{1}{t+1} \\ t^2+1 & \dfrac{1}{t+2} \end{pmatrix}, \quad t \in \mathbb{T}.$$

Prove that

$$(A^\sigma)^{-1}(t) = (A^{-1})^\sigma(t), \quad t \in \mathbb{T}.$$

Theorem 3.1.19 *Let* $m = n$, *A be conformable* Δ*-differentiable at* $t \in \mathbb{T}^\kappa$ *and assume* A^{-1}, $(A^\sigma)^{-1}$ *exist at t. Then*

$$D^\alpha A^{-1}(t) = -A^{-1}(t)(D^\alpha A(t))(A^\sigma)^{-1}(t) + k_1(\alpha, t)(A^\sigma)^{-1}(t),$$

$$D^\alpha A^{-1}(t) = (A^\sigma)^{-1}(t)(D^\alpha A(t))A^{-1}(t) + k_1(\alpha, t)A^{-1}(t).$$

Proof 3.1.20 *We have*

$$I = (AA^{-1})(t).$$

Then

$$O = D^\alpha (AA^{-1})(t)$$

$$= (D^\alpha A(t))A^{-1}(t) + A^\sigma(t)D^\alpha A^{-1}(t) - k_1(\alpha, t)A^\sigma(t)A^{-1}(t)$$

$$= (D^\alpha A(t))(A^\sigma)^{-1}(t) + A(t)D^\alpha A^1(t) - k_1(\alpha, t)A(t)(A^\sigma)^{-1}(t).$$

Hence,

$$A(t)D^\alpha A^{-1}(t) = -(D^\alpha A(t))(A^\sigma)^{-1}(t) + k_1(\alpha, t)A(t)(A^\sigma)^{-1}(t),$$

$$A^\sigma(t)D^\alpha A^{-1}(t) = -(D^\alpha A(t))A^{-1}(t) + k_1(\alpha, t)A^\sigma(t)A^{-1}(t),$$

whereupon we get the desired result. □

Example 3.1.21 *Let* $m = n$, *A and B be conformable* Δ *- differentiable at* $t \in \mathbb{T}^\kappa$. *Then*

$$D^\alpha (AB^{-1})(t) = (D^\alpha A(t))B^{-1}(t) + A^\sigma(t)D^\alpha B^{-1}(t)$$

$$-k_1(\alpha, t)A^\sigma(t)B^{-1}(t)$$

$$= (D^\alpha A(t))B^{-1}(t)$$

$$+A^\sigma(t)\left(-(B^\sigma)^{-1}(t)D^\alpha B(t)B^{-1}(t) + k_1(\alpha, t)B^{-1}(t)\right)$$

$$-k_1(\alpha,t)A^\sigma(t)B^{-1}(t)$$

$$= \left(D^\alpha A(t) - A^\sigma(t)\left(B^\sigma\right)^{-1}(t)D^\alpha B(t)\right)B^{-1}(t).$$

This completes the example.

Definition 3.1.22 *We say that the matrix A is rd-continuous on $\mathbb{T}$ if each entry of A is rd-continuous. The class of such rd-continuous $m \times n$ matrix-valued functions on $\mathbb{T}$ is denoted by*

$$\mathscr{C}_{rd} = \mathscr{C}_{rd}(\mathbb{T}) = \mathscr{C}_{rd}(\mathbb{T},\mathscr{R}^{m\times n}).$$

Below we suppose that A and B are $n \times n$ matrix-valued functions.

Definition 3.1.23 *We say that the matrix A is conformable regressive with respect to $\mathbb{T}$ provided*

$$k_0 I + \mu(A - k_1 I)$$

is invertible for all $t \in \mathbb{T}^\kappa$, and $k_0 - \mu k_1 \neq 0$ on $\mathbb{T}$. The class of such regressive and rd-continuous functions is denoted, similar to the scalar case, by

$$\mathscr{R}_c = \mathscr{R}_c(\mathbb{T}) = \mathscr{R}_c(\mathbb{T},\mathbb{R}^{n\times n}).$$

Theorem 3.1.24 *The matrix-valued function A is conformable regressive if the eigenvalues λ_j of A are conformable regressive for all $1 \leq i \leq n$ and (1.6) holds.*

Proof 3.1.25 *Let $j \in \{1,\ldots,n\}$ be arbitrarily chosen, and let $\lambda_j(t)$ be an eigenvalue corresponding to the eigenfunction $y_j(t)$. Then*

$$(k_0 + \mu(\lambda_j - k_1))y_j = k_0 I y_j + \mu(\lambda_j y_j - k_1 I y_j)$$

$$= k_0 I y_j + \mu(A y_j - k_1 I y_j)$$

$$= (k_0 I + \mu(A_j - k_1 I))y_j,$$

whereupon it follows the desired result. This completes the proof. □

Theorem 3.1.26 *Let A be a 2×2 matrix-valued function. Then A is conformable regressive on $\mathbb{T}$ if and only if*

$$k_0^2 + \mu^2 k_1 + \mu(k_0 - \mu k_1)trA + \mu^2 \det A \neq 0 \quad on \quad \mathbb{T}$$

and (1.6) holds.

Proof 3.1.27 *Let*

$$A = \begin{pmatrix} a_{11} & a_{12} \\ a_{21} & a_{22} \end{pmatrix}.$$

Then

$$\begin{aligned} k_0 I + \mu (A - k_1 I) &= \begin{pmatrix} k_0 & 0 \\ 0 & k_0 \end{pmatrix} + \mu \left(\begin{pmatrix} a_{11} & a_{12} \\ a_{21} & a_{22} \end{pmatrix} - \begin{pmatrix} k_1 & 0 \\ 0 & k_1 \end{pmatrix} \right) \\ &= \begin{pmatrix} k_0 & 0 \\ 0 & k_0 \end{pmatrix} + \mu \begin{pmatrix} a_{11} - k_1 & a_{12} \\ a_{21} & a_{22} - k_1 \end{pmatrix} \\ &= \begin{pmatrix} k_0 + \mu(a_{11} - k_1) & \mu a_{12} \\ \mu a_{21} & k_0 + \mu(a_{22} - k_1) \end{pmatrix}. \end{aligned}$$

Hence,

$$\det\left(k_0 I + \mu(A - k_1 I)\right) = \det \begin{pmatrix} k_0 + \mu(a_{11} - k_1) & \mu a_{12} \\ \mu a_{21} & k_0 + \mu(a_{22} - k_1) \end{pmatrix}$$

$$= \left(k_0 + \mu(a_{11} - k_1)\right)\left(k_0 + \mu(a_{22} - k_1)\right) - \mu^2 a_{21} a_{12}$$

$$= k_0^2 + k_0 \mu(a_{22} - k_1) + k_0 \mu(a_{11} - k_1)$$

$$\quad + \mu^2 (a_{11} - k_1)(a_{22} - k_1) - \mu^2 a_{21} a_{12}$$

$$= k_0^2 + k_0 \mu a_{22} - k_0 k_1 \mu + k_0 \mu a_{11} - k_0 k_1 \mu$$

$$\quad + \mu^2 \left(a_{11} a_{22} - k_1 a_{11} - k_1 a_{22} + k_1^2\right) - \mu^2 a_{21} a_{12}$$

$$= k_0^2 + k_0 \mu(a_{11} + a_{22}) + \mu^2(a_{11} a_{22} - a_{21} a_{12})$$

$$\quad - \mu^2 k_1(a_{11} + a_{22}) + \mu^2 k_1^2$$

$$= k_0^2 + k_0 \mu tr A + \mu^2 det A - \mu^2 k_1 tr A + \mu^2 k_1^2$$

$$= k_0^2 + \mu^2 k_1 + \mu(k_0 - \mu k_1) tr A + \mu^2 det A.$$

This completes the proof. □

Definition 3.1.28 *Let A and B be regressive on* $\mathbb{T}$. *Then we define*

$$A \oplus_c B = (A + B - k_1 I) + \mu k_0^{-1}(A - k_1 I)(B - k_1),$$

$$\ominus_c A \;=\; -k_0 \left(k_0 I + \mu(A - k_1 I)\right)^{-1} (A - k_1 I),$$

$$A \ominus_c B \;=\; A \oplus (\ominus_c B).$$

Theorem 3.1.29 $(\mathscr{R}_c, \oplus_c)$ *is a group.*

Proof 3.1.30 *Let* $A, B < C \in \mathscr{R}_c$. *Then*

$$k_0 - \mu k_1 \neq 0 \quad on \quad \mathbb{T},$$

and

$$\left(k_0 I + \mu(A - k_1 I)\right)^{-1}, \quad \left(k_0 I + \mu(B - k_1 I)\right)^{-1}, \quad \left(k_0 I + \mu(C - k_1 I)\right)^{-1}$$

exist. We have

$$k_0 I + \mu(A \oplus_c B - k_1 I)$$

$$= \; k_0 I + \mu \left(k_0^{-1} \left((A + B - k_1 I)k_0 + \mu(A - k_1 I)(B - k_1 I)\right) - k_1 I \right)$$

$$= \; k_0^{-1} \left(k_0^2 I + \mu \left((A + B - k_1 I)k_0 + \mu(A - k_1 I)(B - k_1 I)\right) - \mu k_0 k_1 I \right)$$

$$= \; k_0^{-1} \left(k_0^2 I + \mu k_0 (A - k_1 I) + \mu k_0 B + \mu(A - k_1 I)(B - k_1 I) - \mu k_0 k_1 I \right)$$

$$= \; k_0^{-1} \left(k_0^2 I + \mu k_0 (A - k_1 I) + \mu k_0 (B - k_1 I) + \mu(A - k_1 I)(B - k_1 I) \right)$$

$$= \; k_0^{-1} \left(k_0 (k_0 I + \mu(A - k_1 I)) + \mu(k_0 I + \mu(A - k_1 I))(B - k_1 I) \right)$$

$$= \; k_0^{-1} \left(k_0 I + \mu(A - k_1 I) \right) \left(k_I + \mu(B - k_1 I) \right),$$

whereupon we conclude that $k_0 I + \mu(A \oplus_c B - k_1 I)$ *is an invertible matrix on* $\mathbb{T}$ *and* $A \oplus_c B \in \mathscr{R}_c$. *Also,*

$$k_0 I + \mu(k_1 I - k_1 I) = k_0 I$$

is invertible and hence, $k_1 I \in \mathscr{R}_c$, *and*

$$(k_1 I) \oplus_c A \;=\; k_0^{-1} \left((k_1 I + A - k_1 I)k_0 + \mu(k_1 I - k_1 I)(A - k_1 I) \right)$$

$$= \; k_0^{-1} k_0 A$$

$$= \; A,$$

$$A \oplus_c (k_1 I) = k_0^{-1} ((A + k_1 I - k_1 I) k_0 + \mu (A - k_1 I)(k_1 I - k_1 I))$$

$$= k_0^{-1} k_0 A$$

$$= A.$$

Consequently, $k_1 I$ is the conformable additive identity for $\oplus_c$. Note that

$$k_0 I + \mu \left(-k_0 (k_0 I + \mu (A - k_1 I))^{-1} (A - k_1 I) + k_1 I - k_1 I \right)$$

$$= k_0 I - \mu k_0 (k_0 I + \mu (A - k_1 I))^{-1} (A - k_1 I)$$

$$= (k_0 I + \mu (A - k_1 I))^{-1} \left(k_0^2 I + \mu k_0 (A - k_1 I) - \mu k_0 (A - k_1 I) \right)$$

$$= k_0^2 (k_0 I + \mu (A - k_1 I))^{-1},$$

i.e.,

$$-k_0 (k_0 I + \mu (A - k_1 I))^{-1} (A - k_1 I) + k_1 I \in \mathscr{R}_c.$$

Next,

$$A \oplus_c \left(-k_0 (k_0 I + \mu (A - k_1 I))^{-1} (A - k_1 I) + k_1 I \right)$$

$${}' = k_0^{-1} \left(k_0 \left(A - k_0 (k_0 I + \mu (A - k_1 I))^{-1} (A - k_1 I) + k_1 I - k_1 I \right) \right.$$

$$\left. + \mu (A - k_1 I) \left(-k_0 (k_0 I + \mu (A - k_1 I))^{-1} (A - k_1 I) + k_1 I - k_1 I \right) \right)$$

$$= k_0^{-1} \left(k_0 \left(A - k_0 (k_0 I + \mu (A - k_1 I))^{-12} (A - k_1 I) \right) \right.$$

$$\left. - \mu k_0 (A - k_1 I) (k_0 I + \mu (A - k_1 I))^{-1} (A - k_1 I) \right)$$

$$= \left(A k_0 (k_0 I + \mu (A - k_1 I))^{-1} (A - k_1 I) \right.$$

$$\left. - \mu (A - k_1 I) (k_0 I + \mu (A - K_1 I))^{-1} (A - k_1 I) \right)$$

$$= A - (k_0 + \mu (A - k_1 I)) (k_0 I + \mu (A - k_1 I))^{-1} (A - k_1 I)$$

$$= A - A + k_1 I$$

$$= k_1 I$$

and

$$\left(-k_0 \left(k_0 I + \mu (A - k_1 I) \right)^{-1} (A - k_1 I) + k_1 I \right) \oplus_c A$$

$$= -k_0 \left(k_0 I + \mu (A - k_1 I) \right)^{-1} (A - k_1 I) + k_1 I + A - k_1 I$$

$$+ \mu k_0^{-1} \left(-k_0 \left(k_0 I + \mu (A - k_1 I) \right)^{-1} (A - k_1 I) + k_1 I + k_1 I - k_1 I \right) (A - k_1 I)$$

$$= -k_0 \left(k_0 I + \mu (A - k_1 I) \right)^{-1} (A - k_1 I) + A$$

$$- \mu \left(k_0 I + \mu (A - k_1 I) \right)^{-1} (A - k_1 I)(A - k_1 I)$$

$$= - \left(k_0 I + \mu (A - k_1 I) \right)^{-1} \left(k_0 I + \mu (A - k_1 I) \right) (A - k_1 I) + A$$

$$= -A + k_1 I + A$$

$$= k_1 I,$$

i.e., the conformable additive inverse of the matrix A is the matrix

$$-k_0 \left(k_0 I + \mu (A - k_1 I) \right)^{-1} (A - k_1 I) + k_1 I.$$

Let

$$P = A \oplus_c B, \quad Q = B \oplus_c C.$$

Then

$$P = A + B - k_1 I + \mu k_0^{-1} (A - k_1 I)(B - k_1 I),$$

$$Q = B + C - k_1 I + \mu k_0^{-1} (B - k_1 I)(C - k_1 I),$$

and

$$P \oplus_c C = P + C - k_1 I + \mu k_0^{-1} (P - k_1 I)(C - k_1 I)$$

$$= A + B - k_1 I + \mu k_0^{-1} (A - k_1 I)(B - k_1 I) + C - k_1 I$$

$$+ \mu k_0^{-1} \left(A + B - k_1 I + \mu k_0^{1} (A - k_1 I)(B - k_1 I) - k_1 I \right) (C - k_1 I)$$

$$= A + B + C - 2k_1I + \mu k_0^{-1}(A - k_1I)(B - k_1I)$$

$$+ \mu k_0^{-1}(A - k_1I)(C - k_1I) + \mu k_0^{-1}(B - k_1I)(C - k_1I)$$

$$+ \left(\mu k_0^{-1}\right)^2 (A - k_1I)(B - k_1I)(C - k_1I),$$

$$A \oplus_c Q = A + Q - k_1I + \mu k_{)}^{-1}(A - k_1I)(Q - k_1I)$$

$$= A + B + C - k_1I + \mu k_0^{-1}(B - k_1I)(C - k_1I) - k_1I$$

$$+ \mu k_{)}^{-1}\left(B + C - k_1I + \mu k_0^{-1}(B - k_1I)(C - k_1I) - k_1I\right)$$

$$= A + B + C - 2k_1I + \mu k_0^{-1}(B - k_1I)(C - k_1I)$$

$$+ \mu k_0^{-1}(A - k_1I)(B - k_1I) + \mu k_0^{-1}(A - k_1I)(C - k_1I)$$

$$+ \left(\mu k_0^{-1}\right)^2 (A - k_1I)(B - k_1I)(C - k_1I).$$

Therefore

$$A \oplus_c Q = P \oplus_c C$$

or

$$A \oplus_c (B \oplus_c C) = (A \oplus_c B) \oplus_c C.$$

This completes the proof. □

Example 3.1.31 *Let* $\mathbb{T} = 2^{\mathbb{N}_0}$,

$$k_1(\alpha, t) = (1 - \alpha)t^\alpha, \quad k_0(\alpha, t) = \alpha t^{1-\alpha}, \quad \alpha \in (0, 1], \quad t \in \mathbb{T},$$

and

$$A(t) = \begin{pmatrix} 1 & t \\ t+1 & 3 \end{pmatrix}, \quad B(t) = \begin{pmatrix} -1 & t \\ t & 2 \end{pmatrix}, \quad {}'t \in \mathbb{T}.$$

We will find

$$(A \oplus_c B)(t), \quad t \in \mathbb{T}.$$

Here

$$\mu(t) = t, \quad t \in \mathbb{T}.$$

Then

$$B(t) - k_1(\alpha, t)I = \begin{pmatrix} -1 & t \\ t & 2 \end{pmatrix} - (1 - \alpha)t^\alpha \begin{pmatrix} 1 & 0 \\ 0 & 1 \end{pmatrix}$$

$$= \begin{pmatrix} -1 & t \\ t & 2 \end{pmatrix} - \begin{pmatrix} (1-\alpha)t^\alpha & 0 \\ 0 & (1-\alpha)t^\alpha \end{pmatrix}$$

$$= \begin{pmatrix} -1-(1-\alpha)t^\alpha & t \\ t & 2-(1-\alpha)t^\alpha \end{pmatrix},$$

$$A(t) - k_1(\alpha,t)I = \begin{pmatrix} 1 & t \\ t+1 & 3 \end{pmatrix} - (1-\alpha)t^\alpha \begin{pmatrix} 1 & 0 \\ 0 & 1 \end{pmatrix}$$

$$= \begin{pmatrix} 1-(1-\alpha)t^\alpha & t \\ t+1 & 3-(1-\alpha)t^\alpha \end{pmatrix},$$

$$(A(t) - k_1(\alpha,t)I)(B(t) - k_1(\alpha,t)I) = \begin{pmatrix} -1-(1-\alpha)t^\alpha & t \\ t & 2-(1-\alpha)t^\alpha \end{pmatrix}$$

$$\times \begin{pmatrix} 1-(1-\alpha)t^\alpha & t \\ t+1 & 3-(1-\alpha)t^\alpha \end{pmatrix}$$

$$= \begin{pmatrix} -1+(1-\alpha)^2t^{2\alpha}+t^2 & 3t-2(1-\alpha)t^{1+\alpha} \\ 2t-1-(1-\alpha)t^\alpha & t^2+t+6-(1-\alpha)t^\alpha-(1-\alpha)^2t^{2\alpha} \end{pmatrix},$$

$$\mu(t)(k_0(\alpha,t))^{-1}(A(t)-k_1(\alpha,t)I)(B(t)-k_1(\alpha,t)I)$$

$$= \frac{t}{\alpha t^{1-\alpha}} \begin{pmatrix} -1+(1-\alpha)^2t^{2\alpha}+t^2 & 3t-2(1-\alpha)t^{1+\alpha} \\ 2t-1-(1-\alpha)t^\alpha & t^2+t+6-(1-\alpha)t^\alpha-(1-\alpha)^2t^{2\alpha} \end{pmatrix}$$

$$= \begin{pmatrix} -\frac{1}{\alpha}t^\alpha+\frac{(1-\alpha)^2}{\alpha}t^{3\alpha}+\frac{1}{\alpha}t^{2+\alpha} & \frac{3}{\alpha}t^{1+\alpha}-\frac{2}{\alpha}(1-\alpha)t^{1+2\alpha} \\ \frac{2}{\alpha}t^{1+\alpha}-\frac{1}{\alpha}t^\alpha-\frac{1}{\alpha}(1-\alpha)t^{2\alpha} & \frac{1}{\alpha}(t^2+t+6)t^\alpha-\frac{1-\alpha}{\alpha}t^\alpha-\frac{1-\alpha}{\alpha}t^{2\alpha}-\frac{(1-\alpha)^2}{\alpha}t^{3\alpha} \end{pmatrix},$$

$$A(t)+B(t)-k_1(\alpha,t)I = \begin{pmatrix} 1 & t \\ t+1 & 3 \end{pmatrix}+\begin{pmatrix} -1 & t \\ t & 2 \end{pmatrix}-(1-\alpha)t^\alpha\begin{pmatrix} 1 & 0 \\ 0 & 1 \end{pmatrix}$$

$$= \begin{pmatrix} 0 & 2t \\ 2t+1 & 5 \end{pmatrix}+\begin{pmatrix} (\alpha-1)t^\alpha & 0 \\ 0 & (\alpha-1)t^\alpha \end{pmatrix}$$

$$= \begin{pmatrix} (\alpha-1)t^\alpha & 2t \\ 2t+1 & 5+(\alpha-1)t^\alpha \end{pmatrix},$$

$$(A\oplus_c B)(t) = \begin{pmatrix} (\alpha-1)t^\alpha & 2t \\ 2t+1 & 5+(\alpha-1)t^\alpha \end{pmatrix}$$

$$+\left(\begin{array}{cc} -\dfrac{1}{\alpha}t^{\alpha}+\dfrac{(1-\alpha)^2}{\alpha}t^{3\alpha}+\dfrac{1}{\alpha}t^{2+\alpha} & \dfrac{3}{\alpha}t^{1+\alpha}-\dfrac{2}{\alpha}(1-\alpha)t^{1+2\alpha} \\[2mm] \dfrac{2}{\alpha}t^{1+\alpha}-\dfrac{1}{\alpha}t^{\alpha}-\dfrac{1}{\alpha}(1-\alpha)t^{2\alpha} & \dfrac{1}{\alpha}\left(t^2+t+6\right)t^{\alpha}-\dfrac{1-\alpha}{\alpha}t^{\alpha}-\dfrac{1-\alpha}{\alpha}t^{2\alpha}-\dfrac{(1-\alpha)^2}{\alpha}t^{3\alpha} \end{array}\right)$$

$$=\left(\begin{array}{cc} \left(\alpha-1-\dfrac{1}{\alpha}\right)t^{\alpha}+\dfrac{(1-\alpha)^2}{\alpha}t^{3\alpha}+\dfrac{1}{\alpha}t^{2+\alpha} & 2t+\dfrac{3}{\alpha}t^{1+\alpha}-\dfrac{2(1-\alpha)}{\alpha}t^{1+2\alpha} \\[2mm] 1+2t+\dfrac{2}{\alpha}t^{1+\alpha}-\dfrac{1}{\alpha}t^{\alpha}-\dfrac{1-\alpha}{\alpha}t^{2\alpha} & \left(\dfrac{t^2+t+6}{\alpha}+\alpha-1\right)t^{\alpha}+5+\dfrac{3}{\alpha}t^{1+\alpha}-\dfrac{2(1-\alpha)}{\alpha}t^{1+2\alpha} \end{array}\right),$$

$\alpha \in (0,1]$, $t \in \mathbb{T}$. *This ends the example.*

Exercise 3.1.32 *Let* $\mathbb{T} = 2^{\mathbb{N}_0}$,

$$k_1(\alpha,t) = (1-\alpha)(1+t)^{\alpha}, \quad k_0(\alpha,t) = \alpha(1+t)^{1-\alpha}, \quad \alpha \in (0,1], \quad t \in \mathbb{T},$$

and

$$A(t) = \left(\begin{array}{cc} 1 & t \\ t & 2 \end{array}\right), \quad B(t) = \left(\begin{array}{cc} t & 3t \\ 1 & 4 \end{array}\right), \quad t \in \mathbb{T}.$$

Find

1. $(A \oplus_c B)(t)$, $t \in \mathbb{T}$,

2. $(\ominus_c A)(t)$, $t \in \mathbb{T}$,

3. $(\ominus_c B)(t)$, $t \in \mathbb{T}$.

With $\overline{A}$ we will denote the conjugate matrix of the matrix A. With A^T we will denote the transpose matrix of the matrix A, and $A^* = \left(\overline{A}\right)^T$ will be the conjugate transpose matrix of the matrix A.

Theorem 3.1.33 *Let* $A \in \mathscr{R}_c$. *Then* $A^* \in \mathscr{R}_c$.

Proof 3.1.34 *We have*

$$(k_0 I + \mu(A - k_1 I))\,(k_0 I + \mu(A - k_1 I))^{-1} = I \quad on \quad \mathbb{T}.$$

Hence,

$$\overline{(k_0 I + \mu(A - k_1 I))\,(k_0 I + \mu(A - k_1 I))^{-1}} = I \quad on \quad \mathbb{T},$$

and

$$\left((k_0 I + \mu(A - k_1 I))^{-1}\right)^*(k_0 I + \mu(A^* - k_1 I)) = I \quad on \quad \mathbb{T},$$

or

$$(k_0 I + \mu(A^* - k_1 I))^{-1}(k_0 I + \mu(A^* - k_1 I)) = I \quad on \quad \mathbb{T}.$$

Therefore

$$k_0 I + \mu(A^* - k_1 I)$$

is invertible on $\mathbb{T}$. *This completes the proof.* □

Theorem 3.1.35 *Let $A \in \mathscr{R}_c$. Then*

$$(\ominus_c A)^* = \ominus_c A^*.$$

Proof 3.1.36 *We have*

$$
\begin{aligned}
(\ominus_c A)^* &= \left(-k_0 \left(k_0 I + \mu (A - k_1 I) \right)^{-1} (A - k_1 I) + k_1 I \right)^* \\[2mm]
&= -k_0 (A^* - k_1 I) \left(k_0 I + \mu (A^* - k_1 I) \right)^{-1} + k_1 I \\[2mm]
&= -k_0 \left(k_0 I + \mu (A^* - k_1 I) \right)^{-1} (A^* - k_1 I) + k_1 I \\[2mm]
&= \ominus_c A^*.
\end{aligned}
$$

This completes the proof. □

Theorem 3.1.37 *Let $A, B \in \mathscr{R}_c$. Then*

$$(A \oplus_c B)^* = B^* \oplus_c A^*.$$

Proof 3.1.38 *We have*

$$
\begin{aligned}
(A \oplus_c B)^* &= \left(A + B - k_1 I + \mu k_0^{-1} (A - k_1 I)(B - k_1 I) \right)^* \\[2mm]
&= B^* + A^* - k_1 I + \mu k_0^{-1} (B^* - k_1 I)(A^* - k_1 I) \\[2mm]
&= B^* \oplus_c A^*.
\end{aligned}
$$

This completes the proof. □

Definition 3.1.39 (Conformable Matrix Exponential Function) *Let $A \in \mathscr{R}_c$ and $t_0 \in \mathbb{T}$. The unique solution of the IVP*

$$D^\alpha Y = A(t)Y, \quad Y(t_0) = I,$$

is called the conformable matrix exponential function. It is denoted by $E_A(\cdot, t_0)$.

Theorem 3.1.40 *Let $A, B \in \mathscr{R}_c$ and $t, s, r \in \mathbb{T}$. Then:*

1. $E_A(t,t) = I$,

2. The matrix composition with sigma is given by

$$E_A(\sigma(t), s) = \left(\left(1 - \mu(t) \frac{k_1(\alpha, t)}{k_0(\alpha, t)} \right) I + \frac{\mu(t)}{k_0(\alpha, t)} A(t) \right) E_A(t, s),$$

3. *The inverse matrix is given by*

$$E_A(s,t) = (E_A(t,s))^{-1} = (E_C(t,s))^*,$$

where

$$C = k_0 \left(-A^* + k_1 I\right) \left(k_0 I + \mu \left(A^* - k_I\right)\right)^{-1},$$

4. $E_A(t,r) = E_A(t,s) E_A(s,r)$,

5. $E_A(t,s) E_B(t,s) = E_{A \oplus_c B}(t,s)$ *if A and B commute.*

Proof 3.1.41 *1. Consider the IVP*

$$D^\alpha Y = A(t) Y, \quad Y(s) = I.$$

By the definition of $E_A(\cdot,s)$ we have

$$E_A(s,s) = I.$$

2. *We have*

$$
\begin{aligned}
E_A(\sigma(t),s) &= \left(1 - \frac{\mu(t)}{k_0(\alpha,t)} k_1(\alpha,t)\right) E_A(t,s) + \frac{\mu(t)}{k_0(\alpha,t)} D^\alpha E_A(t,s) \\
&= \left(1 - \frac{\mu(t)}{k_0(\alpha,t)} k_1(\alpha,t)\right) E_A(t,s) + \frac{\mu(t)}{k_0(\alpha,t)} A(t) E_A(t,s) \\
&= \left(\left(1 - \frac{\mu(t)}{k_0(\alpha,t)} k_1(\alpha,t)\right) I + \frac{\mu(t)}{k_0(\alpha,t)} A(t)\right) E_A(t,s).
\end{aligned}
$$

3. *Let*

$$Z(t) = \left((E_A(t,s))^{-1}\right)^*.$$

Then

$$D^\alpha Z(t) = \left(-(E_A(t,s))^{-1} D^\alpha E_A(t,s) (E_A(\sigma(t),s))^{-1} + k_1(\alpha,t) (E_A(\sigma(t),s))^{-1}\right)^*$$

$$= \left(-(E_A(t,s))^{-1} A(t) E_A(t,s) (E_A(\sigma(t),s))^{-1} + k_1(\alpha,t) (E_A(\sigma(t),s))^{-1}\right)^*$$

$$= \left((-A(t) + k_1(\alpha,t) I) (E_A(\sigma(t),s))^{-1}\right)^*$$

$$= \left((-A(t) + k_1(\alpha,t) I) \left(\frac{1}{k_0(\alpha,t)} (k_0(\alpha,t) I + \mu(t) (A(t) - k_1(\alpha,t) I)) E_A(t,s)\right)^{-1}\right)^*$$

$$= \left(k_0(\alpha,t) (-A(t) + k_1(\alpha,t) I) (k_0(\alpha,t) I + \mu(t) (A(t) - k_1(\alpha,t) I))^{-1} (E_A(t,s))^{-1}\right)^*$$

$$= k_0(\alpha,t) (-A^*(t) + k_1(\alpha,t) I) (k_0(\alpha,t) I + \mu(t) (A^*(t) - k_1(\alpha,t) I))^{-1} \left((E_A(t,s))^{-1}\right)^*$$

$$= k_0(\alpha,t)\left(-A^*(t)+k_1(\alpha,t)I\right)\left(k_0(\alpha,t)I+\mu(t)\left(A^*(t)-k_1(\alpha,t)I\right)\right)^{-1}Z(t).$$

Hence,

$$\left(\left(E_A(t,s)\right)^{-1}\right)^* = E_C(t,s)$$

or

$$\left(E_A(t,s)\right)^{-1} = E_C^*(t,s).$$

4. *Let*

$$Z(t) = E_A(t,s)E_A(s,r).$$

Then

$$
\begin{aligned}
D^\alpha Z(t) &= D^\alpha E_A(t,s)E_A(s,r) \\
&= A(t)E_A(t,s)E_A(s,r) \\
&= A(t)Z(t)
\end{aligned}
$$

and

$$Z(r) = I.$$

Therefore

$$E_A(t,r) = E_A(t,s)E_A(s,r).$$

5. *Let*

$$Z(t) = E_A(t,s)E_B(t,s).$$

Then

$$
\begin{aligned}
D^\alpha Z(t) &= D^\alpha \left(E_A(t,s)E_B(t,s)\right) \\
&= D^\alpha E_A(t,s)E_B(t,s) + E_A(\sigma(t),s)D^\alpha E_B(t,s) \\
&\quad -k_1(\alpha,s)E_A(\sigma(t),s)E_B(t,s) \\
&= A(t)E_A(t,s)E_B(t,s) + E_A(\sigma(t),s)B(t)E_B(t,s) \\
&\quad -k_1(\alpha,s)E_A(\sigma(t),s)E_B(t,s) \\
&= A(t)E_A(t,s)E_B(t,s) \\
&\quad + \left(\left(1-\mu(t)\frac{k_1(\alpha,t)}{k_0(\alpha,t)}\right)I + \frac{\mu(t)}{k_0(\alpha,t)}A(t)\right)B(t)E_A(t,s)E_B(t,s)
\end{aligned}
$$

$$-k_1(\alpha,t)A(t)E_A(t,s)E_B(t,s)$$

$$+\left(\left(1-\mu(t)\frac{k_1(\alpha,t)}{k_0(\alpha,t)}\right)I+\frac{\mu(t)}{k_0(\alpha,t)}A(t)\right)E_A(t,s)E_B(t,s)$$

$$=A(t)E_A(t,s)E_B(t,s)$$

$$+\left(B(t)+\frac{\mu(t)}{k_0(\alpha,t)}\left(A(t)-k_1(\alpha,t)I\right)B(t)\right)E_A(t,s)E_B(t,s)$$

$$-\frac{k_1(\alpha,t)}{k_0(\alpha,t)}\left(k_0(\alpha,t)I+\mu(t)\left(A(t)-k_1(\alpha,t)I\right)\right)E_A(t,s)E_B(t,s)$$

$$=\left(A(t)+B(t)-k_1(\alpha,t)I+\frac{\mu(t)}{k_0(\alpha,t)}\left(A(t)-k_1(\alpha,t)I\right)\left(B(t)-k_1(\alpha,t)I\right)\right)$$

$$\times E_A(t,s)E_B(t,s)$$

$$=(A\oplus_c B)(t)E_A(t,s)E_B(t,s)$$

$$=(A\oplus_c B)(t)Z(t).$$

Therefore

$$E_{A\oplus_c B}(t,s)=E_A(t,s)E_B(t,s).$$

This completes the proof. □

Theorem 3.1.42 *Let $A\in\mathscr{R}_c$ and $s,t\in\mathbb{T}$. Then*

$$D_t^\alpha\left(E_A(s,t)\right)=\left(-A(t)+k_1(\alpha,t)I\right)E_A(s,\sigma(t)),\quad \alpha\in(0,1].$$

Proof 3.1.43 *We have*

$$D_t^\alpha\left(E_A(s,t)\right)\ =\ D^\alpha\left(\left(E_A(t,s)\right)-1\right)=D_t^\alpha\left(E_C(t,s)\right)^*,$$

where

$$C=k_0\left(-A^*+k_1I\right)\left(k_0I+\mu\left(A^*-k_1I\right)\right)^{-1}.$$

Then

$$D_t^\alpha\left(E_A(s,t)\right)\ =\ \left(D_t^\alpha\left(E_C(t,s)\right)\right)^*$$

$$=\ \left(CE_C(t,s)\right)^*$$

$$=\ \left(E_C(t,s)\right)^*C^*$$

$$= (E_C(t,s))^* k_0 (-A(t) + k_1(\alpha,t)I)(k_0(\alpha,t)I + \mu(t)(A(t) - k_1(\alpha,t)I))^{-1}$$

$$= k_0(-A(t) + k_1(\alpha,t)I)(k_0(\alpha,t)I + \mu(t)(A(t) - k_1(\alpha,t)I))^{-1} E_A(s,t)$$

$$= (-A(t) + k_1(\alpha,t)I) E_A(s,\sigma(t)).$$

This completes the proof. $\qquad\qquad\qquad\qquad\qquad\qquad\qquad\qquad\qquad\square$

Theorem 3.1.44 (Variation of Constants) *Let $A \in \mathscr{R}_c$ and $q : \mathbb{T} \to \mathbb{R}^n$ be rd-continuous. Let also, $t_0 \in \mathbb{T}$, $y_0 \in \mathbb{R}^n$ and assume $k_0 I + \mu A$ is invertible on $\mathbb{T}$. Then the IVP*

$$D^\alpha y = A(t)y + q(t), \quad y(t_0) = y_0, \tag{3.1}$$

has a unique solution $y : \mathbb{T} \to \mathbb{R}^n$ and its solution is given by

$$y(t) = y_0 E_A(t,t_0) + \int_{t_0}^t E_B(\sigma(s),t)q(s)\Delta_{\alpha,t}s, \tag{3.2}$$

where

$$B = (\mu k_1 - k_0)A (k_0 I + \mu A)^{-1}.$$

Proof 3.1.45   *1. Let $y : \mathbb{T} \to \mathbb{R}^n$ be given by (3.2). Then*

$$y(t) = y_0 E_A(t,t_0) + E_B(t_0,t)\int_{t_0}^t E_B(\sigma(s),t_0)q(s)\Delta_{\alpha,t}s, \quad t \in \mathbb{T},$$

and

$$D^\alpha y(t) = y_0 D^\alpha E_A(t,t_0) + D^\alpha \left(E_B(t_0,t)\int_{t_0}^t E_B(\sigma(s),t_0)q(s)\Delta_{\alpha,t}s \right)$$

$$= y_0 A(t)E_S(t,t_0)$$

$$+ (D^\alpha E_B(t_0,t))\int_{t_0}^t E_B(\sigma(s),t_0)q(s)\Delta_{\alpha,t}s$$

$$+ E_B(t_0,\sigma(t))E_B(\sigma(t),t_0)q(t)$$

$$- k_1(\alpha,t)E_B(t_0,\sigma(t))\int_{t_0}^t E_B(\sigma(s),t_0)q(s)\Delta_{\alpha,t}s$$

$$= y_0 A(t)E_A(t,t_0)$$

$$+ (-B(t) + k_1(\alpha,t)I) E_B(t_0,\sigma(t))\int_{t_0}^t E_B(\sigma(s),t_0)q(s)\Delta_{\alpha,t}s$$

$$+q(t)$$

$$-k_1(\alpha,t)E_B(t_0,\sigma(t))\int_{t_0}^{t}E_B(\sigma(s),t_0)q(s)\Delta_{\alpha,t}s$$

$$= \; y_0A(t)E_A(t,t_0)$$

$$-B(t)E_B(t_0,\sigma(t))\int_{t_0}^{t}E_B(\sigma(s),t_0)q(s)\Delta_{\alpha,t}s$$

$$+q(t), \quad t\in\mathbb{T}.$$

Note that

$$E_B(t_0,\sigma(t))=k_0(\alpha,t)\left(k_0(\alpha,t)I+\mu(t)(B(t)-k_1(\alpha,t)I)\right)E_B(t_0,t)$$

and

$$(k_0(\alpha,t)I+\mu(t)A(t))B(t)=(\mu(t)k_1(\alpha,t)I-k_0(\alpha,t)I)A(t),$$

whereupon

$$k_0(\alpha,t)B(t) \;=\; A(t)\left(\mu(t)k_1(\alpha,t)I-k_0(\alpha,t)I-\mu(t)B(t)\right)$$

$$=\; -A(t)\left(k_0(\alpha,t)I-\mu(t)k_1(\alpha,t)I+\mu(t)B(t)\right)$$

and

$$k_0(\alpha,t)B(t)\left(k_0(\alpha,t)I-\mu(t)k_1(\alpha,t)I+\mu(t)B(t)\right)^{-1}=-A(t), \quad t\in\mathbb{T}.$$

Hence,

$$-B(t)E_B(t_0,\sigma(t))=A(t)E_B(t_0,t), \quad t\in\mathbb{T}.$$

Therefore

$$D^{\alpha}y(t) \;=\; y_0A(t)E_A(t,t_0)+A(t)E_B(t_0,t)\int_{t_0}^{t}E_B(\sigma(s),t_0)q(s)\Delta_{\alpha,t}s+q(t)$$

$$=\; A(t)\left(y_0E_A(t,t_0)+\int_{t_0}^{t}E_B(\sigma(s),t)q(s)\Delta_{\alpha,t}s\right)+q(t)$$

$$=\; A(t)y(t)+q(t), \quad t\in\mathbb{T}.$$

Also

$$y(t_0)=y_0.$$

Therefore y satisfies (3.1).

2. *Suppose that y_1 and y_2 are two solutions of the IVP (3.1). Let*

$$z = y_1 - y_2 \quad on \quad \mathbb{T}.$$

Then z is the solution of the IVP

$$D^\alpha z = A(t)z, \quad z(t_0) = 0.$$

Therefore $z = 0$ on $\mathbb{T}$, or $y_1 = y_2$ on $\mathbb{T}$. This completes the proof. ☐

Example 3.1.46 *Suppose that $A \in \mathscr{R}_c$, $k_0 I - \mu A$ and $k_0 I - \mu B$ are invertible on $\mathbb{T}$, where*

$$B = (k_0 - \mu = k_1)(k_0 I - \mu A)^{-1} A.$$

Also, assume $q : \mathbb{T} \to \mathbb{R}^n$ is rd-continuous. Consider the IVP

$$D^\alpha y = A(t)y^\sigma + q(t), \quad y(t_0) = y_0.$$

Since

$$y^\sigma(t) = \left(1 - \frac{\mu(t)}{k_0(\alpha,t)}k_1(\alpha,t)\right)y(t) + \frac{\mu(t)}{k_0(\alpha,t)}D^\alpha y(t), \quad t \in \mathbb{T},$$

we get

$$
\begin{aligned}
D^\alpha y(t) &= \left(1 - \frac{\mu(t)}{k_0(\alpha,t)}k_1(\alpha,t)\right)A(t)y(t) + \frac{\mu(t)}{k_0(\alpha,t)}A(t)D^\alpha y(t) \\
&\quad + q(t), \quad t \in \mathbb{T},
\end{aligned}
$$

or

$$(k_0(\alpha,t)I - \mu(t)A(t))D^\alpha y(t) = (k_0(\alpha,t) - \mu(t)k_1(\alpha,t))A(t)y(t) + k_0(\alpha,t)q(t), \quad t \in \mathbb{T},$$

or

$$
\begin{aligned}
D^\alpha y(t) &= (k_0(\alpha,t) - \mu(t)k_1(\alpha,t))(k_0(\alpha,t)I - \mu(t)A(t))^{-1}A(t)y(t) \\
&\quad + k_0(\alpha,t)(k_0(\alpha,t)I - \mu(t)A(t))^{-1}q(t) \\
&= B(t)y(t) + k_0(\alpha,t)(k_0(\alpha,t)I - \mu(t)A(t))^{-1}q(t), \quad t \in \mathbb{T}.
\end{aligned}
$$

Hence, and considering Theorem 3.1.44, we find

$$y(t) = y_0 E_B(t,t_0) + \int_{t_0}^t E_C(\sigma(s),t)p(s)\Delta_{\alpha,t}s, \quad t \in \mathbb{T},$$

where

$$C = (\mu k_1 - k_0)B(k_0 I + \mu B)^{-1} \quad on \quad \mathbb{T}$$

and

$$p = k_0(k_0 I - \mu A)^{-1}q \quad on \quad \mathbb{T}.$$

This ends the example.

Exercise 3.1.47 *Let* $\mathbb{T} = \mathbb{Z}$ *and*

$$k_1(\alpha,t) = (1-\alpha)\left(1+t^2\right)^\alpha, \quad k_0(\alpha,t) = \alpha\left(1+t^2\right)^{1-\alpha}, \quad \alpha \in (0,1], \quad t \in \mathbb{T}.$$

Find the solution of the IVP (3.1), where

$$A(t) = \begin{pmatrix} 1 & t \\ 2 & 1 \end{pmatrix}, \quad q(t) = \begin{pmatrix} t \\ 2t+1 \end{pmatrix}, \quad y_0 = \begin{pmatrix} 1 \\ 0 \end{pmatrix}, \quad t \in \mathbb{T}.$$

Theorem 3.1.48 *Let* $A \in \mathscr{R}_c$ *be a* 2×2 *matrix and assume that* X *is a solution of the system*

$$D^\alpha X = A(t)X.$$

Then X *satisfies Liouville's formula*

$$\det X(t) = E_C(t,t_0)\det X(t_0), \quad t \in \mathbb{T}^\kappa,$$

where

$$\det X(t) = E_C(t,t_0)\det X(t_0), \quad t \in \mathbb{T}^\kappa,$$

where

$$C(t) = trA(t) + \frac{\mu(t)}{k_0(\alpha,t)}\det A(t) - k_1(\alpha,t)\left(1 - \frac{\mu(t)}{k_0(\alpha,t)}k_1(\alpha,t) + \frac{\mu(t)}{k_0(\alpha,t)}trA(t)\right),$$

$t \in \mathbb{T}.$

Proof 3.1.49 *Let*

$$A(t) = \begin{pmatrix} a_{11}(t) & a_{12}(t) \\ a_{21}(t) & a_{22}(t) \end{pmatrix}, \quad X(t) = \begin{pmatrix} x_{11}(t) & x_{12}(t) \\ x_{21}(t) & x_{22}(t) \end{pmatrix}.$$

Then

$$D^\alpha X(t) = \begin{pmatrix} D^\alpha x_{11}(t) & D^\alpha x_{12}(t) \\ D^\alpha x_{21}(t) & D^\alpha x_{22}(t) \end{pmatrix} \quad on \quad \mathbb{T}^\kappa$$

and

$$\begin{pmatrix} D^\alpha x_{11}(t) & D^\alpha x_{12}(t) \\ D^\alpha x_{21}(t) & D^\alpha x_{22}(t) \end{pmatrix} = \begin{pmatrix} a_{11}(t) & a_{12}(t) \\ a_{21}(t) & a_{22}(t) \end{pmatrix}\begin{pmatrix} x_{11}(t) & x_{12}(t) \\ x_{21}(t) & x_{22}(t) \end{pmatrix}$$

on $\mathbb{T}^\kappa$. *Therefore we get the system*

$$\begin{cases} D^\alpha x_{11}(t) = a_{11}(t)x_{11}(t) + a_{12}(t)x_{21}(t) \\[2ex] D^\alpha x_{12}(t) = a_{11}(t)x_{12}(t) + a_{12}(t)x_{22}(t) \\[2ex] D^\alpha x_{21}(t) = a_{21}(t)x_{11}(t) + a_{22}(t)x_{21}(t) \\[2ex] D^\alpha x_{22}(t) = a_{21}(t)x_{12}(t) + a_{22}(t)x_{22}(t), \quad t \in \mathbb{T}^\kappa. \end{cases}$$

Let

$$p_{11}(t) = \left(1 - \frac{\mu(t)}{k_0(\alpha,t)}k_1(\alpha,t)\right)x_{11}(t) + \frac{\mu(t)}{k_0(\alpha,t)}D^\alpha x_{11}(t),$$

$$p_{12}(t) = \left(1 - \frac{\mu(t)}{k_0(\alpha,t)}k_1(\alpha,t)\right)x_{12}(t) + \frac{\mu(t)}{k_0(\alpha,t)}D^\alpha x_{12}(t),$$

$$p_{21}(t) = D^\alpha x_{21}(t),$$

$$p_{22}(t) = D^\alpha x_{22}(t),$$

$$q_{11}(t) = \left(1 - \frac{\mu(t)}{k_0(\alpha,t)}k_1(\alpha,t)\right)x_{11}(t) + \frac{\mu(t)}{k_0(\alpha,t)}D^\alpha x_{11}(t),$$

$$q_{12}(t) = \left(1 - \frac{\mu(t)}{k_0(\alpha,t)}k_1(\alpha,t)\right)x_{12}(t) + \frac{\mu(t)}{k_0(\alpha,t)}D^\alpha x_{12}(t),$$

$$q_{21}(t) = x_{21}(t),$$

$$q_{22}(t) = x_{22}(t), \quad t \in \mathbb{T}^\kappa,$$

and

$$P(t) = \begin{pmatrix} p_{11}(t) & p_{12}(t) \\ p_{21}(t) & p_{22}(t) \end{pmatrix}, \quad Q(t) = \begin{pmatrix} q_{11}(t) & q_{12}(t) \\ q_{21}(t) & q_{22}(t) \end{pmatrix}, \quad t \in \mathbb{T}^\kappa.$$

Next,

$$\det(X(t)) = x_{11}(t)x_{22}(t) - x_{12}(t)x_{21}(t), \quad t \in \mathbb{T},$$

and

$$D^\alpha\left(\det(X(t))\right)$$
$$= D^\alpha(x_{11}(t)x_{22}(t)) - D^\alpha(x_{12}(t)x_{21}(t))$$

$$= D^\alpha x_{11}(t)x_{22}(t) + x_{11}^\sigma(t)D^\alpha x_{22}(t) - k_1(\alpha,t)x_{11}^\sigma(t)x_{22}(t)$$

$$-D^\alpha x_{12}(t)x_{21}(t) - x_{12}^\sigma(t)D^\alpha x_{21}(t) + k_1(\alpha,t)x_{12}^\sigma(t)x_{21}(t)$$

$$= \det\begin{pmatrix} D^\alpha x_{11}(t) & D^\alpha x_{12}(t) \\ x_{21}(t) & x_{22}(t) \end{pmatrix} + \det\begin{pmatrix} x_{11}^\sigma(t) & x_{12}^\sigma(t) \\ D^\alpha x_{21}(t) & D^\alpha x_{22}(t) \end{pmatrix}$$

$$-k_1(\alpha,t)\det\begin{pmatrix} x_{11}^\sigma(t) & x_{12}^\sigma(t) \\ x_{21}(t) & x_{22}(t) \end{pmatrix}$$

$$= \det\begin{pmatrix} a_{11}(t)x_{11}(t) + a_{12}(t)x_{21}(t) & a_{11}(t)x_{12}(t) + a_{12}(t)x_{22}(t) \\ x_{21}(t) & x_{22}(t) \end{pmatrix}$$

$$+\det P(t) - k_1(\alpha,t)\det Q(t)$$

$$= a_{11}(t)\det X(t) + \left(1 - \frac{\mu(t)}{k_0(\alpha,t)}k_1(\alpha,t)\right)\det\begin{pmatrix} x_{11}(t) & x_{12}(t) \\ D^{\alpha}x_{21}(t) & D^{\alpha}x_{22}(t) \end{pmatrix}$$

$$+ \frac{\mu(t)}{k_0(\alpha,t)}\det\begin{pmatrix} D^{\alpha}x_{11}(t) & D^{\alpha}x_{12}(t) \\ D^{\alpha}x_{21}(t) & D^{\alpha}x_{22}(t) \end{pmatrix}$$

$$- k_1(\alpha,t)\left(1 - \frac{\mu(t)}{k_0(\alpha,t)}k_1(\alpha,t)\right)\det X(t)$$

$$- k_1(\alpha,t)\frac{\mu(t)}{k_0(\alpha,t)}\begin{pmatrix} D^{\alpha}x_{11}(t) & D^{\alpha}x_{12}(t) \\ x_{21}(t) & x_{22}(t) \end{pmatrix}$$

$$= \left(a_{11}(t) - k_1(\alpha,t)\left(1 - \frac{\mu(t)}{k_0(\alpha,t)}k_1(\alpha,t)\right)\right)\det X(t)$$

$$+ \left(1 - \frac{\mu(t)}{k_0(\alpha,t)}k_1(\alpha,t)\right)$$

$$\times \det\begin{pmatrix} x_{11}(t) & x_{12}(t) \\ a_{21}(t)x_{11}(t)+a_{22}(t)x_{21}(t) & a_{21}(t)x_{12}(t)+a_{22}(t)x_{22}(t) \end{pmatrix}$$

$$- k_1(\alpha,t)\frac{\mu(t)}{k_0(\alpha,t)}\begin{pmatrix} a_{11}(t)x_{11}(t)+a_{12}(t)x_{21}(t) & a_{11}(t)x_{11}(t)+a_{22}(t)x_{21}(t) \\ x_{21}(t) & x_{22}(t) \end{pmatrix}$$

$$+ \frac{\mu(t)}{k_0(\alpha,t)}\left(D^{\alpha}x_{11}(t)D^{\alpha}x_{22}(t) - D^{\alpha}x_{12}(t)D^{\alpha}x_{21}(t)\right)$$

$$= \left(a_{11}(t) - k_1(\alpha,t)\left(1 - \frac{\mu(t)}{k_0(\alpha,t)}k_1(\alpha,t)\right)\right)\det X(t)$$

$$+ a_{22}(t)\left(1 - \frac{\mu(t)}{k_0(\alpha,t)}k_1(\alpha,t)\right)\det X(t)$$

$$- a_{11}(t)k_1(\alpha,t)\frac{\mu(t)}{k_0(\alpha,t)}\det X(t)$$

$$+ \frac{\mu(t)}{k_0(\alpha,t)}\Big((a_{11}(t)x_{11}(t)+a_{12}(t)x_{21}(t))(a_{21}(t)x_{12}(t)+a_{22}(t)x_{22}(t))$$

$$- (a_{11}(t)x_{12}(t)+a_{12}(t)x_{22}(t))(a_{21}(t)x_{11}(t)+a_{22}(t)x_{21}(t))\Big)$$

$$= \Big(a_{11}(t) + a_{22}(t) - k_1(\alpha,t)\Big(1 - \frac{\mu(t)}{k_0(\alpha,t)}k_1(\alpha,t)$$

$$+ a_{22}(t)\frac{\mu(t)}{k_0(\alpha,t)} + a_{11}(t)\frac{\mu(t)}{k_0(\alpha,t)}\Big)\Big)\det X(t)$$

$$+ \frac{\mu(t)}{k_0(\alpha,t)}\Big(a_{11}(t)a_{21}(t)x_{11}(t)x_{22}(t) + a_{11}(t)a_{22}(t)x_{11}(t)x_{22}(t)$$

$$+a_{12}(t)a_{21}(t)x_{21}(t)x_{12}(t)+a_{12}(t)a_{22}(t)x_{21}(t)x_{22}(t)$$

$$-a_{11}(t)a_{21}(t)x_{11}(t)x_{12}(t)-a_{11}(t)a_{22}(t)x_{12}(t)x_{21}(t)$$

$$-a_{12}(t)a_{21}(t)x_{11}(t)x_{22}(t)-a_{12}(t)a_{22}(t)x_{21}(t)x_{22}(t)\Big)$$

$$= \Bigg(\Big((a_{11}(t)+a_{22}(t)-k_1(\alpha,t)\Big(\big(1-\frac{\mu(t)}{k_0(\alpha,t)}k_1(\alpha,t) \big)$$

$$+a_{22}(t)\frac{\mu(t)}{k_0(\alpha,t)}+a_{11}(t)\frac{\mu(t)}{k_0(\alpha,t)} \Big) \Big) \det X(t)$$

$$+\frac{\mu(t)}{k_0(\alpha,t)}\big(a_{11}(t)a_{22}(t)\det X(t)-a_{12}(t)a_{22}(t)\det X(t) \big)$$

$$= \Big(trA(t)+\frac{\mu(t)}{k_0(\alpha,t)}\det A(t)$$

$$-k_1(\alpha,t) \Big(1-\frac{\mu(t)}{k_0(\alpha,t)}k_1(\alpha,t)+a_{22}(t)\frac{\mu(t)}{k_0(\alpha,t)}+a_{11}(t)\frac{\mu(t)}{k_0(\alpha,t)} \Big) \Big) \det X(t),$$

$t\in\mathbb{T}^\kappa$, *i.e.,*

$$D^\alpha\left(\det X(t)\right)=C(t)\det X(t),\quad t\in\mathbb{T}^\kappa.$$

Therefore

$$\det X(t)=E_C(t,t_0)\det X(t_0),\quad t\in\mathbb{T}^\kappa.$$

This completes the proof. □

3.2 CONSTANT COEFFICIENTS

In this section we suppose that A is an $n\times n$ constant matrix and $A\in\mathscr{R}_c$. Let $t_0\in\mathbb{T}$. Consider the system

$$D^\alpha x=Ax. \tag{3.3}$$

Theorem 3.2.1 *Let u and v be solutions of* (3.3). *Then*

$$x=au+bv,\quad a,b\in\mathbb{C},$$

is a solution of the system (3.3).

Proof 3.2.2 *We have*

$$D^\alpha u(t) = Au(t),$$

$$D^{\alpha} v(t) \quad = \quad Bv(t),$$

$$D^{\alpha} x(t) \quad = \quad D^{\alpha}(au + bv)(t)$$

$$= \quad aD^{\alpha} u(t) + bD^{\alpha} v(t)$$

$$= \quad aAu(t) + bAv(t)$$

$$= \quad A(au(t)) + A(bv(t))$$

$$= \quad A(au(t) + bv(t))$$

$$= \quad Ax(t), \quad t \in \mathbb{T}^{\kappa}.$$

This completes the proof. □

Theorem 3.2.3 *Let λ, ξ be an eigenpair of A. Then*

$$x(t) = E_{\lambda}(t, t_0)\xi, \quad t \in \mathbb{T}^{\kappa},$$

is a solution of the system (3.3).

Proof 3.2.4 *We have*

$$A\xi = \lambda \xi.$$

Then

$$D^{\alpha} x(t) \quad = \quad D^{\alpha} \left(E_{\lambda}(t, t_0)\xi \right)$$

$$= \quad D^{\alpha} \left(E_{\lambda}(t, t_0) \right) \xi$$

$$= \quad \lambda E_{\lambda}(t, t_0)\xi$$

$$= \quad E_{\lambda}(t, t_0)(\lambda \xi)$$

$$= \quad E_{\lambda}(t, t_0)A\xi$$

$$= \quad A \left(E_{\lambda}(t, t_0)\xi \right)$$

$$= \quad Ax(t), \quad t \in \mathbb{T}^{\kappa}.$$

This completes the proof. □

Example 3.2.5 *Consider the system*

$$\begin{cases} D^\alpha x_1(t) &= -3x_1 - 2x_2 \\ D^\alpha x_2(t) &= 3x_1 + 4x_2. \end{cases}$$

Here

$$A = \begin{pmatrix} -3 & -2 \\ 3 & 4 \end{pmatrix}.$$

Then

$$\begin{aligned} 0 &= \det \begin{pmatrix} -3 - \lambda & -2 \\ 3 & 4 - \lambda \end{pmatrix} \\ &= (\lambda - 4)(\lambda + 3) + 6 \\ &= \lambda^2 - \lambda - 6 \end{aligned}$$

and

$$\lambda_1 = 3, \quad \lambda_2 = -2.$$

The considered system is conformable regressive for any time scale for which $-2 \in \mathscr{R}_c$. *Note that*

$$\xi_1 = \begin{pmatrix} 1 \\ -3 \end{pmatrix}, \quad \xi_2 = \begin{pmatrix} -2 \\ 1 \end{pmatrix}$$

are eigenvectors corresponding to λ_1 *and* λ_2, *respectively. Therefore*

$$\begin{aligned} x(t) &= c_1 E_3(t, t_0) \xi_1 + c_2 E_{-2}(t, t_0) \xi_2 \\ &= c_1 E_3(t, t_0) \begin{pmatrix} 1 \\ -3 \end{pmatrix} + c_2 E_{-2}(t, t_0) \begin{pmatrix} -2 \\ 1 \end{pmatrix}, \end{aligned}$$

where c_1 *and* c_2 *are real constants, is a solution of the considered system for any time scale for which* $-2 \in \mathscr{R}_c$.

Example 3.2.6 *Consider the system*

$$\begin{cases} D^\alpha x_1(t) &= x_1(t) - x_2(t) \\ D^\alpha x_2(t) &= -x_1(t) + 2x_2(t) - x_3(t) \\ D^\alpha x_3(t) &= -x_2(t) + x_3(t). \end{cases}$$

Here

$$A = \begin{pmatrix} 1 & -1 & 0 \\ -1 & 2 & -1 \\ 0 & -1 & 1 \end{pmatrix}.$$

Then

$$0 = \det(A - \lambda I)$$

$$= \det \begin{pmatrix} 1-\lambda & -1 & 0 \\ -1 & 2-\lambda & -1 \\ 0 & -1 & 1-\lambda \end{pmatrix}$$

$$= -(\lambda-1)^2(\lambda-2)+(\lambda-1)+(\lambda-1)$$

$$= (\lambda-1)(-(\lambda-1)(\lambda-2)+2)$$

$$= (\lambda-1)(-\lambda^2+3\lambda)$$

$$= -\lambda(\lambda-1)(\lambda-3).$$

Therefore

$$\lambda_1 = 0, \quad \lambda_2 = 1, \quad \lambda_3 = 3.$$

Note that the matrix A is conformable regressive for any time scale and

$$\xi_1 = \begin{pmatrix} 1 \\ 1 \\ 1 \end{pmatrix}, \quad \xi_2 = \begin{pmatrix} 1 \\ 0 \\ -1 \end{pmatrix}, \quad \xi_3 = \begin{pmatrix} 1 \\ -2 \\ 1 \end{pmatrix}$$

are eigenvectors corresponding to λ_1, λ_2 and λ_3, respectively. Consequently,

$$x(t) = c_1 E_0(t,t_0)\xi_1 + c_2 E_1(t,t_0)\xi_2 + c_3 E_3(t,t_0)\xi_3$$

$$= c_1 E_0(t,t_0)\begin{pmatrix} 1 \\ 1 \\ 1 \end{pmatrix} + c_2 E_1(t,t_0)\begin{pmatrix} 1 \\ 0 \\ -1 \end{pmatrix} + c_3 E_3(t,t_0)\begin{pmatrix} 1 \\ -2 \\ 1 \end{pmatrix},$$

where c_1, c_2 and c_3 are constants, is a general solution of the considered system.

Example 3.2.7 *Consider the system*

$$\begin{cases} D^\alpha x_1(t) = -x_1(t) + x_2(t) + x_3(t) \\[2mm] D^\alpha x_2(t) = x_2(t) - x_3(t) + x_4(t) \\[2mm] D^\alpha x_3(t) = 2x_3(t) - 2x_4(t) \\[2mm] D^\alpha x_4(t) = 3x_4(t). \end{cases}$$

Here

$$A = \begin{pmatrix} -1 & 1 & 1 & 0 \\ 0 & 1 & -1 & 1 \\ 0 & 0 & 2 & -2 \\ 0 & 0 & 0 & 3 \end{pmatrix}.$$

Then

$$0 = \det(A - \lambda I)$$

$$= \det \begin{pmatrix} -1-\lambda & 1 & 1 & 0 \\ 0 & 1-\lambda & -1 & 1 \\ 0 & 0 & 2-\lambda & -2 \\ 0 & 0 & 0 & 3-\lambda \end{pmatrix}$$

$$= (\lambda+1)(\lambda-1)(\lambda-2)(\lambda-3)$$

and

$$\lambda_1 = -1, \quad \lambda_2 = 1, \quad \lambda_3 = 2, \quad \lambda_4 = 3.$$

The matrix A is conformable regressive for any time scale for which $-1 \in \mathscr{R}_c$. Note that

$$\xi_1 = \begin{pmatrix} 1 \\ 0 \\ 0 \\ 0 \end{pmatrix}, \quad \xi_2 = \begin{pmatrix} 0 \\ 1 \\ 0 \\ 0 \end{pmatrix}, \quad \xi_3 = \begin{pmatrix} 0 \\ 0 \\ 1 \\ 0 \end{pmatrix} \quad and \quad \xi_4 = \begin{pmatrix} 0 \\ 0 \\ 0 \\ 1 \end{pmatrix}$$

are eigenvectors corresponding to λ_1, λ_2, λ_3 and λ_4, respectively. Consequently,

$$x(t) = c_1 E_{-1}(t,t_0)\xi_1 + c_2 E_1(t,t_0)\xi_2 + c_3 E_2(t,t_0)\xi_3 + c_4 E_5(t,t_0)\xi_4$$

$$= c_1 E_{-1}(t,t_0) \begin{pmatrix} 1 \\ 0 \\ 0 \\ 0 \end{pmatrix} + c_2 E_1(t,t_0) \begin{pmatrix} 0 \\ 1 \\ 0 \\ 0 \end{pmatrix}$$

$$+ c_3 E_2(t,t_0) \begin{pmatrix} 0 \\ 0 \\ 1 \\ 0 \end{pmatrix} + c_4 E_3(t,t_0) \begin{pmatrix} 0 \\ 0 \\ 0 \\ 1 \end{pmatrix},$$

where c_1, c_2, c_3 and c_4 are real constants, is a general solution of the considered system.

Exercise 3.2.8 *Find a general solution of the system*

$$\begin{cases} D^\alpha x_1(t) = x_2(t) \\ \\ D^\alpha x_2(t) = x_1(t). \end{cases}$$

Answer.

$$x(t) = c_1 E_1(t,t_0) \begin{pmatrix} 1 \\ 1 \end{pmatrix} + c_2 E_{-1}(t,t_0) \begin{pmatrix} 1 \\ -1 \end{pmatrix},$$

where $c_1, c_2 \in \mathbb{R}$, for any time scale for which $-1 \in \mathcal{R}_c$.

Theorem 3.2.9 *Assume that $A \in \mathcal{R}_c$. If*

$$x(t) = u(t) + iv(t), \quad t \in \mathbb{T}^\kappa,$$

is a complex vector-valued solution of the system (3.3), where $u(t)$ and $v(t)$ are real vector-valued functions on $\mathbb{T}$, then $u(t)$ and $v(t)$ are real vector-valued solutions of the system (3.3) on $\mathbb{T}$.

Proof 3.2.10 *We have*

$$
\begin{aligned}
D^\alpha x(t) &= A(t)x(t) \\[2mm]
&= A(t)(u(t) + iv(t)) \\[2mm]
&= A(t)u(t) + iA(t)v(t) \\[2mm]
&= D^\alpha u(t) + iD^\alpha v(t), \quad t \in \mathbb{T}^\kappa.
\end{aligned}
$$

Equating real and imaginary parts, we get

$$
\begin{aligned}
D^\alpha u(t) &= A(t)u(t), \\[2mm]
D^\alpha v(t) &= A(t)v(t), \quad t \in \mathbb{T}^\kappa.
\end{aligned}
$$

This completes the proof. □

Example 3.2.11 *Consider the system*

$$
\begin{cases}
D^\alpha x_1(t) &= x_1(t) + x_2(t) \\[2mm]
D^\alpha x_2(t) &= -x_1(t) + x_2(t).
\end{cases}
$$

Here

$$A = \begin{pmatrix} 1 & 1 \\ -1 & 1 \end{pmatrix}.$$

Then

$$
\begin{aligned}
0 &= \det(A - \lambda I) \\[2mm]
&= \det \begin{pmatrix} 1 - \lambda & 1 \\ -1 & 1 - \lambda \end{pmatrix}
\end{aligned}
$$

$$= (\lambda - 1)^2 + 1$$

$$= \lambda^2 - 2\lambda + 1 + 1$$

$$= \lambda^2 - 2\lambda + 2,$$

whereupon

$$\lambda_{1,2} = 1 \pm i.$$

Note that

$$\xi = \begin{pmatrix} 1 \\ i \end{pmatrix}$$

is an eigenvector corresponding to the eigenvalue $\lambda = 1 + i$. *We have*

$$x(t) = E_{1+i}(t, t_0) \begin{pmatrix} 1 \\ i \end{pmatrix}$$

$$= E_1(t, t_0) \left(Cos_{\frac{1}{1+\mu}}(t, t_0) + iSin_{\frac{1}{1+\mu}}(t, t_0) \right) \begin{pmatrix} 1 \\ i \end{pmatrix}$$

$$= E_1(t, t_0) \left(\begin{pmatrix} Cos_{\frac{1}{1+\mu}}(t, t_0) \\ iCos_{\frac{1}{1+\mu}}(t, t_0) \end{pmatrix} + \begin{pmatrix} iSin_{\frac{1}{1+\mu}}(t, t_0) \\ -Sin_{\frac{1}{1+\mu}}(t, t_0) \end{pmatrix} \right)$$

$$= E_1(t, t_0) \begin{pmatrix} Cos_{\frac{1}{1\mu}}(t, t_0) \\ -Sin_{\frac{1}{1\mu}}(t, t_0) \end{pmatrix} + iE_1(t, t_0) \begin{pmatrix} Sin_{\frac{1}{1+\mu}}(t, t_0) \\ Cos_{\frac{1}{1+\mu}}(t, t_0) \end{pmatrix}.$$

Consequently,

$$E_1(t, t_0) \begin{pmatrix} Cos_{\frac{1}{1+\mu}}(t, t_0) \\ -Sin_{\frac{1}{1+\mu}}(t, t_0) \end{pmatrix} \quad and \quad E_1(t, t_0) \begin{pmatrix} Sin_{\frac{1}{1+\mu}}(t, t_0) \\ Cos_{\frac{1}{1+\mu}}(t, t_0) \end{pmatrix}$$

are solutions of the considered system. Therefore

$$x(t) = c_1 E_1(t, t_0) \begin{pmatrix} Cos_{\frac{1}{1+\mu}}(t, t_0) \\ -Sin_{\frac{1}{1+\mu}}(t, t_0) \end{pmatrix} + c_2 E_1(t, t_0) \begin{pmatrix} Sin_{\frac{1}{1+\mu}}(t, t_0) \\ Cos_{\frac{1}{1+\mu}}(t, t_0) \end{pmatrix},$$

where $c_1, c_2 \in \mathbb{R}$, *is a general solution of the considered system.*

Example 3.2.12 *Consider the system*

$$\begin{cases} D^\alpha x_1(t) = x_2(t) \\ \\ D^\alpha x_2(t) = x_3(t) \\ \\ D^\alpha x_3(t) = 2x_1(t) - 4x_2(t) + 3x_3(t). \end{cases}$$

Here

$$A = \begin{pmatrix} 0 & 1 & 0 \\ 0 & 0 & 1 \\ 2 & -4 & 3 \end{pmatrix}.$$

Then

$$\begin{aligned} 0 &= \det(A - \lambda I) \\ &= \det \begin{pmatrix} -\lambda & 1 & 0 \\ 0 & -\lambda & 1 \\ 2 & -4 & 3 - \lambda \end{pmatrix} \\ &= -\lambda^2(\lambda - 3) + 2 - 4\lambda \\ &= -(\lambda^3 - 3\lambda^2 + 4\lambda - 2) \\ &= -(\lambda - 1)(\lambda^2 - 2\lambda + 2), \end{aligned}$$

whereupon

$$\lambda_1 = 1, \quad \lambda_{2,3} = 1 \pm i.$$

Note that

$$\xi_1 = \begin{pmatrix} 1 \\ 1 \\ 1 \end{pmatrix} \quad and \quad \xi_2 = \begin{pmatrix} 1 \\ 1+i \\ 2i \end{pmatrix}$$

are eigenvectors corresponding to the eigenvalues $\lambda_1 = 1$ and $\lambda_2 = 1+i$, respectively. Note that

$$E_{1+i}(t,t_0) \begin{pmatrix} 1 \\ 1+i \\ 2i \end{pmatrix} = E_1(t,t_0)\left(Cos_{\frac{1}{1+\mu}}(t,t_0) + iSin_{\frac{1}{1+\mu}}(t,t_0) \right) \begin{pmatrix} 1 \\ 1+i \\ 2i \end{pmatrix}$$

$$= E_1(t,t_0) \left(\begin{pmatrix} Cos_{\frac{1}{1+\mu}}(t,t_0) \\ (1+i)Cos_{\frac{1}{1+\mu}}(t,t_0) \\ 2iCos_{\frac{1}{1+\mu}}(t,t_0) \end{pmatrix} + i \begin{pmatrix} Sin_{\frac{1}{1+\mu}}(t,t_0) \\ (1+i)Sin_{\frac{1}{1+\mu}}(t,t_0) \\ 2iSin_{\frac{1}{1+\mu}}(t,t_0) \end{pmatrix} \right)$$

$$= E_1(t,t_0) \left(\begin{pmatrix} Cos_{\frac{1}{1+\mu}}(t,t_0) \\ (1+i)Cos_{\frac{1}{1+\mu}}(t,t_0) \\ 2iCos_{\frac{1}{1+\mu}}(t,t_0) \end{pmatrix} + \begin{pmatrix} iSin_{\frac{1}{1+\mu}}(t,t_0) \\ (-1+i)Sin_{\frac{1}{1+\mu}}(t,t_0) \\ -2Sin_{\frac{1}{1+\mu}}(t,t_0) \end{pmatrix} \right)$$

$$= E_1(t,t_0) \left(\begin{pmatrix} Cos_{\frac{1}{1+\mu}}(t,t_0) \\ Cos_{\frac{1}{1+\mu}}(t,t_0) - Sin_{\frac{1}{1+\mu}}(t,t_0) \\ -2Sin_{\frac{1}{1+\mu}}(t,t_0) \end{pmatrix} + i \begin{pmatrix} Sin_{\frac{1}{1+\mu}}(t,t_0) \\ Cos_{\frac{1}{1+\mu}}(t,t_0) + Sin_{\frac{1}{1+\mu}}(t,t_0) \\ 2Cos_{\frac{1}{1+\mu}}(t,t_0) \end{pmatrix} \right).$$

Consequently,

$$x(t) = E_1(t,t_0)\left(c_1\begin{pmatrix}1\\1\\1\end{pmatrix} + c_2\begin{pmatrix} Cos_{\frac{1}{1+\mu}}(t,t_0) \\ Cos_{\frac{1}{1+\mu}}(t,t_0) - Sin_{\frac{1}{1+\mu}}(t,t_0) \\ -2Sin_{\frac{1}{1+\mu}}(t,t_0) \end{pmatrix}\right.$$

$$\left. + c_3\begin{pmatrix} Sin_{\frac{1}{1+\mu}}(t,t_0) \\ Cos_{\frac{1}{1+\mu}}(t,t_0) + Sin_{\frac{1}{1+\mu}}(t,t_0) \\ 2Cos_{\frac{1}{1+\mu}}(t,t_0) \end{pmatrix}\right),$$

where $c_1, c_2, c_3 \in \mathbb{R}$, is a general solution of the considered system.

Exercise 3.2.13 *Find a general solution of the system*

$$\begin{cases} D^\alpha x_1(t) &= x_1(t) - 2x_2(t) + x_3(t) \\[2mm] D^\alpha x_2(t) &= -x_1(t) + x_3(t) \\[2mm] D^\alpha x_3(t) &= x_1(t) - 2x_2(t) + x_3(t). \end{cases}$$

Theorem 3.2.14 (Conformable Putzer Algorithm) *Let $A \in \mathscr{R}_c$ be a constant $n \times n$ matrix and $t_0 \in \mathbb{T}$. If $\lambda_1, \lambda_2, \ldots, \lambda_n$ are the eigenvalues of A, then*

$$E_A(t,t_0) = \sum_{k=0}^{n-1} r_{k+1}(t)P_k,$$

where

$$r(t) = \begin{pmatrix} r_1(t) \\ \vdots \\ r_n(t) \end{pmatrix}$$

is the solution of the IVP

$$D^\alpha r = \begin{pmatrix} \lambda_1 & 0 & 0 & \ldots & 0 \\ 1 & \lambda_2 & 0 & \ldots & 0 \\ 0 & 1 & \lambda_3 & \ldots & 0 \\ \vdots & \vdots & \vdots & \vdots & \vdots \\ 0 & 0 & 0 & \ldots & \lambda_n \end{pmatrix} r, \quad r(t_0) = \begin{pmatrix} 1 \\ 0 \\ 0 \\ \vdots \\ 0 \end{pmatrix}, \tag{3.4}$$

$$P_0 = I,$$

$$P_{k+1} = (A - \lambda_{k+1}I)P_k, \quad 0 \le k \le n-1.$$

Proof 3.2.15 *Since A is conformable regressive, we have that all eigenvalues of A are conformable regressive. Therefore the IVP (3.4) has a unique solution. We set*

$$X(t) = \sum_{k=0}^{n-1} r_{k+1}(t)P_k.$$

We have

$$
\begin{aligned}
P_1 &= (A - \lambda_1 I)P_0 \\
&= (A - \lambda_1 I), \\
P_2 &= (A - \lambda_2 I)P_1 \\
&= (A - \lambda_2 I)(A - \lambda_1 I), \\
&\vdots \\
P_n &= (A - \lambda_n I)P_{n-1} \\
&= (A - \lambda_n I)\ldots(A - \lambda_1 I) \\
&= 0.
\end{aligned}
$$

Therefore

$$
\begin{aligned}
D^\alpha X(t) &= \sum_{k=0}^{n-1} D^\alpha r_{k+1}(t)P_k, \\
D^\alpha X(t) - AX(t) &= \sum_{k=0}^{n-1} D^\alpha r_{k+1}(t)P_k - A\sum_{k=0}^{n-1} r_{k+1}(t)P_k \\
&= D^\alpha r_1(t)P_0 + \sum_{k=1}^{n-1} D^\alpha r_{k+1}(t)P_k \\
&\quad -A\sum_{k=0}^{n-1} r_{k+1}(t)P_k \\
&= \lambda_1 r_1(t)P_0 + \sum_{k=1}^{n-1} \left(r_k(t) + \lambda_{k+1} r_{k+1}(t)\right)P_k \\
&\quad -\sum_{k=1}^{n-1} r_{k+1}(t)AP_k
\end{aligned}
$$

$$
= \sum_{k=1}^{n-1} r_k(t)P_k + \lambda_1 r_1(t)P_0
$$

$$
+ \sum_{k=1}^{n-1} \lambda_{k+1} r_{k+1}(t)P_k - \sum_{k=0}^{n-1} r_{k+1}(t)AP_k
$$

$$
= \sum_{k=1}^{n-1} r_k(t)P_k + \sum_{k=0}^{n-1} \lambda_{k+1} r_{k+1}(t)P_k
$$

$$
- \sum_{k=0}^{n-1} r_{k+1}(t)AP_k
$$

$$
= \sum_{k=1}^{n-1} r_k(t)P_k - \sum_{k=0}^{n-1} (A - \lambda_{k+1}I)r_{k+1}(t)P_k
$$

$$
= \sum_{k=1}^{n-1} r_k(t)P_k - \sum_{k=0}^{n-1} r_{k+1}(t)P_{k+1}
$$

$$
= -r_n(t)P_n
$$

$$
= 0, \quad t \in \mathbb{T}^\kappa.
$$

Also,

$$
X(t_0) = \sum_{k=0}^{n-1} r_{k+1}(t_0)P_k
$$

$$
= r_1(t_0)P_0
$$

$$
= I.
$$

This completes the proof. □

Example 3.2.16 *Consider the system*

$$
\begin{cases}
D^\alpha x_1(t) = 2x_1(t) + x_2(t) + 2x_3(t) \\
\\
D^\alpha x_2(t) = 4x_1(t) + 2x_2(t) + 4x_3(t) \\
\\
D^\alpha x_3(t) = 2x_1(t) + x_2(t) + 2x_3(t), \quad t \in \mathbb{T}^\kappa.
\end{cases}
$$

Here

$$
A = \begin{pmatrix} 2 & 1 & 2 \\ 4 & 2 & 4 \\ 2 & 1 & 2 \end{pmatrix}.
$$

Then

$$
0 = \det(A - \lambda I)
$$

$$= \det \begin{pmatrix} 2-\lambda & 1 & 2 \\ 4 & 2-\lambda & 4 \\ 2 & 1 & 2-\lambda \end{pmatrix}$$

$$= -(\lambda-2)^3 + 8 + 8 + 4(\lambda-2) + 4(\lambda-2) + 4(\lambda-2)$$

$$= -(\lambda-2)^3 + 12(\lambda-2) + 16$$

$$= -\left(\lambda^3 - 6\lambda^2 + 12\lambda - 8 - 12\lambda + 24 - 16\right)$$

$$= -\left(\lambda^3 - 6\lambda^2\right)$$

$$= -\lambda^2(\lambda-6),$$

whereupon

$$\lambda_1 = 0, \quad \lambda_2 = 0, \quad \lambda_3 = 6.$$

Consider the IVPs

$$D^\alpha r_1(t) = 0, \quad r_1(t_0) = 1,$$

$$D^\alpha r_2(t) = r_1(t), \quad r_2(t_0) = 0,$$

$$D^\alpha r_3(t) = r_2(t) + 6r_3(t), \quad r_3(t_0) = 0.$$

We have

$$r_1(t) = E_0(t,t_0), \quad t \in \mathbb{T}^\kappa$$

$$D^\alpha r_2(t) = E_0(t,t_0), \quad r_2(t_0) = 0.$$

Then

$$r_2(t) = \int_{t_0}^t E_f(\sigma(s),t)E_0(s,t_0)\Delta_{\alpha,t}s, \quad t \in \mathbb{T}^\kappa,$$

where

$$f(t) = \frac{k_1(\alpha,t)(k_0(\alpha,t) - \mu(t)k_1(\alpha,t))}{k_0(\alpha,t) - \mu(t)k_1(\alpha,t)}, \quad t \in \mathbb{T},$$

and

$$D^\alpha r_3(t) = r_2(t) + 6r_3(t), \quad r_3(t_0) = 0.$$

Therefore

$$r_3(t) = \int_{t_0}^t E_g(\sigma(s),t)r_2(s)\Delta_{\alpha,t}s, \quad t \in \mathbb{T}^\kappa,$$

where

$$g(t) = -\frac{(6-k_1(\alpha,t))(k_0(\alpha,t)-\mu(t)k_1(\alpha,t))}{k_0(\alpha,t)+\mu(t)(6-k_1(\alpha,t))}, \quad t \in \mathbb{T}.$$

Next,

$$P_0 = \begin{pmatrix} 1 & 0 & 0 \\ 0 & 1 & 0 \\ 0 & 0 & 1 \end{pmatrix},$$

$$P_1 = (A - \lambda_1 I)P_0$$

$$= AP_0$$

$$= A$$

$$= \begin{pmatrix} 2 & 1 & 2 \\ 4 & 2 & 4 \\ 2 & 1 & 2 \end{pmatrix},$$

$$P_2 = (A - \lambda_1 I)(A - \lambda_2 I)$$

$$= A^2 I$$

$$= A^2$$

$$= \begin{pmatrix} 2 & 1 & 2 \\ 4 & 2 & 4 \\ 2 & 1 & 2 \end{pmatrix}\begin{pmatrix} 2 & 1 & 2 \\ 4 & 2 & 4 \\ 2 & 1 & 2 \end{pmatrix}$$

$$= \begin{pmatrix} 12 & 6 & 12 \\ 24 & 12 & 24 \\ 12 & 6 & 12 \end{pmatrix},$$

$$P_3 = 0.$$

Therefore

$$E_A(t,t_0) = r_1(t)P_0 + r_2(t)P_1 + r_3(t)P_2$$

$$= r_1(t)\begin{pmatrix} 1 & 0 & 0 \\ 0 & 1 & 0 \\ 0 & 0 & 1 \end{pmatrix} + r_2(t)\begin{pmatrix} 2 & 1 & 2 \\ 4 & 2 & 4 \\ 2 & 1 & 2 \end{pmatrix}$$

$$+ r_3(t)\begin{pmatrix} 12 & 6 & 12 \\ 24 & 12 & 24 \\ 12 & 6 & 12 \end{pmatrix}, \quad t \in \mathbb{T}^\kappa$$

and

$$\begin{pmatrix} x_1(t) \\ x_2(t) \\ x_3(t) \end{pmatrix} = E_A(t,t_0) \begin{pmatrix} c_1 \\ c_2 \\ c_3 \end{pmatrix}, \quad t \in \mathbb{T}^\kappa,$$

where $c_1, c_2, c_3 \in \mathbb{R}$, is a general solution of the considered system. This ends the example.

Exercise 3.2.17 *Using Putzer's algorithm, find $E_A(t,t_0)$, where*

1.

$$A = \begin{pmatrix} 1 & 2 \\ -1 & 3 \end{pmatrix},$$

2.

$$A = \begin{pmatrix} 1 & -1 & 1 \\ 1 & 0 & 2 \\ -1 & 1 & 1 \end{pmatrix}.$$

3.3   ADVANCED PRACTICAL PROBLEMS

Problem 3.3.1 *Let $\mathbb{T} = 3^{\mathbb{N}_0}$,*

$$k_1(\alpha,t) = (1-\alpha)(1+t)^\alpha, \quad k_0(\alpha,t) = \alpha(1+t)^{1-\alpha}, \quad \alpha \in (0,1], \quad t \in \mathbb{T},$$

$$A(t) = \begin{pmatrix} \dfrac{t+1}{t^2+1} & t^2 - 3t \\ t^3 + t^2 & t^4 - t \end{pmatrix}, \quad t \in \mathbb{T}.$$

Find

$$D^{\frac{1}{2}} A(t), \quad t \in \mathbb{T}.$$

Problem 3.3.2 *Let $\mathbb{T} = 4^{\mathbb{N}_0}$,*

$$k_1(\alpha,t) = (1-\alpha)(1+t^2)^{3\alpha}, \quad k_0(\alpha,t) = \alpha(1+t^2)^{3(1-\alpha)}, \quad \alpha \in (0,1], \quad t \in \mathbb{T},$$

and

$$A(t) = \begin{pmatrix} 2t+5 & t \\ t & t+2 \end{pmatrix}, \quad B(t) = \begin{pmatrix} t^2 - t & t \\ 1 & 2 \end{pmatrix}, \quad t \in \mathbb{T}.$$

Find

$$D^{\frac{1}{4}}(AB)(t), \quad t \in \mathbb{T}.$$

Problem 3.3.3 *Let $\mathbb{T} = 4^{\mathbb{N}_0}$ and*

$$A(t) = \begin{pmatrix} \dfrac{1}{t+2} & \dfrac{1}{2t+3} \\ t+1 & t+2 \end{pmatrix}, \quad t \in \mathbb{T}.$$

Prove that

$$(A^\sigma)^{-1}(t) = (A^{-1})^\sigma(t), \quad t \in \mathbb{T}.$$

Problem 3.3.4 *Let* $\mathbb{T} = 3^{\mathbb{N}_0}$,

$$k_1(\alpha,t) = (1-\alpha)(1+t)^{2\alpha}, \quad k_0(\alpha,t) = \alpha(1+t)^{2(1-\alpha)}, \quad \alpha \in (0,1], \quad t \in \mathbb{T},$$

and

$$A(t) = \begin{pmatrix} 1+t & t^2 \\ t+3 & 2 \end{pmatrix}, \quad B(t) = \begin{pmatrix} t+2 & 3t^2+t+1 \\ t+11 & 4t \end{pmatrix}, \quad t \in \mathbb{T}.$$

Find

1. $(A \oplus_c B)(t)$, $t \in \mathbb{T}$,

2. $(\ominus_c A)(t)$, $t \in \mathbb{T}$,

3. $(\ominus_c B)(t)$, $t \in \mathbb{T}$.

Problem 3.3.5 *Let* $\mathbb{T} = 2^{\mathbb{N}_0}$ *and*

$$k_1(\alpha,t) = (1-\alpha)t^{2\alpha}, \quad k_0(\alpha,t) = \alpha t^{2-2\alpha}, \quad \alpha \in (0,1], \quad t \in \mathbb{T}.$$

Find the solution of the IVP (3.1), *where*

$$A(t) = \begin{pmatrix} t & t^2 \\ 2 & 3 \end{pmatrix}, \quad q(t) = \begin{pmatrix} t+4 \\ t \end{pmatrix}, \quad y_0 = \begin{pmatrix} 2 \\ -1 \end{pmatrix}, \quad t \in \mathbb{T}.$$

Problem 3.3.6 *Find a general solution of the system*

$$\begin{cases} D^\alpha x_1(t) = -x_1(t) + x_2(t) \\ \\ D^\alpha x_2(t) = x_1(t) - 3x_2(t). \end{cases}$$

Problem 3.3.7 *Find a general solution of the system*

$$\begin{cases} D^\alpha x_1(t) = -x_1(t) + 2x_2(t)3x_3(t) + x_4(t) \\ \\ D^\alpha x_2(t) = x_1(t) - 3x_2(t) + x_4(T) \\ \\ D^\alpha x_3(t) = -x_1(t) + x_2(t) + x_3(t) + x_4(t) \\ \\ D^\alpha x_4(t) = x_1(t) + x_4(t). \end{cases}$$

Problem 3.3.8 *Using Putzer's algorithm, find* $E_A(t,t_0)$, *where*

1.

$$A = \begin{pmatrix} -1 & 4 \\ 1 & 3 \end{pmatrix},$$

2.

$$A = \begin{pmatrix} 1 & 1 & 2 \\ -1 & 3 & -2 \\ 4 & -1 & -1 \end{pmatrix}.$$

CHAPTER 4

Linear Conformable Inequalities

Suppose that $\mathbb{T}$ is a time scale with forward jump operator and delta differentiation operator σ and Δ, respectively. Let $\alpha \in (0,1]$, k_0 and k_1 satisfy $(A1)$, and assume (1.6) holds. Also, let $a,b \in \mathbb{T}$, $a < b$, and $J = [a,b]$ be a time scale interval. If $A \subset \mathbb{T}$, we define the sets

$$\mathscr{R}_c^+ = \left\{ f \in \mathscr{R}_c : k_0(\alpha,t) + \mu(t)\left(f(t) - k_1(\alpha,t)\right) > 0 \right.$$

$$\left. for \quad all \quad t \in \mathbb{T}, \quad \alpha \in (0,1] \right\},$$

$$\mathscr{R}_c^+(A) = \left\{ f \in \mathscr{R}(A) : k_0(\alpha,t) + \mu(t)\left(f(t) - k_1(\alpha,t)\right) > 0 \right.$$

$$\left. for \quad all \quad t \in A, \quad \alpha \in (0,1] \right\}.$$

4.1 CONFORMABLE GRONWALL INEQUALITY

Theorem 4.1.1 (Conformable Gronwall Inequality) *Let $f,p,q \in \mathscr{C}_{rd}(J)$ be nonnegative functions such that*

$$k_0(\alpha,t) - \mu(t)(-f(t)p(t) + k_1(\alpha,t)) \neq 0, \quad t \in J,$$

and

$$\frac{k_0(\alpha,t) - \mu(t)k_1(\alpha,t)}{k_0(\alpha,t) - \mu(t)(-p(t)f(t) + k_1(\alpha,t))} > 0 \quad t \in J.$$

Also, let $x \in \mathscr{C}_{rd}(J)$ be a nonnegative function. Then the inequality

$$x(t) \leq f(t) \int_a^t p(s)x(s)\Delta_{\alpha,t}s + q(t), \quad t \in J, \tag{4.1}$$

implies the inequality

$$x(t) \leq f(t) \int_a^t p(s)q(s)E_g(\sigma(s),t)\Delta_{\alpha,t}s + q(t), \quad t \in J,$$

where

$$g(t) = \frac{(-f(t)p(t) + k_1(\alpha,t))(k_0(\alpha,t) - \mu(t)k_1(\alpha,t))}{k_0(\alpha,t) - \mu(t)(-f(t)p(t) + k_1(\alpha,t))}, \quad t \in J.$$

Proof 4.1.2 *Let*

$$y(t) = \int_a^t p(s)x(s)\Delta_{\alpha,t}s, \quad t \in J.$$

Then, using (4.1), we get

$$x(t) \leq f(t)y(t) + q(t), \quad t \in J. \tag{4.2}$$

We have

$$\begin{aligned} D^\alpha y(t) &= p(t)x(t) \\ &\leq p(t)(f(t)y(t) + q(t)) \tag{4.3} \\ &= p(t)f(t)y(t) + p(t)q(t), \quad t \in J, \end{aligned}$$

and

$$y(a) = 0.$$

Note that

$$E_g(t,a) > 0, \quad E_g(\sigma(t),s) > 0, \quad t \in J.$$

We multiply both sides of the inequality (4.3) by $E_g(\sigma(t),a)$ and we get

$$\begin{aligned} E_g(\sigma(t),a)D^\alpha y(t) &\leq p(t)f(t)E_g(\sigma(t),a)y(t) \\ &\quad + p(t)q(t)E_g(\sigma(t),a), \quad t \in J. \end{aligned} \tag{4.4}$$

Observe that

$$g(t)E_g(t,a) - k_1(\alpha,t)E_g(\sigma(t),a)$$

$$= g(t)E_g(t,a) - k_1(\alpha,t)\left(1 + \mu(t)\frac{g(t) - k_1(\alpha,t)}{k_0(\alpha,t)}\right)E_g(t,a)$$

$$= \left(g(t) - k_1(\alpha,t)\left(1 + \mu(t)\frac{g(t) - k_1(\alpha,t)}{k_0(\alpha,t)}\right)\right)E_g(t,a), \quad t \in J,$$

and

$$g(t) - k_1(\alpha,t) = \frac{(-f(t)p(t) + k_1(\alpha,t))(k_0(\alpha,t) - \mu(t)k_1(\alpha,t))}{k_0(\alpha,t) - \mu(t)(-f(t)p(t) + k_1(\alpha,t))} - k_1(\alpha,t)$$

$$
= \frac{1}{k_0(\alpha,t) - \mu(t)(-f(t)p(t) + k_1(\alpha,t))} \Bigg(-f(t)p(t)k_0(\alpha,t)
$$

$$
+ \mu(t)f(t)p(t)k_1(\alpha,t) + k_0(\alpha,t)k_1(\alpha,t) - \mu(t)(k_0(\alpha,t))^2
$$

$$
- k_0(\alpha,t)k_1(\alpha,t) - \mu(t)f(t)p(t)k_1(\alpha,t) + \mu(t)(k_1(\alpha,t))^2 \Bigg)
$$

$$
= -\frac{f(t)p(t)k_0(\alpha,t)}{k_0(\alpha,t) - \mu(t)(-f(t)p(t) + k_1(\alpha,t))}, \quad t \in J,
$$

and

$$
g(t)E_g(t,a) - k_1(\alpha,t)E_g(\sigma(t),a)
$$

$$
= g(t)E_g(t,a) - k_1(\alpha,t)\left(1 + \mu(t)\frac{g(t) - k_1(\alpha,t)}{k_0(\alpha,t)}\right)E_g(t,a)
$$

$$
= \left(g(t) - k_1(\alpha,t) - \mu(t)k_1(\alpha,t)\frac{g(t) - k_1(\alpha,t)}{k_0(\alpha,t)}\right)E_g(t,a)
$$

$$
= (g(t) - k_1(\alpha,t))\frac{k_0(\alpha,t) - \mu(t)k_1(\alpha,t)}{k_0(\alpha,t)}E_g(t,a)
$$

$$
= -\frac{f(t)p(t)k_0(\alpha,t)}{k_0(\alpha,t) - \mu(t)(-f(t)p(t) + k_1(\alpha,t))}\left(\frac{k_0(\alpha,t) - \mu(t)k_1(\alpha,t)}{k_0(\alpha,t)}\right)E_g(t,a)
$$

$$
= -\frac{f(t)p(t)(k_0(\alpha,t) - \mu(t)k_1(\alpha,t))}{k_0(\alpha,t) - \mu(t)(-f(t)p(t) + k_1(\alpha,t))}E_g(t,a), \quad t \in J,
$$

and

$$
-f(t)p(t)E_g(\sigma(t),a) = -f(t)p(t)\left(1 + \mu(t)\frac{g(t) - k_1(\alpha,t)}{k_0(\alpha,t)}\right)E_g(t,a)
$$

$$
= -f(t)p(t)\left(1 - \frac{\mu(t)f(t)p(t)k_0(\alpha,t)}{k_0(\alpha,t)(k_0(\alpha,t) - \mu(t)(-f(t)p(t) + k_1(\alpha,t)))}\right)E_g(t,a)
$$

$$
= -f(t)p(t)\left(1 - \frac{\mu(t)f(t)p(t)}{k_0(\alpha,t) - \mu(t)(-f(t)p(t) + k_1(\alpha,t))}\right)E_g(t,a)
$$

$$
= -f(t)p(t)\frac{k_0(\alpha,t) + \mu(t)f(t)p(t) - \mu(t)k_1(\alpha,t) - \mu(t)f(t)p(t)}{k_0(\alpha,t) - \mu(t)(-f(t)p(t) + k_1(\alpha,t))}E_g(t,a)
$$

$$
= -\frac{f(t)p(t)(k_0(\alpha,t) - \mu(t)k_1(\alpha,t))}{k_0(\alpha,t) - \mu(t)(-f(t)p(t) + k_1(\alpha,t))}E_g(t,a), \quad t \in J.
$$

Therefore

$$
g(t)E_g(t,a) - k_1(\alpha,t)E_g(\sigma(t),a) = -f(t)p(t)E_g(\sigma(t),a), \quad t \in J.
$$

Hence, and considering (4.4), we get

$$E_g(\sigma(t),a)D^\alpha y(t) \;\leq\; (-g(t)E_g(t,a)+k_1(\alpha,t)E_g(\sigma(t),a))\,y(t)$$

$$+p(t)q(t)E_g(\sigma(t),a), \quad t \in J,$$

or

$$E_g(\sigma(t),a)D^\alpha y(t)+g(t)E_g(t,a)y(t)-k_1(\alpha,t)E_g(\sigma(t),a)y(t)$$

$$\leq\; p(t)q(t)E_g(\sigma(t),a), \quad t \in J,$$

or

$$D^\alpha\left(E_g(t,a)y(t)\right) \leq p(t)q(t)E_g(\sigma(t),a), \quad t \in J.$$

From the last inequality, we obtain

$$y(t)E_g(t,a) \leq \int_a^t p(s)q(s)E_g(\sigma(s),a)\Delta_{\alpha,t}s, \quad t \in J,$$

whereupon

$$y(t) \;\leq\; E_g(a,t)\int_a^t p(s)q(s)E_g(\sigma(s),a)\Delta_{\alpha,t}s$$

$$=\; \int_a^t p(s)q(s)E_g(\sigma(s),t)\Delta_{\alpha,t}s, \quad t \in J.$$

Hence, and considering (4.2), we obtain

$$x(t) \;\leq\; f(t)y(t)+q(t)$$

$$\leq\; f(t)\int_a^t p(s)q(s)E_g(\sigma(s),t)\Delta_{\alpha,t}s+q(t), \quad t \in J.$$

This completes the proof. □

Theorem 4.1.3 *Let $f,g,h,p \in \mathscr{C}_{rd}(J)$ be nonnegative functions, $gh \in \mathscr{R}_c(J)$, and*

$$\frac{k_0(\alpha,t)-\mu(t)k_1(\alpha,t)}{k_0(\alpha,t)-\mu(t)(-h(t)g(t)+k_1(\alpha,t))} > 0, \quad t \in J.$$

Then the inequality

$$x(t) \leq f(t)+g(t)\int_a^t (h(s)x(s)+p(s))\Delta_{\alpha,t}s, \quad t \in J,$$

implies the inequality

$$x(t) \leq f(t)+g(t)\int_a^t (h(s)f(s)+p(s))E_q(\sigma(s),t)\Delta_{\alpha,t}s, \quad t \in J,$$

where

$$q(t) = \frac{(-h(t)g(t)+k_1(\alpha,t))(k_0(\alpha,t)-\mu(t)k_1(\alpha,t))}{k_0(\alpha,t)-\mu(t)(-h(t)g(t)+k_1(\alpha,t))}, \quad t \in J.$$

Proof 4.1.4 *Note that*

$$E_q(t,a) > 0, \quad E_q(\sigma(t),a) > 0, \quad t \in J.$$

Let

$$y(t) = \int_a^t (h(s)x(s) + p(s)) \Delta_{\alpha,t}s, \quad t \in J.$$

Then $y(a) = 0$ *and*

$$x(t) \le f(t) + g(t)y(t), \quad t \in J. \tag{4.5}$$

Also,

$$
\begin{aligned}
D^\alpha y(t) &= h(t)x(t) + p(t) \\[2mm]
&\le h(t)(f(t) + g(t)y(t)) + p(t) \\[2mm]
&= h(t)f(t) + p(t) + h(t)g(t)y(t), \quad t \in J,
\end{aligned}
$$

whereupon

$$E_q(\sigma(t),a)D^\alpha y(t) \le (h(t)f(t) + p(t))E_q(\sigma(t),a) + h(t)g(t)y(t)E_q(\sigma(t),a),$$

$t \in J.$ *Observe that*

$$-q(t)E_q(t,a) + k_1(\alpha,t)E_q(\sigma(t),a)$$

$$
\begin{aligned}
&= -q(t)E_q(t,a) + k_1(\alpha,t)\left(1 + \mu(t)\frac{q(t) - k_1(\alpha,t)}{k_0(\alpha,t)}\right)E_q(t,a) \\[2mm]
&= \left(-(q(t) - k_1(\alpha,t)) + \mu(t)k_1(\alpha,t)\frac{q(t) - k_1(\alpha,t)}{k_0(\alpha,t)}\right)E_q(t,a) \\[2mm]
&= (q(t) - k_1(\alpha,t))\left(\frac{\mu(t)k_1(\alpha,t)}{k_0(\alpha,t)} - 1\right)E_q(t,a) \\[2mm]
&= (q(t) - k_1(\alpha,t))\left(\frac{\mu(t)k_1(\alpha,t) - k_0(\alpha,t)}{k_0(\alpha,t)}\right)E_q(t,a), \quad t \in J,
\end{aligned}
$$

and

$$
\begin{aligned}
q(t) - k_1(\alpha,t) &= \frac{(-h(t)g(t) + k_1(\alpha,t))(k_0(\alpha,t) - \mu(t)k_1(\alpha,t))}{k_0(\alpha,t) - \mu(t)(-h(t)g(t) + k_1(\alpha,t))} - k_1(\alpha,t) \\[2mm]
&= \frac{1}{k_0(\alpha,t) - \mu(t)(-h(t)g(t) + k_1(\alpha,t))}\Big(-h(t)g(t)k_0(\alpha,t) \\[2mm]
&\quad + h(t)g(t)\mu(t)k_1(\alpha,t) + k_1(\alpha,t)k_0(\alpha,t) - \mu(t)(k_1(\alpha,t))^2
\end{aligned}
$$

$$\left. -k_1(\alpha,t)k_0(\alpha,t) - \mu(t)k_1(\alpha,t)h(t)g(t) + \mu(t)(k_1(\alpha,t))^2 \right)$$

$$= -\frac{h(t)g(t)k_0(\alpha,t)}{k_0(\alpha,t) - \mu(t)(-h(t)g(t) + k_1(\alpha,t))}, \quad t \in J.$$

Thus,

$$-q(t)E_q(t,a) + k_1(\alpha,t)E_q(\sigma(t),a)$$

$$= -\frac{h(t)g(t)k_0(\alpha,t)}{k_0(\alpha,t) - \mu(t)(-h(t)g(t) + k_1(\alpha,t))} \left(\frac{\mu(t)k_1(\alpha,t) - k_0(\alpha,t)}{k_0(\alpha,t)} \right) E_q(t,a)$$

$$= -\frac{h(t)g(t)(\mu(t)k_1(\alpha,t) - k_0(\alpha,t))}{k_0(\alpha,t) - \mu(t)(-h(t)g(t) + k_1(\alpha,t))} E_q(t,a), \quad t \in J.$$

Next,

$$h(t)g(t)E_q(t,a) = h(t)g(t) \left(1 + \mu(t)\frac{q(t) - k_1(\alpha,t)}{k_0(\alpha,t)} \right) E_q(t,a)$$

$$= h(t)g(t) \left(\frac{k_0(\alpha,t) + \mu(t)q(t) - \mu(t)k_1(\alpha,t)}{k_0(\alpha,t)} \right) E_q(t,a)$$

$$= \frac{h(t)g(t)}{k_0(\alpha,t)} \left(k_0(\alpha,t) - \mu(t)k_1(\alpha,t) \right.$$

$$\left. + \mu(t)\frac{(-h(t)g(t) + k_1(\alpha,t))(k_0(\alpha,t) - \mu(t)k_1(\alpha,t))}{k_0(\alpha,t) - \mu(t)(-h(t)g(t) + k_1(\alpha,t))} \right) E_q(t,a)$$

$$= \frac{h(t)g(t)(k_0(\alpha,t) - \mu(t)k_1(\alpha,t))}{k_0(\alpha,t)(k_0(\alpha,t) - \mu(t)(-h(t)g(t) + k_1(\alpha,t)))} \left(k_0(\alpha,t) \right.$$

$$\left. -\mu(t)(-h(t)g(t) + k_1(\alpha,t)) + \mu(t)(-h(t)g(t) + k_1(\alpha,t)) \right) E_q(t,a)$$

$$= \frac{h(t)g(t)(k_0(\alpha,t) - \mu(t)k_1(\alpha,t))}{(k_0(\alpha,t) - \mu(t)(-h(t)g(t) + k_1(\alpha,t)))} E_q(t,a), \quad t \in J.$$

Therefore

$$-q(t)E_q(t,a) + k_1(\alpha,t)E_q(\sigma(t),a) = h(t)g(t)E_q(\sigma(t),a), \quad t \in J.$$

Hence,

$$E_q(\sigma(t),a)D^\alpha y(t) \le -q(t)y(t)E_q(t,a) + k_1(\alpha,t)E_q(\sigma(t),a) + (h(t)f(t) + p(t))E_q(\sigma(t),a),$$

$t \in J,$ *or*

$$E_q(\sigma(t),a)D^\alpha y(t) + q(t)y(t)E_q(t,a) - k_1(\alpha,t)y(t)$$
$$E_q(\sigma(t),a) \le (h(t)f(t) + p(t))E_q(\sigma(t),a),$$

$t \in J$, or
$$D^\alpha\left(E_q(t,a)y(t)\right) \le (h(t)f(t)+p(t))E_q(\sigma(t),a), \quad t \in J.$$

Therefore
$$E_q(t,a)y(t) \le \int_a^t (h(s)f(s)+p(s))E_q(\sigma(s),a)\Delta_{\alpha,t}s, \quad t \in J,$$

and
$$y(t) \le E_q(a,t)\int_a^t (h(s)f(s)+p(s))E_q(\sigma(s),a)\Delta_{\alpha,t}s$$
$$= \int_a^t (h(s)f(s)+p(s))E_q(\sigma(s),t)\Delta_{\alpha,t}s, \quad t \in J.$$

Hence, and considering (4.5), we get
$$x(t) \le f(t)+g(t)\int_a^t (h(s)f(s)+p(s))E_q(\sigma(s),t)\Delta_{\alpha,t}s, \quad t \in J.$$

This completes the proof. □

Theorem 4.1.5 *Let $0 \in \mathscr{R}_c^+$, assume y, h and v are rd-continuous nonnegative functions on J, and assume k is an rd-continuous positive function on J such that*
$$\frac{h(x)k_0(\alpha,x)}{k_0(\alpha,x)+\mu(x)(k(t)h(x)-k_1(\alpha,x))} - \frac{k_1(\alpha,x)}{k(t)} > 0, \quad a \le x \le t \le b.$$

If
$$y(t) \ge v(x) - \frac{k(t)}{E_0(x,t)}\int_x^t \left(\frac{h(s)}{k_0(\alpha,s)+\mu(s)(k(t)h(s)-k_1(\alpha,s))} - \frac{k_1(\alpha,s)}{k(t)}\right)\Delta_{\alpha,t}s,$$

$a \le x \le t \le b$, *then*
$$y(t) \ge E_{k(t)h}(x,t)v(x), \quad a \le x \le t \le b,$$

where
$$E_{k(t)h}(x,t) = e^{\int_t^x \frac{1}{\mu(s)}\log\left(1+\mu(s)\frac{k(t)h(s)-k_1(\alpha,s)}{k_0(\alpha,s)}\right)\Delta s},$$

$a \le x \le t \le b.$

Proof 4.1.6 *Since $0 \in \mathscr{R}_c^+$, we have*
$$k_0(\alpha,x)-\mu(x)k_1(\alpha,x) > 0, \quad a \le x \le t \le b,$$

whereupon
$$k_0(\alpha,t)+\mu(x)(k(t)h(x)-k_1(\alpha,x)) > 0, \quad a \le x \le t \le b.$$

Let
$$z(x) = y(t) + \frac{k(t)}{E_0(x,t)}\int_x^t \left(\frac{h(s)}{k_0(\alpha,s)+\mu(s)(k(t)h(s)-k_1(\alpha,s))} - \frac{k_1(\alpha,s)}{k(t)}\right)\Delta_{\alpha,t}s,$$

$a \leq x \leq t \leq b$. *Then*

$$z(x) \geq v(x), \quad a \leq x \leq t \leq b,$$

and

$$z(x) = y(t) - k(t) \int_t^x \left(\frac{h(s)}{k_0(\alpha, s) + \mu(s)(k(t)h(s) - k_1(\alpha, s))} - \frac{k_1(\alpha, s)}{k(t)} \right) \Delta_{\alpha, x} s,$$

$a \leq x \leq t \leq b$, *and*

$$
\begin{aligned}
D^\alpha z(x) &= -k(t) \left(\frac{h(x)k_0(\alpha, x)}{k_0(\alpha, x) + \mu(x)(k(t)h(x) - k_1(\alpha, x))} - \frac{k_1(\alpha, x)}{k(t)} \right) v(x) \\
&\geq -k(t) \left(\frac{h(x)k_0(\alpha, x)}{k_0(\alpha, x) + \mu(x)(k(t)h(x) - k_1(\alpha, x))} - \frac{k_1(\alpha, x)}{k(t)} \right) z(x),
\end{aligned}
$$

$a \leq x \leq t \leq b$. *Hence,*

$$
\begin{aligned}
E_{k(t)h}(\sigma(x), t) D^\alpha z(x) &\geq -k(t) \frac{h(x)k_0(\alpha, x)E_{k(t)h}(\sigma(x), t)}{k_0(\alpha, x) + \mu(x)(k(t)h(x) - k_1(\alpha, x))} z(x) \\
&\quad + k_1(\alpha, x)z(x)E_{k(t)h}(\sigma(x), t) \\
&= -k(t)h(x)E_{k(t)h}(x, t)z(x) \\
&\quad + k_1(\alpha, x)E_{k(t)h}(\sigma(x), t)z(x),
\end{aligned}
$$

$a \leq x \leq t \leq b$, *or*

$$E_{k(t)h}(\sigma(x), t) D^\alpha z(x) + k(t)h(x)E_{k(t)h}(x, t)z(x) - k_1(\alpha, x)E_{k(t)h}(\sigma(x), t)z(x) \geq 0,$$

$a \leq x \leq t$. *Therefore*

$$D_x^\alpha \left(E_{k(t)h}(\cdot, t)z(\cdot) \right)(x) \geq 0, \quad a \leq x \leq t \leq b.$$

From the last inequality, we obtain

$$z(t) \geq E_{k(t)h}(x, t)z(x), \quad a \leq x \leq t \leq b.$$

Since

$$z(t) = y(t),$$

$$z(x) \geq v(x), \quad a \leq x \leq t \leq b,$$

we get

$$y(t) \geq E_{k(t)h}(x, t)v(x), \quad a \leq x \leq t \leq b.$$

This completes the proof. □

Theorem 4.1.7 (Gronwall-Type Inequality) *Let* $0 \in \mathscr{R}_c^+$, $y, f, g : J \to \mathbb{R}$ *be nonnegative rd-continuous functions on* J, $-f \in \mathscr{R}_c^+$. *If*

$$D^\alpha y(t) \le f(t) y^\sigma(t) + g(t), \quad t \in J,$$

then

$$y(t) \le y(a) E_{0 \ominus_c (-f)}(t, a) + \int_a^t E_{-f}(s, t) g(s) \Delta_{\alpha, t} s, \quad t \in J.$$

Proof 4.1.8 *We have*

$$E_{-f}(t, a) > 0, \quad t \in J.$$

Then

$$D^\alpha y(t) E_{-f}(t, a) - f(t)(E_{-f}(t, a) y^\sigma(t) \le E_{-f}(t, a) g(t), \quad t \in J,$$

whereupon

$$D^\alpha y(t) E_{-f}(t, a) - f(t)(E_{-f}(t, a) y^\sigma(t) - k_1(\alpha, t) E_{-f}(t, a) y^\sigma(t) \le E_{-f}(t, a) g(t), \quad t \in J,$$

or

$$D^\alpha \left(y E_{-f}(\cdot, a) \right)(t) \le E_{-f}(t, a) g(t), \quad t \in J.$$

Hence,

$$y(t) E_{-f}(t, a) - y(a) E_0(t, a) \le \int_a^t E_{-f}(s, a) g(s) \Delta_{\alpha, t} s, \quad t \in J,$$

or

$$y(t) E_{-f}(t, a) \le y(a) E_0(t, a) + \int_a^t E_{-f}(s, a) g(s) \Delta_{\alpha, t} s, \quad t \in J,$$

and

$$y(t) \le y(a) \frac{E_0(t, a)}{E_{-f}(t, a)} + \int_a^t E_{-f}(a, t) E_{-f}(s, a) g(s) \Delta_{\alpha, t} s, \quad t \in J,$$

or

$$y(t) \le y(a) E_{0 \ominus_c (-f)}(t, a) + \int_a^t E_{-f}(s, t) g(s) \Delta_{\alpha, t} s, \quad t \in J.$$

This completes the proof. □

Theorem 4.1.9 (Gronwall-Type Inequality) *Let* $0 \in \mathscr{R}_c^+$, $y, f, g : J \to \mathbb{R}$ *be nonnegative rd-continuous functions on* J, $-f \in \mathscr{R}_c^+$. *If*

$$D^\alpha y(t) \le f(t) y(t) + g(t), \quad t \in J,$$

then

$$y(t) \le y(a) E_{0 \ominus_c (-h)}(t, a)$$

$$+ \int_a^t \frac{g(s)(k_0(\alpha, s) - \mu(s) k_1(\alpha, s))}{k_0(\alpha, s) - \mu(s)(f(s) + k_1(\alpha, s))} E_{-f}(s, t) \Delta_{\alpha, t} s, \quad t \in J,$$

where

$$h(t) = \frac{k_0(\alpha, t) f(t)}{k_0(\alpha, t) - \mu(t)(f(t) + k_1(\alpha, t))}, \quad t \in J.$$

Proof 4.1.10 *We have*

$$D^\alpha y(t) \leq f(t)y(t) + g(t)$$

$$= \frac{k_0(\alpha,t)f(t)}{k_0(\alpha,t) - \mu(t)k_1(\alpha,t)} y^\sigma(t)$$

$$- \frac{\mu(t)f(t)}{k_0(\alpha,t) - \mu(t)k_1(\alpha,t)} D^\alpha y(t) + g(t), \quad t \in J,$$

whereupon

$$\frac{k_0(\alpha,t) - \mu(t)(f(t) + k_1(\alpha,t))}{k_0(\alpha,t) - \mu(t)k_1(\alpha,t)} D^\alpha y(t) \leq \frac{k_0(\alpha,t)f(t)}{k_0(\alpha,t) - \mu(t)k_1(\alpha,t)} y^\sigma(t) + g(t), \quad t \in J,$$

or

$$D^\alpha y(t) \leq \frac{k_0(\alpha,t)f(t)}{k_0(\alpha,t) - \mu(t)(f(t) + k_1(\alpha,t))} y^\sigma(t)$$

$$+ \frac{g(t)(k_0(\alpha,t) - \mu(t)k_1(\alpha,t))}{k_0(\alpha,t) - \mu(t)(f(t) + k_1(\alpha,t))}$$

$$= h(t)y^\sigma(t)$$

$$+ \frac{g(t)(k_0(\alpha,t) - \mu(t)k_1(\alpha,t))}{k_0(\alpha,t) - \mu(t)(f(t) + k_1(\alpha,t))}, \quad t \in J.$$

From the last inequality and from Theorem 4.1.7, we get the desired result. This completes the proof. □

4.2 CONFORMABLE VOLTERRA-TYPE INTEGRAL INEQUALITIES

Theorem 4.2.1 *Let* $x, f \in \mathscr{C}_{rd}(J)$, $k \in \mathscr{C}_{rd}(J \times J)$ *be nonnegative on* $J \times J$, *and*

$$x(t) \leq f(t) + \int_a^t k(t,s)x(s)\Delta_{\alpha,t}s, \quad t \in J. \tag{4.6}$$

Let also,

$$k_1(t,s) = k(t,s),$$

$$k_l(t,s) = \int_a^t k(t,s_1)k_{l-1}(s_1,s)\Delta s_1, \quad l \in \mathbb{N}, \quad l \geq 2, \quad a \leq s \leq t \leq b.$$

Suppose that

$$H(t,s) = \sum_{n=1}^\infty k_n(t,s)$$

is a uniformly convergent series on $a \leq s \leq t \leq b$. Then

$$x(t) \leq f(t) + \int_a^t H(t,s)f(s)\Delta_{\alpha,t}s, \quad t \in J. \tag{4.7}$$

Proof 4.2.2 *We have*

$$
\begin{aligned}
x(t) &\leq f(t) + \int_a^t k(t,s)x(s)\Delta_{\alpha,t}s \\
&\leq f(t) + \int_a^t k(t,s)\left(f(s) + \int_a^s k(s,s_1)x(s_1)\Delta_{\alpha,s}s_1\right)\Delta_{\alpha,t}s \\
&= f(t) + \int_a^t k(t,s)f(s)\Delta_{\alpha,t}s \\
&\quad + \int_a^t \int_a^s k(t,s)k(s,s_1)x(s_1)\Delta_{\alpha,s}s_1\Delta_{\alpha,t}s \\
&= f(t) + \int_a^t k(t,s)f(s)\Delta_{\alpha,t}s \\
&\quad + \int_a^t \left(\int_a^{s_1} k(t,s)k(s,s_1)\Delta_{\alpha,s_1}s\right)x(s_1)\Delta_{\alpha,t}s_1 \\
&= f(t) + \int_a^t k_1(t,s)f(s)\Delta_{\alpha,t}s + \int_a^t k_2(t,s)x(s)\Delta_{\alpha,t}s, \quad t \in J.
\end{aligned}
$$

Hence, for $t \in J$, we get

$$
\begin{aligned}
x(t) &\leq f(t) + \int_a^t k(t,s)x(s)\Delta_{\alpha,t}s \\
&\leq f(t) + \int_a^t k(t,s)\Big(f(s) + \int_a^s k_1(s,s_1)f(s_1)\Delta_{\alpha,s}s_1 \\
&\quad + \int_a^s k_2(s,s_1)x(s_1)\Delta_{\alpha,t}s_1\Big)\Delta_{\alpha,t}s \\
&= f(t) + \int_a^t k(t,s)f(s)\Delta_{\alpha,t}s \\
&\quad + \int_a^t \int_a^s k(t,s)k_1(s,s_1)f(s_1)\Delta_{\alpha,s}s_1\Delta_{\alpha,t}s \\
&\quad + \int_a^t \int_a^s k(t,s)k_2(s,s_1)x(s_1)\Delta_{\alpha,s}s_1\Delta_{\alpha,t}s \\
&= f(t) + \int_a^t k_1(t,s)f(s)\Delta_{\alpha,t}s
\end{aligned}
$$

$$+ \int_a^t \left(\int_a^{s_1} k(t,s)k_1(s,s_1)\Delta_{\alpha,s_1}s \right) f(s_1)\Delta_{\alpha,t}s_1$$

$$+ \int_a^t \left(\int_a^{s_1} k(t,s)k_2(s,s_1)\Delta_{\alpha,s_1}s \right) x(s_1)\Delta_{\alpha,t}s_1$$

$$= f(t) + \int_a^t k_1(t,s)f(s)\Delta_{\alpha,t}s + \int_a^t k_2(t,s)f(s)\Delta_{\alpha,t}s$$

$$+ \int_a^t k_3(t,s)x(s)\Delta_{\alpha,t}s$$

$$= f(t) + \int_a^t (k_1(t,s) + k_2(t,s)) f(s)\Delta_{\alpha,t}s$$

$$+ \int_a^t k_3(t,s)x(s)\Delta_{\alpha,t}s.$$

Assume that

$$x(t) \le f(t) + \sum_{l=1}^n \int_a^t k_l(t,s)f(s)\Delta_{\alpha,t}s + \int_a^t k_{n+1}(t,s)x(s)\Delta_{\alpha,t}s, \quad t \in J, \qquad (4.8)$$

for some $n \in \mathbb{N}$. We will prove that

$$x(t) \le f(t) + \sum_{l=1}^{n+1} \int_a^t k_l(t,s)f(s)\Delta_{\alpha,t}s + \int_a^t k_{n+2}(t,s)x(s)\Delta_{\alpha,t}s, \quad t \in J.$$

Really, for $t \in J$, we have

$$x(t) \quad \le \quad f(t) + \int_a^t k(t,s)x(s)\Delta_{\alpha,t}s$$

$$\le \quad f(t) + \int_a^t k(t,s)\left(f(s) + \sum_{l=1}^n \int_a^s k_l(s,s_1)f(s_1)\Delta_{\alpha,s}s_1 \right.$$

$$\left. + \int_a^s k_{n+1}(s,s_1)x(s_1)\Delta_{\alpha,s}s_1 \right)\Delta_{\alpha,t}s$$

$$= \quad f(t) + \int_a^t k(t,s)f(s)\Delta_{\alpha,t}s$$

$$+ \sum_{l=1}^n \int_a^t \int_a^s k(t,s)k_l(s,s_1)f(s_1)\Delta_{\alpha,s}s_1\Delta_{\alpha,t}s$$

$$+ \int_a^t \int_a^s k(t,s)k_{n+1}(s,s_1)x(s_1)\Delta_{\alpha,s}s_1\Delta_{\alpha,t}s$$

$$= \quad f(t) + \int_a^t k(t,s)f(s)\Delta_{\alpha,t}s$$

$$+ \sum_{l=1}^{n} \int_a^t \left(\int_a^{s_1} k(t,s) k_l(s,s_1) \Delta_{\alpha,s_1} s \right) f(s_1) \Delta_{\alpha,t} s_1$$

$$+ \int_a^t \left(\int_a^{s_1} k(t,s) k_{n+1}(s,s_1) \Delta_{\alpha,s_1} s \right) x(s_1) \Delta_{\alpha,t} s_1$$

$$= f(t) + \int_a^t k_1(t,s) f(s) \Delta_{\alpha,t} s + \sum_{l=1}^{n} \int_a^t k_{l+1}(t,s) f(s) \Delta_{\alpha,t} s$$

$$+ \int_a^t k_{n+2}(t,s) x(s) \Delta_{\alpha,t} s$$

$$= f(t) + \int_a^t k_1(t,s) f(s) \Delta_{\alpha,t} s$$

$$+ \sum_{l=2}^{n+1} \int_a^t k_l(t,s) f(s) \Delta_{\alpha,t} s$$

$$+ \int_a^t k_{n+2}(t,s) x(s) \Delta_{\alpha,t} s$$

$$= f(t) + \sum_{l=1}^{n+1} \int_a^t k_l(t,s) f(s) \Delta_{\alpha,t} s + \int_a^t k_{n+2}(t,s) x(s) \Delta_{\alpha,t} s.$$

Therefore (4.8) holds for any $n \in \mathbb{N}$. Letting $n \to \infty$ into (4.8) and using that $H(t,s) = \sum_{n=1}^{\infty} k_n(t,s)$ is a uniformly convergent series on $a \le s \le t \le b$, we get the inequality (4.7). This completes the proof. □

Theorem 4.2.3 *Assume that $0 \in \mathscr{R}_c^+$. Let $k : J \times J \to \mathbb{R}$ be a nonnegative continuous function which is nondecreasing with respect to its first argument and $-k \in \mathscr{R}_c^+$. Also, let c be a nonnegative constant. If*

$$y(t) \le c + \int_a^t \frac{k(t,s) k_0(\alpha,s)}{k_0(\alpha,s) - \mu(s)(k(t,s) + k_1(\alpha,s))} y(s) \Delta_{\alpha,t} s, \quad t \in J,$$

then

$$y(t) \le c E_{0 \ominus_c (-k(b,\cdot))}(t,a), \quad t \in J.$$

Proof 4.2.4 *Since k is nondecreasing with respect to its first argument, we have*

$$k(t,s) \le k(b,s),$$

$$k(t,s) + k_1(\alpha,s) \le k(b,s) + k_1(\alpha,s),$$

$$-\mu(s)(k(t,s) + k_1(\alpha,s)) \ge -\mu(s)(k(b,s) + k_1(\alpha,s)),$$

$$k_0(\alpha,s) - \mu(s)(k(t,s) + k_1(\alpha,s)) \geq k_0(\alpha,s) - \mu(s)(k(b,s) + k_1(\alpha,s)),$$

$t, s \in J$, and

$$\frac{k(t,s)k_0(\alpha,s)}{k_0(\alpha,s) - \mu(s)(k(t,s) + k_1(\alpha,s))} \leq \frac{k(b,s)k_0(\alpha,s)}{k_0(\alpha,s) - \mu(s)(k(b,s) + k_1(\alpha,s))}, \quad t, s \in J.$$

From here,

$$y(t) \leq c + \int_a^t \frac{k(b,s)k_0(\alpha,s)}{k_0(\alpha,s) - \mu(s)(k(b,s) + k_1(\alpha,s))} y(s)\Delta_{\alpha,t}s, \quad t \in J.$$

Define the function $z : J \to \mathbb{R}$ as follows.

$$z(t) = c + \int_a^t \frac{k(b,s)k_0(\alpha,s)}{k_0(\alpha,s) - \mu(s)(k(b,s) + k_1(\alpha,s))} y(s)\Delta_{\alpha,t}s, \quad t \in J.$$

Then

$$y(t) \leq z(t), \quad t \in J, \tag{4.9}$$

and

$$D^\alpha z(t) \leq \frac{k(b,t)k_0(\alpha,t)}{k_0(\alpha,t) - \mu(t)(k(b,t) + k_1(\alpha,t))} y(t)$$

$$\leq \frac{k(b,t)k_1(\alpha,t)}{k_0(\alpha,t) - \mu(t)(k(b,t) + k_1(\alpha,t))} z(t), \quad t \in J.$$

Hence,

$$E_{-k(b,\cdot)}(\sigma(t),a)D^\alpha z(t) - \frac{k(b,t)k_0(\alpha,t)E_{-k(b,\cdot)}(\sigma(t),a)}{k_0(\alpha,t) - \mu(t)(k(b,t) + k_1(\alpha,t))} z(t) \leq 0, \quad t \in J,$$

or

$$E_{-k(b,\cdot)}(\sigma(t),a)D^\alpha z(t) - k(b,t)E_{-k(b,\cdot)}(t,a)z(t) \leq 0, \quad t \in J,$$

whereupon

$$E_{-k(b,\cdot)}(\sigma(t),a)D^\alpha z(t) - k(b,t)E_{-k(b,\cdot)}(t,a)z(t) - k_1(\alpha,t)E_{-k(b,\cdot)}(\sigma(t),a)z(t) \leq 0, \quad t \in J.$$

Therefore

$$D^\alpha \left(E_{-k(b,\cdot)}(t,a)z(t) \right) \leq 0, \quad t \in J,$$

and

$$E_{-k(b,\cdot)}(t,a)z(t) - z(a)E_0(t,a) \leq 0, \quad t \in J,$$

or

$$E_{-k(b,\cdot)}(t,a)z(t) \leq cE_0(t,a), \quad t \in J,$$

or

$$z(t) \leq c\frac{E_0(t,a)}{E_{-k(b,\cdot)}(t,a)}, \quad t \in J,$$

or

$$z(t) \leq cE_{0\ominus_c(-k(b,\cdot))}(t,a), \quad t \in J.$$

From the last inequality and from (4.9), we get

$$y(t) \leq cE_{0\ominus_c(-k(b,\cdot))}(t,a), \quad t \in J.$$

This completes the proof. □

Theorem 4.2.5 *Let* $0 \in \mathscr{R}_c^+$, *k be as in Theorem 4.2.3, and* $g : J \to \mathbb{R}$ *be a positive rd-continuous and nondecreasing function on J. If*

$$y(t) \leq g(t) + \int_a^t \frac{k(t,s)k_0(\alpha,s)}{k_0(\alpha,s) - \mu(s)(k(t,s)+k_1(\alpha,s))}y(s)\Delta_{\alpha,t}s, \quad t \in J, \qquad (4.10)$$

then

$$y(t) \leq g(t)E_{0\ominus_c(-k(b,\cdot))}(t,a), \quad t \in J.$$

Proof 4.2.6 *By the inequality (4.10), using that g is a nondecreasing function on J, we obtain*

$$\begin{aligned}
\frac{y(t)}{g(t)} &\leq 1 + \int_a^t \frac{k(t,s)k_0(\alpha,s)}{k_0(\alpha,s) - \mu(s)(k(t,s)+k_1(\alpha,s))}\frac{y(s)}{g(t)}\Delta_{\alpha,t}s \\
&\leq 1 + \int_a^t \frac{k(t,s)k_0(\alpha,s)}{k_0(\alpha,s) - \mu(s)(k(t,s)+k_1(\alpha,s))}\frac{y(s)}{g(s)}\Delta_{\alpha,t}s, \quad t \in J.
\end{aligned}$$

Let

$$x(t) = \frac{y(t)}{g(t)}, \quad t \in J.$$

Then

$$x(t) \leq 1 + \int_a^t \frac{k(t,s)k_0(\alpha,s)}{k_0(\alpha,s) - \mu(s)(k(t,s)+k_1(\alpha,s))}x(s)\Delta_{\alpha,t}s, \quad t \in J.$$

Hence, and considering Theorem 4.2.3, we get

$$x(t) \leq E_{0\ominus_c(-k(b,\cdot))}(t,a), \quad t \in J,$$

whereupon

$$y(t) \leq g(t)E_{0\ominus_c(-k(b,\cdot))}(t,a), \quad t \in J.$$

This completes the proof. □

4.3 CONFORMABLE INEQUALITIES OF GAMIDOV AND RODRIGUES

Theorem 4.3.1 *Let* $0 \in \mathscr{R}_c^+$, f, g_i, h_i, $i \in \{1,\dots,n\}$, *be nonnegative rd-continuous functions on J, and let*

$$g(t) = \sup_{i\in\{1,\dots,n\}} \{g_i(t)\},$$

$$h(t) = \sum_{i=1}^{n} h_i(t), \quad t \in J.$$

Also, let $gh \in \mathscr{R}_c^+$. If

$$y(t) \le f(t) + \sum_{i=1}^{n} g_i(t) \int_a^t h_i(s) y(s) \Delta_{\alpha,t} s, \quad t \in J,$$

then

$$y(t) \le f(t) + g(t) \int_a^t h(s) f(s) E_p(\sigma(s),t) \Delta_{\alpha,t} s, \quad t \in J,$$

where

$$p(t) = \frac{(-g(t)h(t) + k_1(\alpha,t))(k_0(\alpha,t) - \mu(t)k_1(\alpha,t))}{k_0(\alpha,t) + \mu(t)(g(t)h(t) - k_1(\alpha,t))}, \quad t \in J.$$

Proof 4.3.2 *We have*

$$
\begin{aligned}
y(t) &\le f(t) + g(t) \int_a^t \left(\sum_{i=1}^{n} h_i(s) \right) y(s) \Delta_{\alpha,t} s \\
&= f(t) + g(t) \int_a^t h(s) y(s) \Delta_{\alpha,t} s, \quad t \in J.
\end{aligned}
$$

Hence, by Theorem 4.1.1, we get the desired result. This completes the proof. □

Theorem 4.3.3 *Let $0 \in \mathscr{R}_c^+$, y, f, g_1, g_2, h_i, $i \in \{1, \dots, n\}$, be nonnegative rd-continuous functions on J, $h_1 g_1 \in \mathscr{R}_c^+$, $a = t_1 \le t_2 \le \dots \le t_n = b$, c_i, $i \in \{1, \dots, n\}$, be nonnegative constants,*

$$f_1(t) = f(t) + g_2(t) \sum_{i=2}^{n} m_i,$$

$$p(t) = \frac{(-g_1(t)h_1(t) + k_1(\alpha,t))(k_0(\alpha,t) - \mu(t)k_1(\alpha,t))}{k_0(\alpha,t) + \mu(t)(g_1(t)h_1(t) - k_1(\alpha,t))},$$

$$q_1(t) = f_1(t) + g_1(t) \int_{t_1}^t h_1(s) f(s) E_p(\sigma(s),t) \Delta_{\alpha,t} s,$$

$$q_2(t) = \int_{t_1}^t h_1(s) g_2(s) E_p(\sigma(s),t) \Delta_{\alpha,t} s,$$

$$1 > \sum_{i=2}^{n} c_i \int_{t_1}^{t_i} h_i(s) q_2(s) \Delta_{\alpha,t_i} s,$$

$$
\begin{aligned}
M &= \left(1 - \sum_{i=2}^{n} c_i \int_{t_1}^{t_i} h_i(s) q_2(s) \Delta_{\alpha,t_i} s \right)^{-1} \\
&\quad \times \left(\sum_{i=2}^{n} c_i \int_{t_1}^{t_i} h_i(s) q_1(s) \Delta_{\alpha,t} s \right), \quad t \in J.
\end{aligned}
$$

If

$$y(t) \leq f(t) + g_1(t) \int_{t_1}^{t} h_1(s)y(s)\Delta_{\alpha,t}s$$

$$+ g_2(t) \sum_{i=2}^{n} c_i \int_{t_1}^{t_i} h_i(s)y(s)\Delta_{\alpha,t}s, \quad t \in J,$$

then

$$y(t) \leq q_1(t) + Mq_2(t), \quad t \in J.$$

Proof 4.3.4 *Set*

$$m_i = c_i \int_{t_1}^{t_i} h_i(s)y(s)\Delta_{\alpha,t_i}s, \quad i \in \{1,\ldots,n\}, \quad i \in \{1,\ldots,n\}.$$

Then

$$y(t) \leq f(t) + g_2(t) \sum_{i=2}^{n} m_i + g_1(t) \int_{t_1}^{t} h_1(s)y(s)\Delta_{\alpha,t}s$$

$$= f_1(t) + g_1(t) \int_{t_1}^{t} h_1(s)y(s)\Delta_{\alpha,t}s, \quad t \in J.$$

Hence, and considering Theorem 4.1.1, we get

$$y(t) \leq f_1(t) + g_1(t) \int_{t_1}^{t} h_1(s)f_1(s)E_p(\sigma(s),t)\Delta_{\alpha,t}s$$

$$= f_1(t) + g_1(t) \int_{t_1}^{t} h_1(s) \left(f(s) + g_2(s) \sum_{i=2}^{n} m_i \right) E_p(\sigma(s),t)\Delta_{\alpha,t}s$$

$$= f_1(t) + g_1(t) \int_{t_1}^{t} h_1(s)f(s)E_p(\sigma(s),t)\Delta_{\alpha,t}s$$

$$+ g_1(t) \sum_{i=2}^{n} m_i \int_{t_1}^{t} h_1(s)g_2(s)E_p(\sigma(s),t)\Delta_{\alpha,t}s$$

$$= q_1(t) + \sum_{i=2}^{n} m_i q_2(t), \quad t \in J.$$

Since

$$\sum_{i=2}^{n} m_i = \sum_{i=2}^{n} c_i \int_{t_1}^{t_i} h_i(s)y(s)\Delta_{\alpha,t_i}s$$

$$= \sum_{i=2}^{n} c_i \int_{t_1}^{t_i} h_i(s) \left(\sum_{i=2}^{n} m_i q_2(s) + q_1(s) \right) \Delta_{\alpha,t_i}s$$

$$= \sum_{i=2}^{n} c_i \int_{t_1}^{t_i} h_i(s)q_1(s)\Delta_{\alpha,t_i}s$$

$$+ \sum_{i=2}^{n} m_i \left(\sum_{i=2}^{n} c_i \int_{t_1}^{t_i} h_i(s) q_2(s) \Delta_{\alpha,t_i} s \right),$$

whereupon

$$\sum_{i=2}^{n} m_i \left(1 - \sum_{i=2}^{n} c_i \int_{t_1}^{t_i} h_i(s) q_2(s) \Delta_{\alpha,t_i} s \right) = \sum_{i=2}^{n} c_i \int_{t_1}^{t_i} h_i(s) q_1(s) \Delta_{\alpha,t_i} s$$

and

$$\sum_{i=2}^{n} m_i = \left(1 - \sum_{i=2}^{n} c_i \int_{t_1}^{t_i} h_i(s) q_2(s) \Delta_{\alpha,t_i} s \right)^{-1}$$

$$\times \left(\sum_{i=2}^{n} c_i \int_{t_1}^{t_i} h_i(s) q_1(s) \Delta_{\alpha,t} s \right)$$

$$= M.$$

Therefore

$$y(t) \le q_1(t) + M q_2(t), \quad t \in J.$$

This completes the proof. □

4.4 SIMULTANEOUS CONFORMABLE INTEGRAL INEQUALITIES

Theorem 4.4.1 *Let* $0 \in \mathscr{R}_c^+$, $x, y, a, b, p, h_i : J \to \mathbb{R}$, $i \in \{1, 2, 3, 4\}$, *be nonnegative rd-continuous functions on J,*

$$h(t) = \max\{h_1(t) + h_3(t), h_2(t) + h_4(t)\},$$

$$c(t) = a(t) + b(t),$$

$$g(t) = \frac{(-p(t)h(t) + k_1(\alpha,t))(k_0(\alpha,t) - \mu(t)k_1(\alpha,t))}{k_0(\alpha,t) + \mu(t)(p(t)h(t) - k_1(\alpha,t))}, \quad t \in J.$$

If

$$x(t) \le a(t) + p(t) \left(\int_a^t h_1(s) x(s) \Delta_{\alpha,t} s + \int_a^t h_2(s) y(s) \Delta_{\alpha,t} s \right)$$

$$y(t) \le a(t) + p(t) \left(\int_a^t h_3(s) x(s) \Delta_{\alpha,t} s + \int_a^t h_4(s) y(s) \Delta_{\alpha,t} s \right), \quad t \in J,$$

then

$$x(t) \le c(t) + p(t) \int_a^t h(s) c(s) E_g(\sigma(s), t) \Delta_{\alpha,t} s,$$

$$y(t) \le c(t) + p(t) \int_a^t h(s) c(s) E_g(\sigma(s), t) \Delta_{\alpha,t} s, \quad t \in J.$$

Proof 4.4.2 *Let*

$$z(t) = x(t) + y(t), \quad t \in J.$$

Then

$$
\begin{aligned}
z(t) &= x(t) + y(t) \\[2mm]
&\leq a(t) + b(t) + p(t)\left(\int_a^t (h_1(s) + h_3(s))x(s)\Delta_{\alpha,t}s \right. \\[2mm]
&\qquad \left. + \int_a^t (h_2(s) + h_4(s))y(s)\Delta_{\alpha,t}s \right) \\[2mm]
&\leq c(t) + p(t)\left(\int_a^t h(s)x(s)\Delta_{\alpha,t}s + \int_a^t h(s)y(s)\Delta_{\alpha,t}s \right) \\[2mm]
&= c(t) + p(t)\int_a^t h(s)z(s)\Delta_{\alpha,t}s, \quad t \in J.
\end{aligned}
$$

Hence, and considering Theorem 4.1.1, it follows that

$$z(t) \leq c(t) + p(t)\int_a^t h(s)c(s)E_g(\sigma(s),t)\Delta_{\alpha,t}s, \quad t \in J.$$

This completes the proof. □

4.5 CONFORMABLE PACHPATTE'S INEQUALITIES

Theorem 4.5.1 *Let $0 \in \mathscr{R}_c^+$, $y_0 \geq 0$, $y, f, g : J \to \mathbb{R}$ be nonnegative rd-continuous functions on J and $-(f+g) \in \mathscr{R}_c^+$. If*

$$
\begin{aligned}
y(t) &\leq y_0 + \int_a^t f(s)y(s)\Delta_{\alpha,t}s \\[2mm]
&\quad + \int_a^t f(s)\left(\int_a^s g(\tau)y(\tau)\Delta_{\alpha,s}\tau \right)\Delta_{\alpha,t}s, \quad t \in J,
\end{aligned}
$$

then

$$y(t) \leq y_0\left(E_0(t,a) + \int_a^t f(s)E_{0\ominus_c(-h)}(s,a)\Delta_{\alpha,t}s \right), \quad t \in J,$$

where

$$h(t) = \frac{k_0(\alpha,t)(f(t)+g(t))}{k_0(\alpha,t) - \mu(t)(f(t)+g(t)+k_1(\alpha,t))}, \quad t \in J.$$

Proof 4.5.2 *Let*

$$
\begin{aligned}
z(t) &= y_0 + \int_a^t f(s)y(s)\Delta_{\alpha,t}s \\[2mm]
&\quad + \int_a^t f(s)\left(\int_a^s g(\tau)y(\tau)\Delta_{\alpha,s}\tau \right)\Delta_{\alpha,t}s, \quad t \in J.
\end{aligned}
$$

Then

$$y(t) \le z(t), \quad t \in J,$$

and

$$
\begin{aligned}
D^{\alpha}z(t) &= f(t)y(t) + f(t) \int_a^t g(s)y(s)\Delta_{\alpha,t}s \\
&= f(t)\left(y(t) + \int_a^t g(s)y(s)\Delta_{\alpha,t}s\right) \\
&\le f(t)\left(z(t) + \int_a^t g(s)z(s)\Delta_{\alpha,t}s\right), \quad t \in J.
\end{aligned}
$$

Let

$$w(t) = z(t) + \int_a^t g(s)z(s)\Delta_{\alpha,t}s, \quad t \in J.$$

Then

$$
\begin{aligned}
w(a) &= z(a) \\
&= y_0, \\
D^{\alpha}z(t) &\le f(t)w(t), \\
D^{\alpha}w(t) &= D^{\alpha}z(t) + g(t)z(t) \\
&\le f(t)w(t) + g(t)w(t) \\
&= (f(t)+g(t))w(t), \quad t \in J.
\end{aligned}
$$

Hence, by the Gronwall inequality and Theorem 4.1.7, we get

$$
\begin{aligned}
w(t) &\le w(a)E_{0\ominus_c(-h)}(t,a) \\
&= y_0 E_{0\ominus_c(-h)}(t,a), \quad t \in J.
\end{aligned}
$$

Then

$$
\begin{aligned}
D^{\alpha}z(t) &\le f(t)w(t) \\
&\le y_0 f(t)E_{0\ominus_c(-h)}(t,a), \quad t \in J.
\end{aligned}
$$

Now we integrate the last inequality from a to t and we find

$$z(t) \le z(a)E_0(t,a) + y_0 \int_a^t f(s)E_{0\ominus_c(-h)}(s,a)\Delta_{\alpha,t}s$$

$$= y_0 E_0(t,a) + y_0 \int_a^t f(s) E_{0 \ominus_c(-h)}(s,a) \Delta_{\alpha,t} s, \quad t \in J,$$

and

$$y(t) \leq y_0 \left(1 + \int_a^t f(s) E_{0 \ominus_c(-h)}(s,a) \Delta_{\alpha,t} s \right), \quad t \in J.$$

This completes the proof. □

Theorem 4.5.3 *Let* $0 \in \mathscr{R}_c^+$, $f,g,h,y : J \to \mathbb{R}$ *be nonnegative rd-continuous functions on* J, $y_0 \geq 0$ *and* $-(f+g+h) \in \mathscr{R}_c^+$. *If*

$$y(t) \leq y_0 + \int_a^t f(s) y(s) \Delta_{\alpha,t} s$$

$$+ \int_a^t f(s) \left(\int_a^s g(\tau) y(\tau) \Delta_{\alpha,s} \tau \right) \Delta_{\alpha,s} t$$

$$+ \int_a^t f(s) \left(\int_a^s g(\tau) \left(\int_a^\tau h(l) y(l) \Delta_{\alpha,\tau} l \right) \Delta_{\alpha,s} \tau \right) \Delta_{\alpha,t} s, \quad t \in J,$$

then

$$y(t) \leq y_0 \left(E_0(t,a) + \int_a^t f(s) E_{0 \ominus_c(-p)}(s,a) \Delta_{\alpha,t} s \right), \quad t \in J,$$

where

$$p(t) = f(t) + g(t) + h(t), \quad t \in J.$$

Proof 4.5.4 *Let*

$$z(t) = y_0 + \int_a^t f(s) y(s) \Delta_{\alpha,t} s$$

$$+ \int_a^t f(s) \left(\int_a^s g(\tau) y(\tau) \Delta_{\alpha,s} \tau \right) \Delta_{\alpha,s} t$$

$$+ \int_a^t f(s) \left(\int_a^s g(\tau) \left(\int_a^\tau h(l) y(l) \Delta_{\alpha,\tau} l \right) \Delta_{\alpha,s} \tau \right) \Delta_{\alpha,t} s, \quad t \in J.$$

Then

$$y(t) \leq z(t), \quad t \in J,$$

$$z(a) = y_0,$$

and

$$D^\alpha z(t) = f(t) y(t) + f(t) \int_a^t g(s) y(s) \Delta_{\alpha,t} s$$

$$+f(t) \int_a^t g(s) \left(\int_a^s h(\tau)y(\tau)\Delta_{\alpha,s}\tau \right) \Delta_{\alpha,s}t$$

$$= f(t) \left(y(t) + \int_a^t g(s)y(s)\Delta_{\alpha,t}s \right.$$

$$+ \left. \int_a^t g(s) \left(\int_a^s h(\tau)y(\tau)\Delta_{\alpha,s}\tau \right) \Delta_{\alpha,s}t \right)$$

$$\leq f(t) \left(z(t) + \int_a^t g(s)z(s)\Delta_{\alpha,t}s \right.$$

$$+ \left. \int_a^t g(s) \left(\int_a^s h(\tau)z(\tau)\Delta_{\alpha,s}\tau \right) \Delta_{\alpha,s}t \right), \quad t \in J.$$

Let

$$w(t) = z(t) + \int_a^t g(s)z(s)\Delta_{\alpha,t}s$$

$$+ \int_a^t g(s) \left(\int_a^s h(\tau)z(\tau)\Delta_{\alpha,s}\tau \right) \Delta_{\alpha,s}t, \quad t \in J.$$

Then

$$w(a) = z(a)$$

$$= y_0,$$

$$z(t) \leq w(t),$$

$$D^\alpha z(t) \leq f(t)w(t),$$

$$D^\alpha w(t) = D^\alpha z(t) + g(t)z(t)$$

$$+ g(t) \int_a^t h(s)z(s)\Delta_{\alpha,t}s$$

$$\leq f(t)w(t) + g(t)w(t)$$

$$+ g(t) \int_a^t h(s)w(s)\Delta_{\alpha,t}s$$

$$\leq (f(t) + g(t)) \left(w(t) + \int_a^t h(s)w(s)\Delta_{\alpha,t}s \right), \quad t \in J.$$

Let

$$x(t) = w(t) + \int_a^t h(s)w(s)\Delta_{\alpha,t}s, \quad t \in J.$$

Then

$$x(a) = w(a)$$

$$= y_0,$$

$$w(t) \leq x(t),$$

$$D^\alpha w(t) \leq (f(t) + g(t))x(t),$$

$$D^\alpha x(t) = D^\alpha w(t) + h(t)w(t)$$

$$\leq (f(t) + g(t))w(t) + h(t)w(t)$$

$$= (f(t) + g(t) + h(t))w(t), \quad t \in J.$$

Consequently, by the Gronwall inequality and Theorem 4.1.9, we get

$$x(t) \leq y_0 E_{0\ominus_c(-p)}(t, a), \quad t \in J.$$

Next,

$$w(t) \leq x(t)$$

$$\leq y_0 E_{0\ominus_c(-p)}(t, a), \quad t \in J,$$

and

$$D^\alpha z(t) \leq h(t)w(t)$$

$$\leq y_0 f(t) E_{0\ominus_c(-p)}(t, a), \quad t \in J.$$

We integrate the last inequality from a to t and we obtain

$$z(t) \leq z(a)E_0(t, a) + y_0 \int_a^t f(s)E_{0\ominus_c(-p)}(s, a)\Delta_{\alpha,t}s$$

$$= y_0\left(E_0(t, a) + \int_a^t f(s)E_{0\ominus_c(-p)}(s, a)\Delta_{\alpha,t}s\right), \quad t \in J,$$

and

$$y(t) \leq z(t)$$

$$= y_0\left(E_0(t, a) + \int_a^t f(s)E_{0\ominus_c(-p)}(s, a)\Delta_{\alpha,t}s\right), \quad t \in J.$$

This completes the proof. □

4.6 A CONFORMABLE INTEGRO-DYNAMIC INEQUALITY

Theorem 4.6.1 *Let $a_1, b_1, c_1, y, D^\alpha y : J \to \mathbb{R}$ be nonnegative rd-continuous functions on J, $b_1 \geq 1$ on J, $-b_1(c_1 + 1) \in \mathscr{R}_c^+$,*

$$f(t) = c_1(t)\left(y(a)E_0(t,a) + \int_a^t a_1(s)\Delta_{\alpha,t}s + a_1(t)\right),$$

$$h(t) = \frac{k_0(\alpha,t)b_1(t)(c_1(t)+1)}{k_0(\alpha,t) - \mu(t)(b_1(t)(c_1(t)+1) + k_1(\alpha,t))}, \quad t \in J.$$

If

$$D^\alpha y(t) \leq a_1(t) + b_1(t)\int_a^t a_1(s)\,(y(s) + D^\alpha y(s))\Delta_{\alpha,t}s, \quad t \in J,$$

then

$$D^\alpha y(t) \leq a_1(t)$$

$$+ b_1(t)\int_a^t \frac{f(s)(k_0(\alpha,s) - \mu(s)k_1(\alpha,s))}{k_0(\alpha,s) - \mu(s)(b_1(s)(c_1(s)+1) + k_1(\alpha,s))}E_{-h}(s,t)\Delta_{\alpha,t}s,$$

$$y(t) \leq y(a)E_0(t,a) + \int_a^t a_1(s)\Delta_{\alpha,t}s$$

$$+ \int_a^t b_1(\tau)\int_a^\tau \frac{f(s)(k_0(\alpha,s) - \mu(s)k_1(\alpha,s))}{k_0(\alpha,s) - \mu(s)(b_1(s)(c_1(s)+1) + k_1(\alpha,s))}E_{-h}(s,\tau)\Delta_{\alpha,\tau}s\Delta_{\alpha,s}\tau,$$

$t \in J$.

Proof 4.6.2 *Let*
$$z(t) = \int_a^t c_1(s)\,(y(s) + D^\alpha y(s))\Delta_{\alpha,t}s, \quad t \in J. \tag{4.11}$$

Then

$$z(a) = 0,$$

$$D^\alpha y(t) \leq a_1(t) + b_1(t)z(t), \quad t \in J.$$

Hence, integrating from a to t, we find

$$y(t) \leq y(a)E_0(t,a) + \int_a^t a_1(s)\Delta_{\alpha,t}s$$

$$+ \int_a^t b_1(s)z(s)\Delta_{\alpha,t}s, \quad t \in J.$$

By (4.11), we get

$$D^\alpha z(t) \leq c_1(t)\,(y(t) + D^\alpha y(t))$$

$$\leq c_1(t)\left(y(a)E_0(t,a) + \int_a^t a_1(s)\Delta_{\alpha,t}s\right.$$

$$\left. + \int_a^t b_1(s)z(s)\Delta_{\alpha,t}s + a_1(t) + b_1(t)z(t)\right)$$

$$= c_1(t)\left(y(a)E_0(t,a) + \int_a^t a_1(s)\Delta_{\alpha,t}s + a_1(t)\right)$$

$$+ c_1(t)\left(b_1(t)z(t) + \int_a^t b_1(s)z(s)\Delta_{\alpha,t}s\right)$$

$$= f(t) + c_1(t)\left(b_1(t)z(t) + \int_a^t b_1(s)z(s)\Delta_{\alpha,t}s\right), \quad t \in J.$$

Set

$$w(t) = z(t) + \int_a^t b_1(s)z(s)\Delta_{\alpha,t}s, \quad t \in J.$$

Then

$$z(t) \leq w(t),$$

$$D^\alpha z(t) \leq f(t) + c_1(t)b_1(t)w(t), \quad t \in J,$$

$$w(a) = z(a)$$

$$= 0$$

and

$$D^\alpha w(t) = D^\alpha z(t) + b_1(t)z(t)$$

$$\leq f(t) + c_1(t)b_1(t)w(t) + b(t)w(t)$$

$$= f(t) + b_1(t)(c_1(t) + 1)w(t), \quad t \in J.$$

Thus, by the Gronwall inequality and Theorem 4.1.9, it follows that

$$w(t) \leq \int_a^t \frac{f(s)(k_0(\alpha,s) - \mu(s)k_1(\alpha,s))}{k_0(\alpha,s) - \mu(s)(b_1(s)(c_1(s)+1) + k_1(\alpha,s))}E_{-h}(s,t)\Delta_{\alpha,t}s,$$

$t \in J.$ *Then*

$$z(t) \leq w(t)$$

$$\leq \int_a^t \frac{f(s)(k_0(\alpha,s) - \mu(s)k_1(\alpha,s))}{k_0(\alpha,s) - \mu(s)(b_1(s)(c_1(s)+1) + k_1(\alpha,s))} E_{-h}(s,t)\Delta_{\alpha,t}s,$$

$$D^{\alpha}y(t) \leq a(t)$$

$$+ \int_a^t \frac{f(s)(k_0(\alpha,s) - \mu(s)k_1(\alpha,s))}{k_0(\alpha,s) - \mu(s)(b_1(s)(c_1(s)+1) + k_1(\alpha,s))} E_{-h}(s,t)\Delta_{\alpha,t}s, \quad t \in J,$$

and

$$y(t) \leq y(a)E_0(t,a) + \int_a^t a_1(s)\Delta_{\alpha,t}s$$

$$+ \int_a^t b_1(\tau) \int_a^\tau \frac{f(s)(k_0(\alpha,s) - \mu(s)k_1(\alpha,s))}{k_0(\alpha,s) - \mu(s)(b_1(s)(c_1(s)+1) + k_1(\alpha,s))} E_{-h}(s,\tau)\Delta_{\alpha,\tau}s\Delta_{\alpha,s}\tau,$$

$t \in J$. This completes the proof. □

CHAPTER 5

Cauchy-Type Problem for a Class of Nonlinear Conformable Dynamic Equations

Throughout this chapter, suppose that $\mathbb{T}$ is a time scale with forward jump operator and delta differentiation operator σ and Δ, respectively. Let $\alpha \in (0,1]$, k_0 and k_1 satisfy $(A1)$, and assume

$$k_0(\alpha,t) - \mu(t)k_1(\alpha,t) > 0, \quad \alpha \in (0,1], \quad t \in \mathbb{T},$$

holds. Let $t_0 \in \mathbb{T}$ and $y_0 \in \mathbb{R}$.

5.1   EXISTENCE AND UNIQUENESS OF SOLUTIONS

In this section we will investigate the Cauchy problem

$$D^\alpha y(t) = f(t,y(t)), \quad t > t_0, \tag{5.1}$$

$$y(t_0) = y_0, \tag{5.2}$$

where

(H) $f : \mathbb{T} \times \mathbb{R} \to \mathbb{R}$, $f \in \mathscr{C}(\mathbb{T} \times \mathbb{R})$, $\int_{t_0}^{t} f(s,z)\Delta_{\alpha,t}s$ exists for any $z \in \mathbb{R}$ and for any $t \geq t_0$,

$|f(t,s)| \leq M$ for any $t \geq t_0$ and for any $s \in \mathbb{R}$ and for some positive constant M,

$$|f(t,y_1) - f(t,y_2)| \leq L|y_1 - y_2|$$

for any $t \geq t_0$ and for any $y_1, y_2 \in \mathbb{R}$.

Firstly, we will note that the IVP (5.1), (5.2) is equivalent to the equation

$$y(t) = y_0 E_0(t,t_0) + \int_{t_0}^{t} f(s,y(s))\Delta_{\alpha,t}s, \quad t \geq t_0.$$

Theorem 5.1.1 *Suppose that f satisfies (H). Then the problem (5.1), (5.2) has a unique solution $y \in \mathscr{C}^1([t_0, \infty))$.*

Proof 5.1.2 *Let $a \in \mathbb{T}$, $a > t_0$, $a < \infty$, be arbitrarily chosen and fixed. Let K be a positive constant such that*

$$h_l(t, t_0) \leq K \frac{(t - t_0)^l}{l!}, \quad l \in \mathbb{N}, \quad t \in [t_0, a].$$

Consider the IVP

$$D^\alpha y(t) = f(t, y(t)), \quad t \in [t_0, a], \tag{5.3}$$

$$y(t_0) = y_0. \tag{5.4}$$

We define the sequence $\{y_l(t)\}_{l \in \mathbb{N}}$, $t \in [t_0, a]$, as follows

$$y_0(t) = y_0 E_0(t, t_0),$$

$$y_l(t) = y_0 E_0(t, t_0) + \int_{t_0}^t f(s, y_{l-1}(s)) \Delta_{\alpha, t} s, \quad t \in [t_0, a], \quad l \in \mathbb{N}.$$

We have

$$
\begin{aligned}
|y_1(t) - y_0(t)| &= \left| y_0 E_0(t, t_0) + \int_{t_0}^t f(s, y_0(s)) \Delta_{\alpha, t} s - y_0 E_0(t, t_0) \right| \\
&= \left| \int_{t_0}^t d(s, y_0(s)) \Delta_{\alpha, t} s \right| \\
&\leq \int_{t_0}^t |f(s, y_0(s))| \Delta_{\alpha, t} s \\
&\leq M \int_{t_0}^t \Delta_{\alpha, t} s \\
&= M h_1(t, t_0), \quad t \in [t_0, a].
\end{aligned}
$$

Assume that

$$|y_{l-1}(t) - y_{l-2}(t)| \leq M L^{l-2} h_{l-1}(t, t_0), \quad t \in [t_0, a], \quad l \in \mathbb{N},$$

for some $l \in \mathbb{N}$, $l \geq 2$. We will prove that

$$|y_l(t) - y_{l-1}(t)| \leq M L^{l-1} h_l(t, t_0), \quad t \in [t_0, a]. \tag{5.5}$$

Really,

$$
\begin{aligned}
|y_l(t) - y_{l-1}(t)| &= \left| y_0 E_0(t, t_0) + \int_{t_0}^t f(s, y_{l-1}(s)) \Delta_{\alpha, t} s \right. \\
&\quad \left. - y_0 E_0(t, t_0) - \int_{t_0}^t f(s, y_{l-2}(s)) \Delta_{\alpha, t} s \right|
\end{aligned}
$$

$$= \left| \int_{t_0}^{t} f(s,y_{l-1}(s)) \Delta_{\alpha,t} s - \int_{t_0}^{t} f(s,y_{l-2}(s)) \Delta_{\alpha,t} s \right|$$

$$= \left| \int_{t_0}^{t} \left(f(s,y_{l-1}(s)) - f(s,y_{l-2}(s)) \right) \Delta_{\alpha,t} s \right|$$

$$\leq \int_{t_0}^{t} |f(s,y_{l-1}(s)) - f(s,y_{l-2}(s))| \Delta_{\alpha,t} s$$

$$\leq L \int_{t_0}^{t} |y_{l-1}(s) - y_{l-2}(s)| \Delta_{\alpha,t} s$$

$$\leq ML^{l-1} \int_{t_0}^{t} h_{l-1}(s,t_0) \Delta_{\alpha,t} s$$

$$\leq ML^{l-1} h_l(t,t_0), \quad t \in [t_0,a].$$

Therefore (5.5) holds for any $l \in \mathbb{N}$, $l \geq 1$, and for any $t \in [t_0,a]$. Note that

$$\left| \lim_{l\to\infty} (y_l(t) - y_0(t)) \right| = \left| \sum_{l=1}^{\infty} (y_l(t) - y_{l-1}(t)) \right|$$

$$\leq \sum_{l=1}^{\infty} |y_l(t) - y_{l-1}(t)|$$

$$\leq \frac{M}{L} \sum_{l=1}^{\infty} L^l h_l(t,t_0)$$

$$\leq \frac{M}{L} \sum_{l=1}^{\infty} L^l h_l(a,t_0)$$

$$\leq K \frac{M}{L} \sum_{l=1}^{\infty} \frac{(L(a-t_0))^l}{l!}$$

$$\leq K \frac{M}{L} e^{L(a-t_0)}$$

$$< \infty, \quad t \in [t_0,a].$$

Consequently,

$$\sum_{l=1}^{\infty} (y_l(t) - y_{l-1}(t))$$

is uniformly convergent on $[t_0,a]$. Hence, there exists

$$\lim_{l\to\infty} (y_l(t) - y_0(t)) = \sum_{l=1}^{\infty} (y_l(t) - y_{l-1}(t)), \quad t \in [t_0,a].$$

From here, there exists

$$y(t) = \lim_{l\to\infty} y_l(t), \quad t \in [t_0,a],$$

and y satisfies (5.3), (5.4). Now we assume that the problem (5.3), (5.4) has two solutions y and z. We have

$$z(t) = y_0 E_0(t,t_0) + \int_{t_0}^t f(s,z(s))\Delta_{\alpha,t} s, \quad t \in [t_0,a].$$

Note that

$$\begin{aligned}
|y_0(t) - z(t)| &= \left| y_0 E_0(t,t_0) - y_0 E_0(t,t_0) \right. \\
&\quad \left. - \int_{t_0}^t f(s,z(s))\Delta_{\alpha,t} s \right| \\
&= \left| \int_{t_0}^t f(s,z(s))\Delta_{\alpha,t} s \right| \\
&\leq \int_{t_0}^t |f(s,z(s))|\Delta_{\alpha,t} s \\
&\leq bMh_1(t,t_0), \quad t \in [t_0,a].
\end{aligned}$$

Assume that

$$|y_{l-1}(t) - z(t)| \leq ML^{l-1}h_l(t,t_0), \quad t \in [t_0,a],$$

for some $l \in \mathbb{N}$. We will prove that

$$|y_l(t) - z(t)| \leq ML^l h_{l+1}(t,t_0), \quad t \in [t_0,a]. \tag{5.6}$$

Really, we have

$$\begin{aligned}
|y_l(t) - z(t)| &= \left| y_0 E_0(t,t_0) + \int_{t_0}^t f(s,y_{l-1}(s))\Delta_{\alpha,t} s \right. \\
&\quad \left. - 0 y_0 E_0(t,t_0) - \int_{t_0}^t f(s,z(s))\Delta_{\alpha,t} s \right| \\
&= \left| \int_{t_0}^t (f(s,y_{l-1}(s)) - f(s,z(s)))\Delta_{\alpha,t} s \right| \\
&\leq \int_{t_0}^t |f(s,y_{l-1}(s)) - f(s,z(s))|\Delta_{\alpha,t} s \\
&\leq L \int_{t_0}^t |y_{l-1}(s) - z(s)|\Delta_{\alpha,t} s \\
&\leq ML^l \int_{t_0}^t h_l(s,t_0)\Delta_{\alpha,t} s \\
&= ML^l h_{l+1}(t,t_0), \quad t \in [t_0,a].
\end{aligned}$$

Therefore (5.6) holds for any $l \in \mathbb{N}$. Because

$$\lim_{l \to \infty} ML^l h_{l+1}(t,t_0) = 0, \quad t \in [t_0, a],$$

we get

$$\lim_{l \to \infty} (y_l(t) - z(t)) = 0, \quad t \in [t_0, a].$$

From here

$$y(t) = z(t), \quad t \in [t_0, a].$$

In this way we get that the problem (5.3), (5.4) has a unique solution in $\mathscr{C}^1([t_0, a])$, which we will denote by y^1. Now we consider the problem

$$D^\alpha y(t) = f(t, y(t)), \quad t \in [a, 2a], \tag{5.7}$$

$$y(a) = y^1(a). \tag{5.8}$$

As in the above, we get that the problem (5.7), (5.8) has a unique solution $y^2 \in \mathscr{C}^1([a, 2a])$, and so on. Then

$$y(t) = \begin{cases} y^1(t) & if \quad t \in [t_0, a] \\ y^2(t) & if \quad t \in [a, 2a] \\ \vdots \end{cases}$$

is the unique solution of the problem (5.1), (5.2) that belongs to $\mathscr{C}^1([t_0, \infty))$. This completes the proof. □

5.2   THE DEPENDENCY OF THE SOLUTION UPON THE INITIAL DATA

Let $z_0 \in \mathbb{R}$. Consider the IVPs

$$D^\alpha y(t) = f(t, y(t)), \quad t \in [t_0, a],$$

$$y(t_0) = y_0 \tag{5.9}$$

and

$$D^\alpha z(t) = f(t, z(t)), \quad t \in [t_0, a],$$

$$z(t_0) = z_0, \tag{5.10}$$

where f satisfies $(H1)$ and

$$k_0(\alpha, t) - \mu(t)(k_1(\alpha, t) - L) > 0, \quad t \in [t_0, a].$$

Let P be a positive constant such that

$$E_0(t, t_0) \le P, \quad t \in [t_0, a].$$

We have

$$y(t) = y_0 E_0(t,t_0) + \int_{t_0}^t f(s,y(s))\Delta_{\alpha,t}s,$$

$$z(t) = y_0 E_0(t,t_0) + \int_{t_0}^t f(s,z(s))\Delta_{\alpha,t}s, \quad t \in [t_0,a].$$

Then

$$|y(t) - z(t)| = \left| y_0 E_0(t,t_0) + \int_{t_0}^t f(s,y(s))\Delta_{\alpha,t}s \right.$$

$$\left. -z_0 E_0(t,t_0) - \int_{t_0}^t f(s,z(s))\Delta_{\alpha,t}s \right|$$

$$= \left| (y_0 - z_0)E_0(t,t_0) + \int_{t_0}^t (f(s,y(s)) - f(s,z(s)))\Delta_{\alpha,t}s \right|$$

$$\leq |y_0 - z_0|E_0(t,t_0) + \left| \int_{t_0}^t (f(s,y(s)) - f(s,z(s)))\Delta_{\alpha,t}s \right|$$

$$\leq |y_0 - z_0|E_0(t,t_0) + \int_{t_0}^t |f(s,y(s)) - f(s,z(s))|\Delta_{\alpha,t}s$$

$$\leq P|y_0 - z_0| + L\int_{t_0}^t |y(s) - z(s)|\Delta_{\alpha,t}s, \quad t \in [t_0,a].$$

From the last inequality and from Theorem 4.1.1, we get

$$|y(t) - z(t)| \leq P|y_0 - z_0|\left(L\int_{t_0}^t E_g(\sigma(s),t)\Delta_{\alpha,t}s + 1 \right), \quad t \in [t_0,a],$$

where

$$g(t) = \frac{(-L+k_1(\alpha,t))(k_0(\alpha,t) - \mu(t)k_1(\alpha,t))}{k_0(\alpha,t) - \mu(t)(k_1(\alpha,t) - L)}, \quad t \in [t_0,a].$$

5.3 LYAPUNOV FUNCTIONS

Let $0 \in \mathscr{R}_c$. Suppose that x is a solution of equation (5.3) and let $V : \mathbb{R} \to \mathbb{R}$ be a continuously differentiable function. By Pötzsche's chain rule (see the appendix of this book), we get

$$(V(x(t)))^\Delta = \int_0^1 V'\left(x(t) + h\mu(t)x^\Delta(t)\right)dh\,x^\Delta(t)$$

$$= \int_0^1 V'\left(x(t) + h\mu(t)\left(\frac{1}{k_0(\alpha,t)}D^\alpha x(t) - \frac{k_1(\alpha,t)}{k_0(\alpha,t)}x(t)\right)\right)dh$$

$$\times \left(\frac{1}{k_0(\alpha,t)}D^\alpha x(t) - \frac{k_1(\alpha,t)}{k_0(\alpha,t)}x(t)\right)$$

$$= \int_0^1 V'\left(\frac{k_0(\alpha,t) - h\mu(t)k_1(\alpha,t)}{k_0(\alpha,t)}x(t) + \frac{h\mu(t)}{k_0(\alpha,t)}D^\alpha x(t)\right) dh$$

$$\times \left(\frac{1}{k_0(\alpha,t)}D^\alpha x(t) - \frac{k_1(\alpha,t)}{k_0(\alpha,t)}x(t)\right), \quad t \in \mathbb{T}^\kappa,$$

whereupon

$$k_0(\alpha,t)(V(x(t)))^\Delta = \int_0^1 V'\left(\frac{k_0(\alpha,t) - h\mu(t)k_1(\alpha,t)}{k_0(\alpha,t)}x(t) + \frac{h\mu(t)}{k_0(\alpha,t)}D^\alpha x(t)\right) dh$$

$$\times \left(D^\alpha x(t) - \frac{k_1(\alpha,t)}{k_0(\alpha,t)}x(t)\right)$$

$$= \int_0^1 V'\left(\frac{k_0(\alpha,t) - h\mu(t)k_1(\alpha,t)}{k_0(\alpha,t)}x(t) + \frac{h\mu(t)}{k_0(\alpha,t)}f(t,x(t))\right) dh$$

$$\times \left(f(t,x(t)) - \frac{k_1(\alpha,t)}{k_0(\alpha,t)}x(t)\right), \quad t \in \mathbb{T}^\kappa,$$

and

$$D^\alpha(V(x(t))) = \int_0^1 V'\left(\frac{k_0(\alpha,t) - h\mu(t)k_1(\alpha,t)}{k_0(\alpha,t)}x(t) + \frac{h\mu(t)}{k_0(\alpha,t)}f(t,x(t))\right) dh$$

$$\times \left(f(t,x(t)) - \frac{k_1(\alpha,t)}{k_0(\alpha,t)}x(t)\right) + k_1(\alpha,t)V(x(t)), \quad t \in \mathbb{T}^\kappa.$$

This motivates us to define $D^\alpha V : \mathbb{T} \times \mathbb{R} \to \mathbb{R}$ as follows

$$D^\alpha(V(t,x)) = \int_0^1 V'\left(\frac{k_0(\alpha,t) - h\mu(t)k_1(\alpha,t)}{k_0(\alpha,t)}x(t) + \frac{h\mu(t)}{k_0(\alpha,t)}f(t,x)\right) dh$$

$$\times \left(f(t,x) - \frac{k_1(\alpha,t)}{k_0(\alpha,t)}x(t)\right) + k_1(\alpha,t)V(x), \quad t \in \mathbb{T}^\kappa.$$

Let $t \in \mathbb{T}^\kappa$. If $\mu(t) = 0$, then

$$D^\alpha(V(t,x)) = V'\left(\frac{k_0(\alpha,t) - h\mu(t)k_1(\alpha,t)}{k_0(\alpha,t)}x(t)\right) dh$$

$$\times \left(f(t,x) - \frac{k_1(\alpha,t)}{k_0(\alpha,t)}x(t)\right) + k_1(\alpha,t)V(x).$$

If $\mu(t) \neq 0$ and $f(t,x) \neq 0$, then

$$D^\alpha(V(t,x)) = \frac{k_0(\alpha,t)}{\mu(t)f(t,x)}\left(V\left(\frac{k_0(\alpha,t) - h\mu(t)k_1(\alpha,t)}{k_0(\alpha,t)}x(t) + \frac{h\mu(t)}{k_0(\alpha,t)}f(t,x)\right)\right.$$

$$\left. -V\left(\frac{k_0(\alpha,t) - h\mu(t)k_1(\alpha,t)}{k_0(\alpha,t)}x\right)\right)\left(f(t,x) - \frac{k_1(\alpha,t)}{k_0(\alpha,t)}x\right)$$

$$+k_1(\alpha,t)V(x).$$

Example 5.3.1 *Let $V(x) = x$. Then*

$$D^{\alpha}(V(t,x)) = \left(f(t,x) - \frac{k_1(\alpha,t)}{k_0(\alpha,t)}x \right) + k_1(\alpha,t)x.$$

Exercise 5.3.2 *Let $V(x) = x^2$. Find $D^{\alpha}(V(t,x))$.*

5.4 BOUNDEDNESS OF SOLUTIONS

Definition 5.4.1 *We say that a solution y of the IVP (5.3), (5.4) is uniformly bounded if there exists a constant $C = C(y_0)$, which does not depend on t_0, such that*

$$\|y\| = \sup_{t \in [t_0,\infty)} |y(t)| \leq C.$$

Theorem 5.4.2 *Let $0,1 \in \mathcal{R}_c^+$, and assume there exists a Lyapunov function $V : \mathbb{R} \to [0,\infty)$ such that for all $(t,y) \in [t_0,\infty) \times \mathbb{R}$ we have*

$$V(y) \quad \to \quad \infty, \quad as \quad \|y\|\infty,$$

$$V(y) \quad \leq \quad \phi(\|y\|),$$

$$D^{\alpha}(V(t,y)) \quad \leq \quad \psi(\|y\|) + L$$

and

$$\left(\psi\left(\phi^{-1}(V(y))\right) + L \right) \frac{k_0(\alpha,t) + \mu(t)(1 - k_1(\alpha,t))}{k_0(\alpha,t)}$$

$$+ V(y) \left(1 - k_1(\alpha,t) \frac{k_0(\alpha,t) + \mu(t)(1 - k_1(\alpha,t))}{k_0(\alpha,t)} \right) \leq \gamma,$$

where ϕ, ψ are functions such that $\phi : [0,\infty) \to [0,\infty)$, $\psi : [0,\infty) \to (-\infty,0]$, ψ is nonincreasing and ϕ^{-1} exists, L and γ are nonnegative constants. Then any solution of the IVP (5.3), (5.4), defined on $[t_0,\infty)$, is uniformly bounded.

Proof 5.4.3 *Let y be a solution of the IVP (5.3), (5.4), defined on $[t_0,\infty)$. Then*

$$D^{\alpha}\left(V(t,y)E_1(t,t_0)\right) \quad = \quad D^{\alpha}(V(t,y))E_1^{\sigma}(t,t_0) + V(y(t))E_1(t,t_0)$$

$$- k_1(\alpha,t)V(y(t))E_1^{\sigma}(t,t_0)$$

$$= \quad D^{\alpha}(V(t,y))\frac{k_0(\alpha,t) + \mu(t)(1 - k_1(\alpha,t))}{k_0(\alpha,t)}E_1(t,t_0)$$

$$+ V(y(t))E_1(t,t_0)$$

$$-\frac{k_1(\alpha,t)(k_0(\alpha,t)+\mu(t)(1-k_1(\alpha,t)))}{k_0(\alpha,t)}V(y(t))E_1(t,t_0)$$

$$\leq (\psi(\|y\|)+L)\frac{k_0(\alpha,t)+\mu(t)(1-k_1(\alpha,t))}{k_0(\alpha,t)}E_1(t,t_0)$$

$$+V(y(t))E_1(t,t_0)$$

$$-k_1(\alpha,t)\frac{k_0(\alpha,t)+\mu(t)(1-k_1(\alpha,t))}{k_0(\alpha,t)}V(y(t))E_1(t,t_0)$$

$$\leq \left(\psi\left(\phi^{-1}(V(y(t)))\right)+L\right)\frac{k_0(\alpha,t)+\mu(t)(1-k_1(\alpha,t))}{k_0(\alpha,t)}E_1(t,t_0)$$

$$+V(y(t))E_1(t,t_0)$$

$$-k_1(\alpha,t)\frac{k_0(\alpha,t)+\mu(t)(1-k_1(\alpha,t))}{k_0(\alpha,t)}V(y(t))E_1(t,t_0)$$

$$= \left(\left(\psi\left(\phi^{-1}(V(y(t)))\right)+L\right)\frac{k_0(\alpha,t)+\mu(t)(1-k_1(\alpha,t))}{k_0(\alpha,t)}\right.$$

$$\left.+V(y(t))\left(1-k_1(\alpha,t)\frac{k_0(\alpha,t)+\mu(t)(1-k_1(\alpha,t))}{k_0(\alpha,t)}\right)\right)E_1(t,t_0)$$

$$\leq \gamma E_1(t,t_0), \quad t \in [t_0,\infty).$$

Hence,

$$V(y(t))E_1(t,t_0) \leq V(y_0)E_0(t,t_0)+\gamma E_1(t,t_0), \quad t \in [t_0,\infty),$$

and

$$V(y(t)) \leq V(y_0)\frac{E_0(t,t_0)}{E_1(t,t_0)}+\gamma, \quad t \in [t_0,\infty), \quad t \in [t_0,\infty). \tag{5.11}$$

Since $0,1 \in \mathscr{R}_c^+$, we have

$$\frac{E_0(t,t_0)}{E_1(t,t_0)} = \frac{e_{-\frac{k_1}{k_0}}(t,t_0)}{e_{\frac{1-k_1}{k_0}}(t,t_0)}$$

$$= \frac{e^{\int_{t_0}^{t}\frac{1}{\mu(\tau)}\log\left(1-\mu(\tau)\frac{k_1(\alpha,\tau)}{k_0(\alpha,\tau)}\right)\Delta\tau}}{e^{\int_{t_0}^{t}\frac{1}{\mu(\tau)}\log\left(1+\mu(\tau)\frac{1-k_1(\alpha,\tau)}{k_0(\alpha,\tau)}\right)\Delta\tau}}$$

$$= e^{\int_{t_0}^{t}\frac{1}{\mu(\tau)}\log\left(\frac{k_0(\alpha,\tau)-\mu(\tau)k_1(\alpha,\tau)}{k_0(\alpha,\tau)+\mu(\tau)(1-k_1(\alpha,\tau))}\right)\Delta\tau}$$

$$\leq 1, \quad t \in [t_0,\infty).$$

Then, by (5.11), we get

$$V(y(t)) \leq V(y_0) + \gamma, \quad t \in [t_0, \infty).$$

Because $V(y) \to \infty$, as $\|y\| \to \infty$, by the last inequality, we get that there exists a constant $C > 0$, depending on $V(y_0)$, γ and L only, such that

$$\|y\| \leq C.$$

This completes the proof. □

Theorem 5.4.4 *Let $0, 1 \in \mathscr{R}_c^+$, and assume there exists a Lyapunov function $V : \mathbb{R} \to [0, \infty)$ such that for all $(t, y) \in [t_0, \infty) \times \mathbb{R}$, we have*

$$V(y) \to \infty, \quad as \quad \|y\| \to \infty, \tag{5.12}$$

$$D^\alpha(V(t,y)) \leq -\lambda_1 \|y\|^r + L,$$

$$V(y) \leq \lambda_2 \|y\|^q,$$

and

$$\left(-\frac{\lambda_1}{\lambda_2^{\frac{r}{q}}} (V(y))^{\frac{r}{q}} + L \right) \frac{k_0(\alpha,t) + \mu(t)(1 - k_1(\alpha,t))}{k_0(\alpha,t)}$$

$$+ V(y)\left(1 - k_1(\alpha,t) \frac{k_0(\alpha,t) + \mu(t)(1 - k_1(\alpha,t))}{k_0(\alpha,t)} \right) \leq \gamma,$$

where λ_1, λ_2, r, q are positive constants, L and γ are nonnegative constants. Then any solution of the IVP (5.3), (5.4), defined on $[t_0, \infty)$, is uniformly bounded.

Proof 5.4.5 *Let y be a solution of the IVP (5.3), (5.4), defined on $[t_0, \infty)$. We have*

$$D^\alpha(V(t,y)E_1(t,t_0)) = D^\alpha(V(t,y))E_1^\sigma(t,t_0) + V(y(t))E_1(t,t_0)$$

$$-k_1(\alpha,t)V(y(t))E_1^\sigma(t,t_0)$$

$$= D^\alpha(V(t,y))\frac{k_0(\alpha,t) + \mu(t)(1 - k_1(\alpha,t))}{k_0(\alpha,t)}E_1(t,t_0)$$

$$+V(y(t))E_1(t,t_0)$$

$$-k_1(\alpha,t)\frac{k_0(\alpha,t) + \mu(t)(1 - k_1(\alpha,t))}{k_0(\alpha,t)}V(y(t))E_1(t,t_0)$$

$$\leq (-\lambda_1 \|y\|^r + L)\frac{k_0(\alpha,t) + \mu(t)(1 - k_1(\alpha,t))}{k_0(\alpha,t)}E_1(t,t_0)$$

$$+V(y(t))\left(1-k_1(\alpha,t)\frac{k_0(\alpha,t)+\mu(t)(1-k_1(\alpha,t))}{k_0(\alpha,t)}\right)E_1(t,t_0)$$

$$\leq \left(-\frac{\lambda_1}{\lambda_2^{\frac{r}{q}}}(V(y(t)))^{\frac{r}{q}}+L\right)\frac{k_0(\alpha,t)+\mu(t)(1-k_1(\alpha,t))}{k_0(\alpha,t)}E_1(t,t_0)$$

$$+V(y(t))\left(1-k_1(\alpha,t)\frac{k_0(\alpha,t)+\mu(t)(1-k_1(\alpha,t))}{k_0(\alpha,t)}\right)E_1(t,t_0)$$

$$\leq \left(\left(-\frac{\lambda_1}{\lambda_2^{\frac{r}{q}}}(V(y(t)))^{\frac{r}{q}}+L\right)\frac{k_0(\alpha,t)+\mu(t)(1-k_1(\alpha,t))}{k_0(\alpha,t)}\right.$$

$$\left.+V(y(t))\left(1-k_1(\alpha,t)\frac{k_0(\alpha,t)+\mu(t)(1-k_1(\alpha,t))}{k_0(\alpha,t)}\right)\right)E_1(t,t_0)$$

$$\leq \gamma E_1(t,t_0), \quad t\in[t_0,\infty).$$

Now, as in the proof of Theorem 5.4.2, we get

$$V(y(t)) \leq V(y_0)+\gamma, \quad t\in[t_0,\infty), \tag{5.13}$$

and there exists a constant $C>0$, depending on $V(y_0)$, γ and L only, such that

$$\|y\| \leq C.$$

This completes the proof. □

Theorem 5.4.6 *Assume that all conditions of Theorem 5.4.4 are satisfied with (5.12) replaced by*

$$\lambda_3\|y\|^p \leq V(y), \tag{5.14}$$

where λ_3 and p are positive constants. Then any solution of the IVP (5.3), (5.4), defined on $[t_0,\infty)$, is uniformly bounded and

$$\|y\| \leq \left(\frac{V(y_0)+\gamma}{\lambda_3}\right)^{\frac{1}{p}}.$$

Proof 5.4.7 *Let y be a solution of the IVP (5.3), (5.4), defined on $[t_0,\infty)$. By (5.13), we get*

$$V(y(t)) \leq V(y_0)+\gamma, \quad t\in[t_0,\infty).$$

Hence, and considering (5.14), we arrive at

$$\lambda_3\|y\|^p \leq V(y_0)+\gamma,$$

whereupon

$$\|y\|^p \leq \frac{V(y_0)+\gamma}{\lambda_3}$$

and

$$\|y\| \leq \left(\frac{V(y_0) + \gamma}{\lambda_3} \right)^{\frac{1}{p}}.$$

This completes the proof. □

Theorem 5.4.8 *Let* $0, \lambda_1 \in \mathscr{R}_c^+$ *and there exists a Lyapunov function* $V : \mathbb{R} \to [0, \infty)$ *such that for all* $(t, y) \in [t_0, \infty) \times \mathbb{R}$ *we have*

$$V(y) \;\to\; \infty, \quad as \quad \|y\|\infty,$$

$$D^\alpha(V(t, y)) \;\leq\; -\lambda_1 V(y) + L$$

and

$$(-\lambda_1 V(y(t))) + L) \frac{k_0(\alpha, t) + \mu(t)(\lambda_1 - k_1(\alpha, t))}{k_0(\alpha, t)}$$

$$+ V(y(t)) \left(\lambda_1 - k_1(\alpha, t) \frac{k_0(\alpha, t) + \mu(t)(\lambda_1 - k_1(\alpha, t))}{k_0(\alpha, t)} \right)$$

where L *and* γ *are nonnegative constants. Then any solution of the IVP* (5.3), (5.4)*, defined on* $[t_0, \infty)$*, is uniformly bounded.*

Proof 5.4.9 *Let* y *be a solution of the IVP* (5.3), (5.4)*, defined on* $[t_0, \infty)$*. Then*

$$D^\alpha \left(V(t, y) E_{\lambda_1}(t, t_0) \right) \;=\; D^\alpha(V(t, y)) E_{\lambda_1}^\sigma(t, t_0) + \lambda_1 V(y(t)) E_{\lambda_1}(t, t_0)$$

$$- k_1(\alpha, t) V(y(t)) E_{\lambda_1}^\sigma(t, t_0)$$

$$= \; D^\alpha(V(t, y)) \frac{k_0(\alpha, t) + \mu(t)(\lambda_1 - k_1(\alpha, t))}{k_0(\alpha, t)} E_{\lambda_1}(t, t_0)$$

$$+ \lambda_1 V(y(t)) E_{\lambda_1}(t, t_0)$$

$$- \frac{k_1(\alpha, t)(k_0(\alpha, t) + \mu(t)(\lambda_1 - k_1(\alpha, t)))}{k_0(\alpha, t)} V(y(t)) E_{\lambda_1}(t, t_0)$$

$$\leq \; (-\lambda_1 V(y(t)) + L) \frac{k_0(\alpha, t) + \mu(t)(\lambda_1 - k_1(\alpha, t))}{k_0(\alpha, t)} E_{\lambda_1}(t, t_0)$$

$$+ \lambda_1 V(y(t)) E_{\lambda_1}(t, t_0)$$

$$- k_1(\alpha, t) \frac{k_0(\alpha, t) + \mu(t)(\lambda_1 - k_1(\alpha, t))}{k_0(\alpha, t)} V(y(t)) E_{\lambda_1}(t, t_0)$$

$$= \; \left((-\lambda_1 V(y(t)) + L) \frac{k_0(\alpha, t) + \mu(t)(\lambda_1 - k_1(\alpha, t))}{k_0(\alpha, t)} \right.$$

$$+V(y(t)) \left(\lambda_1 - k_1(\alpha,t) \frac{k_0(\alpha,t) + \mu(t)(\lambda_1 - k_1(\alpha,t))}{k_0(\alpha,t)} \right) \right) E_{\lambda_1}(t,t_0)$$

$$\leq \gamma E_{\lambda_1}(t,t_0), \quad t \in [t_0,\infty).$$

Consequently,

$$V(y(t))E_{\lambda_1}(t,t_0) \leq V(y_0)E_0(t,t_0) + \gamma E_{\lambda_1}(t,t_0), \quad t \in [t_0,\infty),$$

and

$$V(y(t)) \leq V(y_0) \frac{E_0(t,t_0)}{E_{\lambda_1}(t,t_0)} + \gamma, \quad t \in [t_0,\infty), \quad t \in [t_0,\infty). \tag{5.15}$$

Now, using that $0, \lambda_1 \in \mathscr{R}_c^+$, *we obtain*

$$\frac{E_0(t,t_0)}{E_{\lambda_1}(t,t_0)} = \frac{e_{-\frac{k_1}{k_0}}(t,t_0)}{e_{\frac{\lambda_1-k_1}{k_0}}(t,t_0)}$$

$$= \frac{e^{\int_{t_0}^t \frac{1}{\mu(\tau)} \log\left(1-\mu(\tau)\frac{k_1(\alpha,\tau)}{k_0(\alpha,\tau)}\right)\Delta\tau}}{e^{\int_{t_0}^t \frac{1}{\mu(\tau)} \log\left(1+\mu(\tau)\frac{\lambda_1-k_1(\alpha,\tau)}{k_0(\alpha,\tau)}\right)\Delta\tau}}$$

$$= e^{\int_{t_0}^t \frac{1}{\mu(\tau)} \log\left(\frac{k_0(\alpha,\tau)-\mu(\tau)k_1(\alpha,\tau)}{k_0(\alpha,\tau)+\mu(\tau)(\lambda_1-k_1(\alpha,\tau))}\right)\Delta\tau}$$

$$\leq 1, \quad t \in [t_0,\infty).$$

From here and (5.15), we get

$$V(y(t)) \leq V(y_0) + \gamma, \quad t \in [t_0,\infty). \tag{5.16}$$

Since $V(y) \to \infty$, *as* $\|y\| \to \infty$, *by the last inequality, we conclude that there exists a constant* $C > 0$, *depending on* $V(y_0)$, γ *and* L *only, such that*

$$\|y\| \leq C.$$

This completes the proof. □

Theorem 5.4.10 *Assume that all conditions of Theorem 5.4.8 are satisfied with*

$$V(y) \to \infty, \quad as \quad \|y\| \to \infty$$

replaced by

$$\lambda_2 \|y\|^p \leq V(y), \tag{5.17}$$

where λ_2 *and* p *are positive constants. Then any solution of the IVP (5.3), (5.4), defined on* $[t_0,\infty)$, *is uniformly bounded and*

$$\|y\| \leq \left(\frac{V(y_0)+\gamma}{\lambda_2}\right)^{\frac{1}{p}}.$$

Proof 5.4.11 *Let y be a solution of the IVP (5.3), (5.4), defined on $[t_0,\infty)$. By (5.16), we get*

$$V(y(t)) \leq V(y_0) + \gamma, \quad t \in [t_0,\infty).$$

Hence, and considering (5.17), we arrive at

$$\lambda_2 \|y\|^p \leq V(y_0) + \gamma,$$

whereupon

$$\|y\|^p \leq \frac{V(y_0) + \gamma}{\lambda_2}$$

and

$$\|y\| \leq \left(\frac{V(y_0) + \gamma}{\lambda_2}\right)^{\frac{1}{p}}.$$

This completes the proof. □

Exercise 5.4.12 *Let $\mathbb{T} = 2\mathbb{Z}$,*

$$k_1(\alpha,t) = (1-\alpha)(1+t^2)^{4\alpha}, \quad k_0(\alpha,t) = \alpha(1+t^2)^{4(1-\alpha)}, \quad \alpha \in (0,1], \quad t \in \mathbb{T}.$$

Consider the IVP

$$D^{\frac{1}{4}}y = ay^{\frac{1}{3}} + by, \quad t > 0,$$

$$y(0) = 2,$$

where a and b are constants. If

1. *if $V(x) = x$,*

2. *$V(x) = x^2$,*

find conditions for the constants a and b so that any solution of the considered IVP, defined on $[0,\infty)$, is uniformly bounded.

5.5 EXPONENTIAL STABILITY

In this section, in addition we suppose that $f(t,0) = 0$, $t \in [t_0,\infty)$.

Definition 5.5.1 *We say that the trivial solution of equation (5.3) is exponentially stable if there exist positive constants d and M and a nonnegative constant C such that for any solution of the IVP (5.3), (5.4) we have*

$$|y(t)| \leq C(y_0,t_0)(E_{\ominus_c M}(t,t_0))^d, \quad t \in [t_0,\infty).$$

Theorem 5.5.2 *Let* $0, M \in \mathcal{R}_c^+$, $M > 0$, *and assume there exists a Lyapunov function* $V : \mathbb{R} \to [0, \infty)$ *such that for all* $(t, y) \in [t_0, \infty) \times \mathbb{R}$ *we have*

$$W(\|y\|) \leq V(y) \leq \phi(\|y\|),$$

$$D^\alpha(V(t,y)) \leq \psi(\|y\|) + L$$

and

$$\left(\psi\left(\phi^{-1}(V(y))\right) + L\right) \frac{k_0(\alpha,t) + \mu(t)(M \oplus_c 0 - k_1(\alpha,t))}{k_0(\alpha,t)}$$

$$+ V(y)\left(1 - k_1(\alpha,t)\frac{k_0(\alpha,t) + \mu(t)(M \oplus_c 0 - k_1(\alpha,t))}{k_0(\alpha,t)}\right) \leq -(M \oplus_c 0),$$

where ϕ, ψ *are functions such that* $W, \phi : [0, \infty) \to [0, \infty)$, $\psi : [0, \infty) \to (-\infty, 0]$, ψ *is nonincreasing and* ϕ *and* W *are strictly increasing,* L *is a nonnegative constant. Then any solution of the IVP (5.3) satisfies*

$$\|y\| \leq W^{-1}\left((V(y_0) + 1)E_{\ominus_c M}(t, t_0)\right), \quad t \in [t_0, \infty).$$

Proof 5.5.3 *Let* y *be a solution of the IVP (5.3), (5.4), defined on* $[t_0, \infty)$. *Then*

$$D^\alpha\left(V(t,y)E_{M \oplus_c 0}(t,t_0)\right) = D^\alpha(V(t,y))E_{M \oplus_c 0}^\sigma(t,t_0) + V(y(t))E_{M \oplus_c 0}(t,t_0)$$

$$- k_1(\alpha,t)V(y(t))E_{M \oplus_c 0}^\sigma(t,t_0)$$

$$= D^\alpha(V(t,y))\frac{k_0(\alpha,t) + \mu(t)(M \oplus_c 0 - k_1(\alpha,t))}{k_0(\alpha,t)}E_{M \oplus_c 0}(t,t_0)$$

$$+ (M \oplus_c 0)V(y(t))E_{M \oplus_c 0}(t,t_0)$$

$$- \frac{k_1(\alpha,t)(k_0(\alpha,t) + \mu(t)(M \oplus_c 0 - k_1(\alpha,t)))}{k_0(\alpha,t)}V(y(t))E_{M \oplus_c 0}(t,t_0)$$

$$\leq (\psi(\|y\|) + L)\frac{k_0(\alpha,t) + \mu(t)(M \oplus_c 0 - k_1(\alpha,t))}{k_0(\alpha,t)}E_{M \oplus_c 0}(t,t_0)$$

$$+ (M \oplus_c 0)V(y(t))E_{M \oplus_c 0}(t,t_0)$$

$$- k_1(\alpha,t)\frac{k_0(\alpha,t) + \mu(t)(M \oplus_c 0 - k_1(\alpha,t))}{k_0(\alpha,t)}V(y(t))E_{M \oplus_c 0}(t,t_0)$$

$$\leq \left(\psi\left(\phi^{-1}(V(y(t)))\right) + L\right)\frac{k_0(\alpha,t) + \mu(t)(M \oplus_c 0 - k_1(\alpha,t))}{k_0(\alpha,t)}E_{M \oplus_c 0}(t,t_0)$$

$$+(M\oplus_c 0)V(y(t))E_{M\oplus_c 0}(t,t_0)$$

$$-k_1(\alpha,t)\frac{k_0(\alpha,t)+\mu(t)(M\oplus_c 0-k_1(\alpha,t))}{k_0(\alpha,t)}V(y(t))E_{M\oplus_c 0}(t,t_0)$$

$$= \left(\left(\psi\left(\phi^{-1}(V(y(t)))\right)+L\right)\frac{k_0(\alpha,t)+\mu(t)(M\oplus_c 0-k_1(\alpha,t))}{k_0(\alpha,t)}\right.$$

$$\left.+V(y(t))\left(M\oplus_c 0-k_1(\alpha,t)\frac{k_0(\alpha,t)+\mu(t)(M\oplus_c 0-k_1(\alpha,t))}{k_0(\alpha,t)}\right)\right)E_{M\oplus_c 0}(t,t_0)$$

$$\leq -(M\oplus_c 0)E_{M\oplus_c 0}(t,t_0), \quad t\in[t_0,\infty).$$

Hence,

$$V(y(t))E_{M\oplus_c 0}(t,t_0) \leq V(y_0)E_0(t,t_0)-E_{M\oplus_c 0}(t,t_0)+E_0(t,t_0)$$

$$\leq (V(y_0)+1)E_0(t,t_0), \quad t\in[t_0,\infty),$$

and

$$V(y(t))\leq (V(y_0)+1)E_{\ominus_c M}(t,t_0), \quad t\in[t_0,\infty), \quad t\in[t_0,\infty).$$

Now, using that $V(y(t))\geq W(\|y\|)$, we get

$$\|y\|\leq W^{-1}\left((V(y_0)+1)E_{\ominus_c M}(t,t_0)\right), \quad t\in[t_0,\infty).$$

This completes the proof. □

Theorem 5.5.4 *Let $0,M\in\mathscr{R}_c^+$, and assume there exists a Lyapunov function $V:\mathbb{R}\to[0,\infty)$ such that for all $(t,y)\in[t_0,\infty)\times\mathbb{R}$, we have*

$$D^\alpha(V(t,y)) \leq -\lambda_1\|y\|^r+L,$$

$$\lambda_3\|y\|^p \leq V(y)\leq\lambda_2\|y\|^q,$$

and

$$\left(-\frac{\lambda_1}{\lambda_2^{\frac{r}{q}}}(V(y))^{\frac{r}{q}}+L\right)\frac{k_0(\alpha,t)+\mu(t)(M\oplus_c 0-k_1(\alpha,t))}{k_0(\alpha,t)}$$

$$+V(y)\left(M\oplus_c 0-k_1(\alpha,t)\frac{k_0(\alpha,t)+\mu(t)(M\oplus_c -k_1(\alpha,t))}{k_0(\alpha,t)}\right)\leq -(M\oplus_c 0),$$

where λ_1, λ_2, r, p, q are positive constants, L is a nonnegative constant. Then the trivial solution of equation (5.3) is exponentially stable.

Proof 5.5.5 *Let y be a solution of the IVP (5.3), (5.4), defined on* $[t_0, \infty)$. *We have*

$$D^\alpha(V(t,y)E_{M\oplus_c0}(t,t_0)) = D^\alpha(V(t,y))E^\sigma_{M\oplus_c0}(t,t_0) + V(y(t))E_{M\oplus_c0}(t,t_0)$$

$$-k_1(\alpha,t)V(y(t))E^\sigma_{M\oplus_c0}(t,t_0)$$

$$= D^\alpha(V(t,y))\frac{k_0(\alpha,t)+\mu(t)(M\oplus_c0-k_1(\alpha,t))}{k_0(\alpha,t)}E_{M\oplus_c0}(t,t_0)$$

$$+(M\oplus_c0)V(y(t))E_{M\oplus_c0}(t,t_0)$$

$$-k_1(\alpha,t)\frac{k_0(\alpha,t)+\mu(t)(M\oplus_c0-k_1(\alpha,t))}{k_0(\alpha,t)}V(y(t))E_{M\oplus_c0}(t,t_0)$$

$$\leq (-\lambda_1\|y\|^r+L)\frac{k_0(\alpha,t)+\mu(t)(M\oplus_c0-k_1(\alpha,t))}{k_0(\alpha,t)}E_{M\oplus_c0}(t,t_0)$$

$$+V(y(t))\left(M\oplus_c0-k_1(\alpha,t)\frac{k_0(\alpha,t)+\mu(t)(1-k_1(\alpha,t))}{k_0(\alpha,t)}\right)E_{M\oplus_c0}(t,t_0)$$

$$\leq \left(-\frac{\lambda_1}{\lambda_2^{\frac{r}{q}}}(V(y(t)))^{\frac{r}{q}}+L\right)\frac{k_0(\alpha,t)+\mu(t)(M\oplus_c0-k_1(\alpha,t))}{k_0(\alpha,t)}E_{M\oplus_c0}(t,t_0)$$

$$+V(y(t))\left(M\oplus_c0-k_1(\alpha,t)\frac{k_0(\alpha,t)+\mu(t)(M\oplus_c0-k_1(\alpha,t))}{k_0(\alpha,t)}\right)E_{M\oplus_c0}(t,t_0)$$

$$\leq \left(\left(-\frac{\lambda_1}{\lambda_2^{\frac{r}{q}}}(V(y(t)))^{\frac{r}{q}}+L\right)\frac{k_0(\alpha,t)+\mu(t)(M\oplus_c0-k_1(\alpha,t))}{k_0(\alpha,t)}\right.$$

$$\left.+V(y(t))\left(M\oplus_c0-k_1(\alpha,t)\frac{k_0(\alpha,t)+\mu(t)(M\oplus_c0-k_1(\alpha,t))}{k_0(\alpha,t)}\right)\right)E_{M\oplus_c0}(t,t_0)$$

$$\leq -(M\oplus_c0)E_{M\oplus_c0}(t,t_0), \quad t\in[t_0,\infty).$$

Now, as in the proof of Theorem 5.5.2, we get

$$V(y(t)) \leq (V(y_0)+1)e_{\ominus_cM}(t,t_0), \quad t\in[t_0,\infty),$$

and

$$y(t) \leq \left(\frac{V(y_0)+1}{\lambda_3}e_{\ominus_cM}(t,t_0)\right)^{\frac{1}{p}}, \quad t\in[t_0,\infty).$$

This completes the proof. □

Theorem 5.5.6 *Let* $0, \lambda_1 \in \mathcal{R}_c^+$, *and assume there exists a Lyapunov function* $V : \mathbb{R} \to [0,\infty)$ *such that for all* $(t,y) \in [t_0,\infty) \times \mathbb{R}$ *we have*

$$\lambda\|y\|^p \leq V(y),$$

$$D^\alpha(V(t,y)) \leq -\lambda_1 V(y) + L$$

and

$$(-\lambda_1 V(y(t)) + L)\frac{k_0(\alpha,t) + \mu(t)(\lambda_1 - k_1(\alpha,t))}{k_0(\alpha,t)}$$

$$+ V(y(t))\left(\lambda_1 - k_1(\alpha,t)\frac{k_0(\alpha,t) + \mu(t)(\lambda_1 - k_1(\alpha,t))}{k_0(\alpha,t)}\right) \leq -(\lambda_1 \oplus_c 0)$$

where L is a nonnegative constant. Then the trivial solution of equation (5.3) is exponentially stable.

Proof 5.5.7 *Let y be a solution of the IVP (5.3), (5.4), defined on $[t_0, \infty)$. Then*

$$D^\alpha\left(V(t,y)E_{\lambda_1 \oplus_c 0}(t,t_0)\right) = D^\alpha(V(t,y))E^\sigma_{\lambda_1 \oplus_c 0}(t,t_0) + (\lambda_1 \oplus_c 0)V(y(t))E_{\lambda_1 \oplus_c 0}(t,t_0)$$

$$- k_1(\alpha,t)V(y(t))E^\sigma_{\lambda_1 \oplus_c 0}(t,t_0)$$

$$= D^\alpha(V(t,y))\frac{k_0(\alpha,t) + \mu(t)(\lambda_1 \oplus_c 0 - k_1(\alpha,t))}{k_0(\alpha,t)}E_{\lambda_1 \oplus_c 0}(t,t_0)$$

$$+ (\lambda_1 \oplus_c 0)V(y(t))E_{\lambda_1 \oplus_c 0}(t,t_0)$$

$$- \frac{k_1(\alpha,t)(k_0(\alpha,t) + \mu(t)(\lambda_1 \oplus_c 0 - k_1(\alpha,t)))}{k_0(\alpha,t)}V(y(t))E_{\lambda_1 \oplus_c 0}(t,t_0)$$

$$\leq (-\lambda_1 V(y(t)) + L)\frac{k_0(\alpha,t) + \mu(t)(\lambda_1 \oplus_c 0 - k_1(\alpha,t))}{k_0(\alpha,t)}E_{\lambda_1 \oplus_c 0}(t,t_0)$$

$$+ (\lambda_1 \oplus_c 0)V(y(t))E_{\lambda_1 \oplus_c 0}(t,t_0)$$

$$- k_1(\alpha,t)\frac{k_0(\alpha,t) + \mu(t)(\lambda_1 \oplus_c 0 - k_1(\alpha,t))}{k_0(\alpha,t)}V(y(t))E_{\lambda_1 \oplus_c 0}(t,t_0)$$

$$= \left((-\lambda_1 V(y(t)) + L)\frac{k_0(\alpha,t) + \mu(t)(\lambda_1 \oplus_c 0 - k_1(\alpha,t))}{k_0(\alpha,t)}\right.$$

$$\left. + V(y(t))\left(\lambda_1 \oplus_c 0 - k_1(\alpha,t)\frac{k_0(\alpha,t) + \mu(t)(\lambda_1 \oplus_c 0 - k_1(\alpha,t))}{k_0(\alpha,t)}\right)\right)E_{\lambda_1 \oplus_c 0}(t,t_0)$$

$$\leq -(\lambda_1 \oplus_c 0)E_{\lambda_1 \oplus_c 0}(t,t_0), \quad t \in [t_0,\infty).$$

Consequently,

$$V(y(t)) \leq (V(y_0) + 1)E_{\ominus_c \lambda_1}(t,t_0), \quad t \in [t_0,\infty),$$

and

$$|y(t)| \leq \left(\frac{V(0)+1}{\lambda} E_{\ominus_c \lambda_1}(t,t_0) \right)^{\frac{1}{p}}, \quad t \in [t_0, \infty).$$

This completes the proof. □

Exercise 5.5.8 *Let* $\mathbb{T} = 4^{\mathbb{N}_0}$,

$$k_1(\alpha,t) = (1-\alpha)t^{8\alpha}, \quad k_0(\alpha,t) = \alpha t^{8(1-\alpha)}, \quad \alpha \in (0,1], \quad t \in \mathbb{T}.$$

Consider the IVP

$$D^{\frac{1}{7}}y = t + y^{\frac{1}{3}} + \frac{at+b}{1+y^8}, \quad t > 1,$$

$$y(1) = 1,$$

where a,b *are constants. If*

1. $V(x) = x$,

2. $V(x) = x^2$,

find conditions for the constants a,b *so that the trivial solution is exponentially stable.*

5.6 ADVANCED PRACTICAL PROBLEMS

Problem 5.6.1 *Let* $V(x) = ax^4$, *where* $a > 0$. *Find* $D^{\alpha}(V(t,x))$.

Problem 5.6.2 *Let*

$$V(x) = \int_0^x p(s)ds,$$

where $p : \mathbb{R} \to [0,\infty)$ *is a continuous function. Find* $D^{\alpha}(V(t,x))$.

Problem 5.6.3 *Let* $\mathbb{T} = 2^{\mathbb{N}_0}$,

$$k_1(\alpha,t) = (1-\alpha)(1+t+t^2)^{2\alpha}, \quad k_0(\alpha,t) = \alpha(1+t+t^2)^{2(1-\alpha)}, \quad \alpha \in (0,1], \quad t \in \mathbb{T}.$$

Consider the IVP

$$D^{\frac{1}{6}}y = 1 + t + (a+b)ty^{\frac{1}{7}} + (ct+d)y^{\frac{1}{3}}, \quad t > 1,$$

$$y(1) = 1,$$

where a,b,c,d *are constants. If*

1. $V(x) = x$,

2. $V(x) = x^2$,

find conditions for the constants a, b, c, d so that any solution of the considered IVP, defined on $[1, \infty)$, is uniformly bounded.

Problem 5.6.4 *Let* $\mathbb{T} = 3^{\mathbb{N}_0}$,

$$k_1(\alpha, t) = (1 - \alpha)(1 + t^4)^{6\alpha}, \quad k_0(\alpha, t) = \alpha(1 + t^4)^{6(1-\alpha)}, \quad \alpha \in (0, 1], \quad t \in \mathbb{T}.$$

Consider the IVP

$$D^{\frac{1}{9}}y = at + \frac{b}{\sqrt{1+y^2}}, \quad t > 1,$$

$$y(1) = 4,$$

where a, b are constants. If

1. $V(x) = x,$

2. $V(x) = x^2,$

find conditions for the constants a, b so that the trivial solution is exponentially stable.

CHAPTER 6

Higher-Order Linear Conformable Dynamic Equations with Constant Coefficients

Suppose that $\mathbb{T}$ is a time scale with forward jump operator and delta differentiation operator σ and Δ, respectively. Let $\alpha \in (0,1]$, $n \in \mathbb{N}$, k_0 and k_1 satisfy $(A1)$, $k_0(\alpha, \cdot)$, $k_1(\alpha, \cdot) \in \mathscr{C}_{rd}^{n-1}(\mathbb{T})$ for any $\alpha \in (0,1]$. Also, assume (1.6) holds and $t_0 \in \mathbb{T}$.

6.1 HOMOGENEOUS HIGHER-ORDER LINEAR CONFORMABLE DYNAMIC EQUATIONS WITH CONSTANT COEFFICIENTS

Consider the equation

$$(D^\alpha)^n y + a_1 (D^\alpha)^{n-1} y + \cdots + a_{n-1} D^\alpha y + a_n y = 0 \quad \text{on} \quad \mathbb{T}^{\kappa^n} \tag{6.1}$$

where $a_l \in \mathbb{R}$, $l \in \{1, \ldots, n\}$.

Definition 6.1.1 *The equation* (6.1) *is called a homogeneous nth-order linear conformable dynamic equation with constant coefficients.*

Definition 6.1.2 *A function $y \in \mathscr{C}_{rd}^n(\mathbb{T})$ that satisfies equation* (6.1) *will be called a solution of equation* (6.1).

Theorem 6.1.3 *Let $y_1, y_2 \in \mathscr{C}_{rd}^n(\mathbb{T})$ be two solutions of equation* (6.1). *Then*

$$b_1 y_1 + b_2 y_2$$

is a solution of equation (6.1) *for any $b_1, b_2 \in \mathbb{R}$.*

Proof 6.1.4 *We have*

$$(D^\alpha)^n (b_1 y_1 + b_2 y_2) + a_1 (D^\alpha)^{n-1} (b_1 y_1 + b_2 y_2) + \cdots + a_{n-1} D^\alpha (b_1 y_1 + b_2 y_2)$$

$$+ a_n(b_1 y_1 + b_2 y_2)$$

$$= \left((D^\alpha)^n (b_1 y_1) + (D^\alpha)^n (b_2 y_2) \right) + a_1 \left((D^\alpha)^{n-1} (b_1 y_1) + (D^\alpha)^{n-1} (b_2 y_2) \right)$$

$$+ \cdots + a_{n-1} \left(D^\alpha (b_1 y_1) + D^\alpha (b_2 y_2) \right) + a_n (b_1 y_1 + b_2 y_2)$$

$$= \left(b_1 (D^\alpha)^n y_1 + b_2 (D^\alpha)^n y_2 \right) + a_1 \left(b_1 (D^\alpha)^{n-1} y_1 + b_2 (D^\alpha)^{n-1} y_2 \right)$$

$$+ \cdots + a_{n-1} \left(b_1 D^\alpha y_1 + b_2 D^\alpha y_2 \right) + a_n (b_1 y_1 + b_2 y_2)$$

$$= b_1 \left((D^\alpha)^n y_1 + a_1 (D^\alpha)^{n-1} y_1 + \cdots + a_{n-1} D^\alpha y_1 + a_n y_1 \right)$$

$$+ b_2 \left((D^\alpha)^n y_2 + a_1 (D^\alpha)^{n-1} y_2 + \cdots + a_{n-1} D^\alpha y_2 + a_n y_2 \right)$$

$$= 0 \quad on \quad \mathbb{T}^{\kappa^n}.$$

This completes the proof. □

Now, we will search a solution of equation (6.1) in the form

$$y(t) = E_\lambda(t, t_0), \quad t \in \mathbb{T}^{\kappa^n},$$

where λ is a constant. We have

$$D^\alpha y(t) = D^\alpha E_\lambda(t, t_0)$$

$$= \lambda E_\lambda(t, t_0),$$

$$(D^\alpha)^2 y(t) = (D^\alpha)^2 E_\lambda(t, t_0)$$

$$= \lambda D^\alpha E_\lambda(t, t_0)$$

$$= \lambda^2 E_\lambda(t, t_0),$$

$$\vdots$$

$$(D^\alpha)^n y(t) = (D^\alpha)^n E_\lambda(t, t_0)$$

$$= \lambda^{n-1} D^\alpha E_\lambda(t, t_0)$$

$$= \lambda^n E_\lambda(t, t_0), \quad t \in \mathbb{T}^{\kappa^n}.$$

Thus, using (6.1), we get

$$\lambda^n E_\lambda(t,t_0) + a_1 \lambda^{n-1} E_\lambda(t,t_0) + \cdots + a_{n-1}\lambda E_\lambda(t,t_0) + a_n E_\lambda(t,t_0) = 0, \quad t \in \mathbb{T}^{\kappa^n},$$

or

$$\left(\lambda^n + a_1\lambda^{n-1} + \cdots + a_{n-1}\lambda + a_n\right) E_\lambda(t,t_0) = 0, \quad t \in \mathbb{T}^{\kappa^n}.$$

This leads to the following definition.

Definition 6.1.5 *The equation*

$$\lambda^n + a_1\lambda^{n-1} + \cdots + a_{n-1}\lambda + a_n = 0 \tag{6.2}$$

is called the characteristic equation of equation (6.1).

Example 6.1.6 *The characteristic equation of the equation*

$$(D^\alpha)^3 y + 3(D^\alpha)^2 y + 2D^\alpha y + y = 0, \quad t \in \mathbb{T}^{\kappa^3},$$

is

$$\lambda^3 + 3\lambda^2 + 2\lambda + 1 = 0.$$

Exercise 6.1.7 *Find the characteristic equations for each of the following equations.*

1.
$$(D^\alpha)^4 y - 2(D^\alpha)^3 y + D^\alpha y = 0, \quad t \in \mathbb{T}^{\kappa^4},$$

2.
$$(D^\alpha)^5 y - y = 0, \quad t \in \mathbb{T}^{\kappa^3},$$

3.
$$(D^\alpha)^3 y - 2(D^\alpha)^2 y - D^\alpha y - 10y = 0, \quad t \in \mathbb{T}^{\kappa^3},$$

4.
$$(D^\alpha)^3 y + 7(D^\alpha)^2 y + 12D^\alpha y - 13y = 0, \quad t \in \mathbb{T}^{\kappa^3},$$

5.
$$(D^\alpha)^4 y - (D^\alpha)^2 y - 5D^\alpha y - 18y = 0, \quad t \in \mathbb{T}^{\kappa^4}.$$

Let $\lambda_1, \lambda_2, \ldots, \lambda_n$ be the roots of the characteristic equation (6.2). Then equation (6.1) can be rewritten in the following way

$$(D^\alpha - \lambda_1)(D^\alpha - \lambda_2)\ldots(D^\alpha - \lambda_n)y = 0 \quad \text{on} \quad \mathbb{T}^{\kappa^n}.$$

We set

$$(D^\alpha - \lambda_2)\ldots(D^\alpha - \lambda_n)y = y_1 \quad \text{on} \quad \mathbb{T}^{\kappa^n}.$$

Then we get

$$(D^\alpha - \lambda_1)y_1 = 0 \quad \text{on} \quad \mathbb{T}^{\kappa^n},$$

which is a first-order linear conformable equation. Let y_1 be its solution. Then we set

$$(D^\alpha - \lambda_3)\dots(D^\alpha - \lambda_n)y = y_2 \quad \text{on} \quad \mathbb{T}^{\kappa^n}$$

and we obtain

$$(D^\alpha - \lambda_2)y_2 = y_1 \quad \text{on} \quad \mathbb{T}^{\kappa^n}.$$

We solve the last equation and so on.

Example 6.1.8 *Let $\mathbb{T} = 2^{\mathbb{N}_0}$,*

$$k_1(\alpha,t) = (1 - \alpha)t^{4\alpha}, \quad k_0(\alpha,t) = \alpha t^{4(1-\alpha)}, \quad \alpha \in (0,1], \quad t \in \mathbb{T}.$$

Consider the equation

$$\left(D^{\frac{1}{4}}\right)^2 y - y = 0.$$

The characteristic equation is

$$\lambda^2 - 1 = 0.$$

Its roots are

$$\lambda_1 = 1, \quad \lambda_2 = -1.$$

The given equation we can rewritten in the form

$$\left(D^{\frac{1}{4}} - 1\right)\left(D^{\frac{1}{4}} + 1\right)y = 0.$$

We set

$$\left(D^{\frac{1}{4}} + 1\right)y = y_1.$$

Thus we get

$$D^{\frac{1}{4}}y_1 - y_1 = 0.$$

Its solution is given by

$$y_1(t) = c_1 E_1(t,t_0), \quad t \in \mathbb{T}.$$

Here c_1 is a constant. Hence,

$$D^{\frac{1}{4}}y + y = c_1 E_1(t,t_0),$$

or

$$D^{\frac{1}{4}}y = -y + c_1 E_1(t,t_0), \quad t \in \mathbb{T}. \tag{6.3}$$

Note that

$$\sigma(t) = 2t,$$

$$\mu(t) = t,$$

$$k_0\left(\frac{1}{4},t\right) = \frac{1}{4}t^3,$$

$$k_1\left(\frac{1}{4},t\right) = \frac{3}{4}t, \quad t \in \mathbb{T}.$$

The equation (6.3) is a first-order linear conformable dynamic equation. Here

$$p(t) = -1,$$

$$q(t) = c_1 E_1(t,t_0), \quad t \in \mathbb{T}.$$

Then

$$g(t) = \frac{\left(p(t) - k_1\left(\frac{1}{4},t\right)\right)\left(\mu(t)k_1\left(\frac{1}{4},t\right) - k_0\left(\frac{1}{4},t\right)\right)}{k_0\left(\frac{1}{4},t\right) + \mu(t)\left(p(t) - k_1\left(\frac{1}{4},t\right)\right)}$$

$$= \frac{\left(-1 - \frac{3}{4}t\right)\left(\frac{3}{4}t^2 - \frac{1}{4}t^3\right)}{\frac{1}{4}t^3 + t\left(-1 - \frac{3}{4}t\right)}$$

$$= -\frac{(4+3t)(3t^2 - t^3)}{4(t^3 - 4t - 3t^2)}$$

$$= -\frac{(3t+4)(3t-t^2)}{4(t^2 - 3t - 4)}, \quad t \in \mathbb{T}, \quad t \neq 4.$$

Therefore

$$y(t) = c_2 E_p(t,t_0) + c_1 \int_{t_0}^t E_1(s,t_0) E_g(\sigma(s),t) \Delta_{\frac{1}{4},t} s, \quad t \in \mathbb{T}, \quad t \neq 4,$$

is a solution of the considered equation. Here c_2 is a constant. This ends the example.

Exercise 6.1.9 *Let $\mathbb{T} = 3\mathbb{Z}$,*

$$k_1(\alpha,t) = (1-\alpha)\left(1+t^6\right)^\alpha, \quad k_0(\alpha,t) = \alpha\left(1+t^6\right)^{1-\alpha}, \quad \alpha \in (0,1], \quad t \in \mathbb{T}.$$

Find a solution for each of the following equations.

1.
$$\left(D^{\frac{1}{4}}\right)^3 y - y = 0, \quad t \in \mathbb{T},$$

2.
$$\left(D^{\frac{1}{2}}\right)^2 y - 3D^{\frac{1}{2}}y + 2y = 0, \quad t \in \mathbb{T},$$

3.
$$\left(D^{\frac{1}{3}}\right)^2 y - 6D^{\frac{1}{3}}y + 5y = 0, \quad t \in \mathbb{T},$$

4.
$$\left(D^{\frac{1}{2}}\right)^4 y - y = 0, \quad t \in \mathbb{T},$$

5.
$$\left(D^{\frac{1}{3}}\right)^3 y - D^{\frac{1}{3}}y = 0, \quad t \in \mathbb{T}.$$

6.2 NONHOMOGENEOUS HIGHER-ORDER LINEAR CONFORMABLE DYNAMIC EQUATIONS WITH CONSTANT COEFFICIENTS

Consider the equation

$$(D^\alpha)^n y + a_1 (D^\alpha)^{n-1} y + \cdots + a_{n-1} D^\alpha y + a_n y = f(t), \quad t \in \mathbb{T}^{\kappa^n}, \tag{6.4}$$

where $a_l \in \mathbb{R}$, $l \in \{1, \ldots, n\}$, $f \in \mathscr{C}_{rd}(\mathbb{T})$.

Theorem 6.2.1 *Let*

$$f(t) = g(t) + ih(t), \quad t \in \mathbb{T},$$

where $g, h \in \mathscr{C}_{rd}(\mathbb{T})$, $g, h : \mathbb{T} \to \mathbb{R}$. *Suppose that*

$$y(t) = u(t) + iv(t), \quad t \in \mathbb{T}^{\kappa^n},$$

where $u, v \in \mathscr{C}_{rd}^n(\mathbb{T})$, $u, v : \mathbb{T} \to \mathbb{R}$, *is a solution of equation* (6.4). *Then u and v are solutions of the equations*

$$(D^\alpha)^n y + a_1 (D^\alpha)^{n-1} y + \cdots + a_{n-1} D^\alpha y + a_n y = g(t), \quad t \in \mathbb{T}^{\kappa^n}.$$

$$(D^\alpha)^n y + a_1 (D^\alpha)^{n-1} y + \cdots + a_{n-1} D^\alpha y + a_n y = h(t), \quad t \in \mathbb{T}^{\kappa^n},$$

respectively.

Proof 6.2.2 *We have*

$$
\begin{aligned}
g(t) + ih(t) &= (D^\alpha)^n (u + iv)(t) + a_1 (D^\alpha)^{n-1} (u + iv)(t) \\
&\quad + \cdots + a_{n-1} D^\alpha (u + iv)(t) + a_n (u + iv)(t) \\
&= (D^\alpha) (u(t) + iv(t)) + a_1 \left((D^\alpha)^{n-1} u(t) + i (D^\alpha)^{n-1} v(t) \right) \\
&\quad + \cdots + a_{n-1} D^\alpha (u(t) + iv(t)) + a_n (u(t) + iv(t)) \\
&= (D^\alpha)^n u(t) + a_1 (D^\alpha)^{n-1} u(t) + \cdots + a_{n-1} D^\alpha u(t) + a_n u(t) \\
&\quad + i \left((D^\alpha)^n v(t) + a_1 (D^\alpha)^{n-1} v(t) + \cdots + a_{n-1} D^\alpha v(t) + a_n v(t) \right),
\end{aligned}
$$

$t \in \mathbb{T}^{\kappa^n}$. *Hence, equating the real and imaginary parts of both sides of the last equation, we get*

$$(D^\alpha)^n u(t) + a_1 (D^\alpha)^{n-1} u(t) + \cdots + a_{n-1} D^\alpha u(t) + a_n u(t) = g(t), \quad t \in \mathbb{T}^{\kappa^n}.$$

$$(D^\alpha)^n v(t) + a_1 (D^\alpha)^{n-1} v(t) + \cdots + a_{n-1} D^\alpha v(t) + a_n v(t) = h(t), \quad t \in \mathbb{T}^{\kappa^n}.$$

This completes the proof. □

Let $\lambda_1, \ldots, \lambda_n$ be the roots of equation (6.2). Then we can rewrite equation (6.4) in the form

$$(D^\alpha - \lambda_1)(D^\alpha - \lambda_2)\ldots(D^\alpha - \lambda_n)y = f(t), \quad t \in \mathbb{T}^{\kappa^n}.$$

Let y be a solution of equation (6.4). We set

$$y_1 = (D^\alpha - \lambda_2)\ldots(D^\alpha - \lambda_n)y, \quad t \in \mathbb{T}^{\kappa^n}.$$

Then we get

$$D^\alpha y_1 - \lambda_1 y_1 = f(t), \quad t \in \mathbb{T}^{\kappa^n},$$

which is a first-order linear conformable dynamic equation. Let y_1 be its solution. Then we set

$$y_2 = (D^\alpha - \lambda_3)\ldots(D^\alpha - \lambda_n)y, \quad t \in \mathbb{T}^{\kappa^n}$$

and we obtain

$$D^\alpha y_2 - \lambda_2 y_2 = y_1, \quad t \in \mathbb{T}^{\kappa^n},$$

and so on.

Example 6.2.3 *Let* $\mathbb{T} = \mathbb{Z}$,

$$k_1(\alpha, t) = (1 - \alpha)(1 + t^2)^\alpha, \quad k_0(\alpha, t) = \alpha(1 + t^2)^{1-\alpha}, \quad \alpha \in (0, 1], \quad t \in \mathbb{T}.$$

Consider the equation

$$\left(D^{\frac{1}{4}}\right)^3 y + 2\left(D^{\frac{1}{4}}\right)^2 y - D^{\frac{1}{4}}y - 2y = t, \quad t \in \mathbb{T}.$$

The characteristic equation is

$$\lambda^3 + 2\lambda^2 - \lambda - 2 = 0$$

or

$$\lambda^2(\lambda + 2) - (\lambda + 2) = 0,$$

or

$$(\lambda^2 - 1)(\lambda + 2) = 0,$$

or

$$(\lambda - 1)(\lambda + 1)(\lambda + 2) = 0.$$

Therefore

$$\lambda_1 = 1, \quad \lambda_2 = -1, \quad \lambda_3 = -2.$$

The considered equation can be rewritten in the following form.

$$\left(D^{\frac{1}{4}} - 1\right)\left(D^{\frac{1}{4}} + 1\right)\left(D^{\frac{1}{4}} + 2\right)y = t, \quad t \in \mathbb{T}.$$

Let

$$y_1 = \left(D^{\frac{1}{4}} + 1\right)\left(D^{\frac{1}{4}} + 2\right)y.$$

Hence, we get the equation

$$D^{\frac{1}{4}}y_1 - y_1 = t, \quad t \in \mathbb{T},$$

or

$$D^{\frac{1}{4}}y_1 = y_1 + t, \quad t \in \mathbb{T}.$$

Here

$$\mu(t) = 1,$$

$$\sigma(t) = t+1,$$

$$p(t) = 1,$$

$$q(t) = t,$$

$$k_0\left(\frac{1}{4},t\right) = \frac{1}{4}(1+t^2)^{\frac{3}{4}},$$

$$k_1\left(\frac{1}{4},t\right) = \frac{3}{4}(1+t^2)^{\frac{1}{4}},$$

$$\begin{aligned} g(t) &= \frac{\left(p(t)-k_1\left(\frac{1}{4},t\right)\right)\left(\mu(t)k_1\left(\frac{1}{4},t\right)-k_0\left(\frac{1}{4},t\right)\right)}{k_0\left(\frac{1}{4},t\right)+\mu(t)\left(p(t)-k_1\left(\frac{1}{4},t\right)\right)} \\ &= \frac{\left(1-\frac{3}{4}(1+t^2)^{\frac{1}{4}}\right)\left(\frac{3}{4}(1+t^2)^{\frac{1}{4}}-\frac{1}{4}(1+t^2)^{\frac{3}{4}}\right)}{\frac{1}{4}(1+t^2)^{\frac{3}{4}}+1-\frac{3}{4}(1+t^2)^{\frac{1}{4}}}, \quad t \in \mathbb{T}. \end{aligned}$$

Therefore

$$y_1(t) = c_1 E_p(t,t_0) + \int_{t_0}^t sE_g(\sigma(s),t)\Delta_{\frac{1}{4},t}s, \quad t \in \mathbb{T},$$

where c_1 is a constant. Now we set

$$\left(D^{\frac{1}{4}}+2\right)y = y_2.$$

Then

$$D^{\frac{1}{4}}y_2 + y_2 = y_1(t), \quad t \in \mathbb{T},$$

or

$$D^{\frac{1}{4}}y_2 = -y_2 + y_1(t), \quad t \in \mathbb{T}.$$

Here

$$p_1(t) = -1,$$

$$q_1(t) = y_1(t),$$

$$
\begin{aligned}
g_1(t) &= \frac{\left(p_1(t) - k_1\left(\frac{1}{4},t\right)\right)\left(\mu(t)k_1\left(\frac{1}{4},t\right) - k_0\left(\frac{1}{4},t\right)\right)}{k_0\left(\frac{1}{4},t\right) + \mu(t)\left(p_1(t) - k_1\left(\frac{1}{4},t\right)\right)} \\
&= \frac{\left(-1 - \frac{3}{4}(1+t^2)^{\frac{1}{4}}\right)\left(\frac{3}{4}(1+t^2)^{\frac{1}{4}} - \frac{1}{4}(1+t^2)^{\frac{3}{4}}\right)}{\frac{1}{4}(1+t^2)^{\frac{3}{4}} - 1 - \frac{3}{4}(1+t^2)^{\frac{1}{4}}}, \quad t \in \mathbb{T}.
\end{aligned}
$$

Therefore

$$y_2(t) = c_2 E_{p_1}(t,t_0) + \int_{t_0}^{t} y_1(s) E_{g_1}(\sigma(s),t)\Delta_{\frac{1}{4},t}s, \quad t \in \mathbb{T},$$

where c_2 is a constant. Now we get the equation

$$D^{\frac{1}{4}}y = -2y + y_2(t), \quad t \in \mathbb{T}.$$

Here

$$p_2(t) = -2,$$

$$q_2(t) = y_2(t),$$

$$
\begin{aligned}
g_2(t) &= \frac{\left(p_2(t) - k_1\left(\frac{1}{4},t\right)\right)\left(\mu(t)k_1\left(\frac{1}{4},t\right) - k_0\left(\frac{1}{4},t\right)\right)}{k_0\left(\frac{1}{4},t\right) + \mu(t)\left(p_2(t) - k_1\left(\frac{1}{4},t\right)\right)} \\
&= \frac{\left(-2 - \frac{3}{4}(1+t^2)^{\frac{1}{4}}\right)\left(\frac{3}{4}(1+t^2)^{\frac{1}{4}} - \frac{1}{4}(1+t^2)^{\frac{3}{4}}\right)}{\frac{1}{4}(1+t^2)^{\frac{3}{4}} - 2 - \frac{3}{4}(1+t^2)^{\frac{1}{4}}}, \quad t \in \mathbb{T}.
\end{aligned}
$$

Consequently,

$$y(t) = c_3 E_{p_2}(t,t_0) + \int_{t_0}^{t} y_2(s) E_{g_2}(\sigma(s),t)\Delta_{\frac{1}{4},t}s, \quad t \in \mathbb{T},$$

where c_3 is a constant, is a solution. This ends the example.

Exercise 6.2.4 *Let $\mathbb{T} = 4\mathbb{Z}$,*

$$k_1(\alpha,t) = (1-\alpha)\left(1+t^4\right)^{4\alpha}, \quad k_0(\alpha,t) = \alpha(1+t^4)^{4(1-\alpha)}, \quad \alpha \in (0,1], \quad t \in \mathbb{T}.$$

Find a solution for each of the following equations.

1.

$$\left(D^{\frac{1}{4}}\right)^3 y + y = t + 1, \quad t \in \mathbb{T},$$

2.

$$\left(D^{\frac{1}{2}}\right)^2 y - 4D^{\frac{1}{2}}y + 3y = t^2, \quad t \in \mathbb{T},$$

3.
$$\left(D^{\frac{1}{3}}\right)^2 y - 8D^{\frac{1}{3}}y + 7y = t + 2, \quad t \in \mathbb{T},$$

4.
$$\left(D^{\frac{1}{2}}\right)^4 y - 16y = t^2 - t, \quad t \in \mathbb{T},$$

5.
$$\left(D^{\frac{1}{3}}\right)^3 y - 4\left(D^{\frac{1}{3}}\right)^2 y + 3D^{\frac{1}{3}}y = t, \quad t \in \mathbb{T}.$$

6.3   ADVANCED PRACTICAL PROBLEMS

Problem 6.3.1 *Find the characteristic equations for each of the following equations.*

1.
$$(D^\alpha)^3 y - (D^\alpha)^2 y - 2D^\alpha y - 3y = 0,$$

2.
$$(D^\alpha)^4 y - D^\alpha y = 0,$$

3.
$$(D^\alpha)^{10} y - (D^\alpha)^8 y - (D^\alpha)^4 y - y = 0.$$

4.
$$(D^\alpha)^5 y - 3(D^\alpha)^3 y - y = 0,$$

5.
$$(D^\alpha)^2 y - 4D^\alpha y + y = 0.$$

Problem 6.3.2 *Let* $\mathbb{T} = 3^{\mathbb{N}_0}$,

$$k_1(\alpha, t) = (1 - \alpha)t^{6\alpha}, \quad k_0(\alpha, t) = \alpha t^{6(1-\alpha)}, \quad \alpha \in (0, 1], \quad t \in \mathbb{T}.$$

Find a solution for each of the following equations.

1.
$$\left(D^{\frac{1}{4}}\right)^3 y - y = 0, \quad t \in \mathbb{T},$$

2.
$$\left(D^{\frac{1}{2}}\right)^2 y - 3D^{\frac{1}{2}}y + 2y = 0, \quad t \in \mathbb{T},$$

3.
$$\left(D^{\frac{1}{3}}\right)^2 y - 6D^{\frac{1}{3}}y + 5y = 0, \quad t \in \mathbb{T},$$

4.
$$\left(D^{\frac{1}{2}}\right)^4 y - y = 0, \quad t \in \mathbb{T},$$

5.
$$\left(D^{\frac{1}{3}}\right)^{3}y - D^{\frac{1}{3}}y = 0, \quad t \in \mathbb{T}.$$

Problem 6.3.3 *Let* $\mathbb{T} = 3^{\mathbb{N}_0}$,

$$k_1(\alpha,t) = (1-\alpha)t^{3\alpha}, \quad k_0(\alpha,t) = \alpha t^{3(1-\alpha)}, \quad \alpha \in (0,1], \quad t \in \mathbb{T}.$$

Find a solution for each of the following equations.

1.
$$\left(D^{\frac{1}{4}}\right)^{3}y + y = t + 1, \quad t \in \mathbb{T},$$

2.
$$\left(D^{\frac{1}{2}}\right)^{2}y - 4D^{\frac{1}{2}}y + 3y = t^{2}, \quad t \in \mathbb{T},$$

3.
$$\left(D^{\frac{1}{3}}\right)^{2}y - 8D^{\frac{1}{3}}y + 7y = t + 2, \quad t \in \mathbb{T},$$

4.
$$\left(D^{\frac{1}{2}}\right)^{4}y - 16y = t^{2} - t, \quad t \in \mathbb{T},$$

5.
$$\left(D^{\frac{1}{3}}\right)^{3}y - 4\left(D^{\frac{1}{3}}\right)^{2}y + 3D^{\frac{1}{3}}y = t, \quad t \in \mathbb{T}.$$

CHAPTER 7

Second-Order Conformable Dynamic Equations

Suppose that $\mathbb{T}$ is a time scale with forward jump operator and delta differentiation operator σ and Δ, respectively. Let $\alpha \in (0,1]$, k_0 and k_1 satisfy $(A1)$, $k_0(\alpha,\cdot)$, $k_1(\alpha,\cdot) \in \mathscr{C}^1_{rd}(\mathbb{T}^\kappa)$. Also, assume (1.6) holds and $t_0 \in \mathbb{T}$.

7.1 HOMOGENEOUS SECOND-ORDER LINEAR CONFORMABLE DYNAMIC EQUATIONS

Consider the equation

$$(D^\alpha)^2 y + a(t)D^\alpha y + b(t)y = 0, \quad t \in \mathbb{T}^{\kappa^2}, \tag{7.1}$$

where $a,b \in \mathscr{C}_{rd}(\mathbb{T})$.

Definition 7.1.1 *A function $y \in \mathscr{C}^2_{rd}(\mathbb{T})$ that satisfies equation (7.1) will be called a solution of equation (7.1).*

Theorem 7.1.2 *Let y_1 and y_2 be solutions of equation (7.1). Then*

$$py_1 + qy_2$$

is a solution of equation (7.1) for any $p,q \in \mathbb{R}$.

Proof 7.1.3 *We have*

$$(D^\alpha)^2 y_1 + a(t)D^\alpha y_1 + b(t)y_1 = 0,$$

$$(D^\alpha)^2 y_2 + a(t)D^\alpha y_2 + b(t)y_2 = 0, \quad t \in \mathbb{T}^{\kappa^2}.$$

Hence,

$$(D^\alpha)^2 (py_1 + qy_2) + a(t)D^\alpha(py_1 + qy_2) + b(t)(py_1 + qy_2)$$

$$= p(D^{\alpha})^2 y_1 + q(D^{\alpha})^2 y_2 + pa(t)D^{\alpha}y_1 + qa(t)D^{\alpha}y_2$$

$$+ pb(t)y_1 + qb(t)y_2$$

$$= p\left((D^{\alpha})^2 y_1 + a(t)D^{\alpha}y_1 + b(t)y_1\right)$$

$$+ q\left((D^{\alpha})^2 y_2 + a(t)D^{\alpha}y_2 + b(t)y_2\right)$$

$$= 0, \quad t \in \mathbb{T}^{\kappa^2}.$$

This completes the proof. □

Definition 7.1.4 *For any two functions y_1, $y_2 \in \mathscr{C}^1_{rd}(\mathbb{T})$ we define the conformable Wronskian by*

$$W(y_1, y_2) = \det \begin{pmatrix} y_1 & y_2 \\ D^{\alpha}y_1 & D^{\alpha}y_2 \end{pmatrix}.$$

Definition 7.1.5 *We say that two solutions y_1 and y_2 of equation (7.1) form a fundamental set of solutions for (7.1) if*

$$W(y_1, y_2)(t) \neq 0 \quad for \quad any \quad t \in \mathbb{T}^{\kappa}.$$

Definition 7.1.6 *With $\mathscr{R}^+_c$ we will denote the set of all functions $f : \mathbb{T} \to \mathbb{R}$ such that*

$$k_0 + \mu(f - k_1) > 0, \quad k_0 - \mu k_1 \neq 0 \quad on \quad \mathbb{T}.$$

Example 7.1.7 *Let $a, b \in \mathbb{R}$ be such that*

$$\frac{-a \pm \sqrt{a^2 - 4b}}{2} \in \mathscr{R}^+_c, \quad a^2 - 4b \neq 0 \quad on \quad \mathbb{T}.$$

Such $\mathbb{T}$, k_0 and k_1 exist. Really, let $\mathbb{T} = \mathbb{Z}$,

$$k_0(\alpha, t) = \alpha, \quad k_1(\alpha, t) = 1 - \alpha, \quad \alpha \in (0, 1], \quad t \in \mathbb{T},$$

and

$$a = \frac{1}{4}, \quad b = 0, \quad \alpha = \frac{3}{4}.$$

Then

$$k_0\left(\frac{3}{4}, t\right) = \frac{3}{4},$$

$$k_1\left(\frac{3}{4},t\right) = 1 - \frac{3}{4}$$

$$= \frac{1}{4},$$

$$k_0\left(\frac{3}{4},t\right) + \mu(t)\left(\frac{-a - \sqrt{a^2 - 4b}}{2} - k_1\left(\frac{3}{4},t\right)\right) = \frac{3}{4} + \left(\frac{-\frac{1}{4} - \frac{1}{4}}{2} - \frac{1}{4}\right)$$

$$= \frac{3}{4} - \left(\frac{1}{4} + \frac{1}{4}\right)$$

$$= \frac{3}{4} - \frac{1}{2}$$

$$= \frac{1}{4},$$

$$k_0\left(\frac{3}{4},t\right) + \mu(t)\left(\frac{-a + \sqrt{a^2 - 4b}}{2} - k_1\left(\frac{3}{4},t\right)\right) = \frac{3}{4} - \frac{1}{4}$$

$$= \frac{1}{2}, \quad t \in \mathbb{T}.$$

We set

$$g_1 = \frac{-a + \sqrt{a^2 - 4b}}{2}, \quad g_2 = \frac{-a - \sqrt{a^2 - 4b}}{2}.$$

Note that

$$(D^\alpha)^2 E_{g_1} + aD^\alpha E_{g_1} + bE_{g_1}$$

$$= \left(\left(\frac{-a + \sqrt{a^2 - 4b}}{2}\right)^2 + a\frac{-a + \sqrt{a^2 - 4b}}{2} + b\right)E_{g_1}$$

$$= \left(\frac{a^2 - 2a\sqrt{a^2 - 4b} + a^2 - 4b}{4} + a\frac{-a + \sqrt{a^2 - 4b}}{2} + b\right)E_{g_1}$$

$$= \left(\frac{a^2 - a\sqrt{a^2 - 4b} - 2b}{2} + \frac{-a^2 + a\sqrt{a^2 - 4b}}{2} + b\right)E_{g_1}$$

$$= 0 \quad on \quad \mathbb{T}^{\kappa^2},$$

and

$$(D^\alpha)^2 E_{g_2} + aD^\alpha E_{g_2} + bE_{g_2}$$

$$= \left(\left(\frac{-a - \sqrt{a^2 - 4b}}{2} \right)^2 + a \frac{-a - \sqrt{a^2 - 4b}}{2} + b \right) E_{g_2}$$

$$= \left(\frac{a^2 + 2a\sqrt{a^2 - 4b} + a^2 - 4b}{4} + a \frac{-a - \sqrt{a^2 - 4b}}{2} + b \right) E_{g_2}$$

$$= \left(\frac{a^2 + a\sqrt{a^2 - 4b} - 2b}{2} + \frac{-a^2 - a\sqrt{a^2 - 4b}}{2} + b \right) E_{g_2}$$

$$= 0 \quad on \quad \mathbb{T}^{\kappa^2},$$

i.e., E_{g_1} and E_{g_2} are solutions of (7.1). Next,

$$W(y_1, y_2)(t) = \det \begin{pmatrix} E_{g_1} & E_{g_2} \\ g_1 E_{g_1} & g_2 E_{g_2} \end{pmatrix}$$

$$= (g_2 - g_1) E_{g_1} E_{g_2}$$

$$= -\sqrt{a^2 - 4b} E_{g_1} E_{g_2}$$

$$\neq 0 \quad on \quad \mathbb{T}^{\kappa}.$$

Therefore E_{g_1} and E_{g_2} form a fundamental system for (7.1). This ends the example.

Theorem 7.1.8 Let y_1 and y_2 be solutions of equation (7.1). Then

$$\frac{k_0 - \mu k_1}{k_0} W(y_1, y_2) = \det \begin{pmatrix} y_1^{\sigma} & y_2^{\sigma} \\ D^{\alpha} y_1 & D^{\alpha} y_2 \end{pmatrix}.$$

Proof 7.1.9 We have

$$\det \begin{pmatrix} y_1^{\sigma} & y_2^{\sigma} \\ D^{\alpha} y_1 & D^{\alpha} y_2 \end{pmatrix} = \frac{1}{k_0} \det \begin{pmatrix} k_0 y_1^{\sigma} & k_0 y_2^{\sigma} \\ D^{\alpha} y_1 & D^{\alpha} y_2 \end{pmatrix}$$

$$= \frac{1}{k_0} \det \begin{pmatrix} k_0 y_1 + k_0 \mu y_1^{\Delta} & k_0 y_2 + k_0 \mu y_2^{\Delta} \\ D^{\alpha} y_1 & D^{\alpha} y_2 \end{pmatrix}$$

$$= \frac{1}{k_0} \det \begin{pmatrix} (k_0 - k_1 \mu) y_1 + \mu D^{\alpha} y_1 & (k_0 - k_1 \mu) y_2 + \mu D^{\alpha} y_2 \\ D^{\alpha} y_1 & D^{\alpha} y_2 \end{pmatrix}$$

$$= \frac{1}{k_0} \det \begin{pmatrix} (k_0 - k_1 \mu) y_1 & (k_0 - k_1 \mu) y_2 \\ D^{\alpha} y_1 & D^{\alpha} y_2 \end{pmatrix}$$

$$= \frac{k_0 - k_1 \mu}{k_0} \det \begin{pmatrix} y_1 & y_2 \\ D^{\alpha} y_1 & D^{\alpha} y_2 \end{pmatrix}$$

$$= \frac{k_0 - k_1 \mu}{k_0} W(y_1, y_2).$$

This completes the proof. □

Theorem 7.1.10 *Let y_1 and y_2 be solutions of equation (7.1). Then*

$$D^\alpha W(y_1, y_2) = \det \begin{pmatrix} y_1^\sigma & y_2^\sigma \\ (D^\alpha)^2 y_1 & (D^\alpha)^2 y_2 \end{pmatrix} - \frac{k_1(k_0 - \mu k_1)}{k_0} W(y_1, y_2).$$

Proof 7.1.11 *We have*

$$
\begin{aligned}
D^\alpha W(y_1, y_2) &= D^\alpha \left(\det \begin{pmatrix} y_1 & y_2 \\ D^\alpha y_1 & D^\alpha y_2 \end{pmatrix} \right) \\[2mm]
&= D^\alpha \left(y_1 D^\alpha y_2 - y_2 D^\alpha y_1 \right) \\[2mm]
&= D^\alpha \left(y_1 D^\alpha y_2 \right) - D^\alpha \left(y_2 D^\alpha y_1 \right) \\[2mm]
&= (D^\alpha y_1)(D^\alpha y_2) + y_1^\sigma (D^\alpha)^2 y_2 - k_1 y_1^\sigma D^\alpha y_2 \\[2mm]
&\quad - (D^\alpha y_2)(D^\alpha y_1) - y_2^\sigma (D^\alpha)^2 y_1 + k_1 y_2^\sigma D^\alpha y_1 \\[2mm]
&= \det \begin{pmatrix} y_1^\sigma & y_2^\sigma \\ (D^\alpha)^2 y_1 & (D^\alpha)^2 y_2 \end{pmatrix} - k_1 \det \begin{pmatrix} y_1^\sigma & y_2^\sigma \\ D^\alpha y_1 & D^\alpha y_2 \end{pmatrix} \\[2mm]
&= \det \begin{pmatrix} y_1^\sigma & y_2^\sigma \\ (D^\alpha)^2 y_1 & (D^\alpha)^2 y_2 \end{pmatrix} - \frac{k_1(k_0 - \mu k_1)}{k_0} W(y_1, y_2).
\end{aligned}
$$

This completes the proof. □

Theorem 7.1.12 *Let y_1 and y_2 be solutions of equation (7.1). Then*

$$D^\alpha W(y_1, y_2) = - \left(a \frac{k_0 - \mu k_1}{k_0} + \frac{k_1(k_0 - \mu k_1)}{k_0} - \frac{b\mu}{k_0} \right) W(y_1, y_2).$$

Proof 7.1.13 *Note that*

$$(D^\alpha)^2 y_1 = -a D^\alpha y_1 - b y_1,$$

$$(D^\alpha)^2 y_2 = -a D^\alpha y_2 - b y_2.$$

Then, using Theorem 7.1.8 and Theorem 7.1.10, we get

$$
\begin{aligned}
D^\alpha W(y_1, y_2) &= \det \begin{pmatrix} y_1^\sigma & y_2^\sigma \\ (D^\alpha)^2 y_1 & (D^\alpha)^2 y_2 \end{pmatrix} - \frac{k_1(k_0 - \mu k_1)}{k_0} W(y_1, y_2) \\[2mm]
&= \det \begin{pmatrix} y_1^\sigma & y_2^\sigma \\ -a D^\alpha y_1 - b y_1 & -a D^\alpha y_2 - b y_2 \end{pmatrix} - \frac{k_1(k_0 - \mu k_1)}{k_0} W(y_1, y_2) \\[2mm]
&= \det \begin{pmatrix} y_1^\sigma & y_2^\sigma \\ -a D^\alpha y_1 & -a D^\alpha y_2 \end{pmatrix} + \det \begin{pmatrix} y_1^\sigma & y_2^\sigma \\ -b y_1 & -b y_2 \end{pmatrix}
\end{aligned}
$$

$$-\frac{k_1(k_0-\mu k_1)}{k_0}W(y_1,y_2)$$

$$= -a\det\begin{pmatrix} y_1^{\sigma} & y_2^{\sigma} \\ D^{\alpha}y_1 & D^{\alpha}y_2 \end{pmatrix} - b\det\begin{pmatrix} y_1+\mu y_1^{\Delta} & y_2+\mu y_2^{\Delta} \\ y_1 & y_2 \end{pmatrix}$$

$$-\frac{k_1(k_0-\mu k_1)}{k_0}W(y_1,y_2)$$

$$= -a\frac{k_0-\mu k_1}{k_0}W(y_1,y_2) - \frac{b\mu}{k_0}\det\begin{pmatrix} k_0 y_1^{\Delta} & k_0 y_2^{\Delta} \\ y_1 & y_2 \end{pmatrix}$$

$$-\frac{k_1(k_0-\mu k_1)}{k_0}W(y_1,y_2)$$

$$= -\left(a\frac{k_0-\mu k_1}{k_0} + \frac{k_1(k_0-\mu k_1)}{k_0}\right)W(y_1,y_2)$$

$$-\frac{b\mu}{k_0}\det\begin{pmatrix} D^{\alpha}y_1 & D^{\alpha}y_2 \\ y_1 & y_2 \end{pmatrix}$$

$$= -\left(a\frac{k_0-\mu k_1}{k_0} + \frac{k_1(k_0-\mu k_1)}{k_0}\right)W(y_1,y_2) + \frac{b\mu}{k_0}W(y_1,y_2)$$

$$= -\left(a\frac{k_0-\mu k_1}{k_0} + \frac{k_1(k_0-\mu k_1)}{k_0} - \frac{b\mu}{k_0}\right)W(y_1,y_2).$$

This completes the proof. □

Theorem 7.1.14 (Abel's Formula) *Let y_1 and y_2 be solutions of equation (7.1). Then*

$$W(y_1,y_2)(t) = E_p(t,t_0)W(y_1,y_2)(t_0),$$

where

$$p = -\left(a\frac{k_0-\mu k_1}{k_0} + \frac{k_1(k_0-\mu k_1)}{k_0} - \frac{b\mu}{k_0}\right).$$

Proof 7.1.15 *By Theorem 7.1.12, we obtain*

$$D^{\alpha}W(y_1,y_2) = pW(y_1,y_2).$$

Hence, we get the assertion. This completes the proof. □

We will now introduce hyperbolic functions. Consider the equation

$$(D^{\alpha})^2 - \gamma^2 y = 0, \quad t \in \mathbb{T}^{\kappa^2}, \tag{7.2}$$

where $\gamma \in \mathbb{R}$, $\gamma > 0$ and $\pm\gamma \in \mathscr{R}_c$. Let

$$y_1(t) = Cosh_\gamma(t,t_0), \quad y_2(t) = Sinh_\gamma(t,t_0). \quad t \in \mathbb{T}^{\kappa^2}. \tag{7.3}$$

Then

$$D^\alpha y_1(t) \;=\; \gamma Sinh_\gamma(t,t_0),$$

$$(D^\alpha)^2 y_1(t) \;=\; \gamma^2 Cosh_\gamma(t,t_0),$$

$$D^\alpha y_2(t) \;=\; \gamma Cosh_\gamma(t,t_0),$$

$$(D^\alpha)^2 y_2(t) \;=\; \gamma^2 Sinh_\gamma(t,t_0), \quad t \in \mathbb{T}^{\kappa^2}.$$

Hence,

$$(D^\alpha)^2 y_1(t) - \gamma^2 y_1(t) \;=\; \gamma^2 Cosh_\gamma(t,t_0) - \gamma^2 Cosh_\gamma(t,t_0)$$

$$\;=\; 0, \quad t \in \mathbb{T}^{\kappa^2},$$

and

$$(D^\alpha)^2 y_2(t) - \gamma^2 y_2(t) \;=\; \gamma^2 Sinh_\gamma(t,t_0) - \gamma^2 Sinh_\gamma(t,t_0)$$

$$\;=\; 0, \quad t \in \mathbb{T}^{\kappa^2}.$$

Consequently, y_1 and y_2 are solutions of equation (7.2). Note that

$$\gamma \oplus_c (-\gamma) \;=\; \frac{(\gamma - \gamma - k_1)k_0 + \mu(\gamma - k_1)(-\gamma - k_1)}{k_0}$$

$$\;=\; \frac{-k_0 k_1 - \mu(\gamma - k_1)(\gamma + k_1)}{k_0}$$

$$\;=\; -k_1 - \frac{\mu}{k_0}(\gamma - k_1)(\gamma + k_1),$$

$$k_0 + \mu\,(\gamma \oplus_c (-\gamma) - k_1) \;=\; k_0 + \mu\left(-k_1 - \frac{\mu}{k_0}(\gamma - k_1)(\gamma + k_1) - k_1\right)$$

$$\;=\; k_0 - 2\mu k_1 - \frac{\mu^2}{k_0}(\gamma - k_1)(\gamma + k_1)$$

$$\;=\; k_0 - 2\mu k_1 - \frac{\mu^2}{k_0}\gamma^2 + \frac{\mu^2}{k_0}k_1^2$$

$$\;=\; \frac{k_0^2 - 2\mu k_0 k_1 - \mu^2 \gamma^2 + \mu^2 k_1^2}{k_0}$$

$$= \frac{(k_0 - \mu k_1)^2 - \mu^2 \gamma^2}{k_0}$$

$$= \frac{(k_0 - \mu k_1 - \mu \gamma)(k_0 - \mu k_1 + \mu \gamma)}{k_0}$$

$$\neq 0, \quad \text{on} \quad \mathbb{T}.$$

Hence,

$$W(y_1, y_2)(t) = \det \begin{pmatrix} y_1(t) & y_2(t) \\ D^\alpha y_1(t) & D^\alpha y_2(t) \end{pmatrix}$$

$$= \det \begin{pmatrix} Cosh_\gamma(t,t_0) & Sinh_\gamma(t,t_0) \\ \gamma Sinh_\gamma(t,t_0) & \gamma Cosh_\gamma(t,t_0) \end{pmatrix}$$

$$= \gamma \left(Cosh_\gamma(t,t_0) \right)^2 - \gamma \left(Sinh_\gamma(t,t_0) \right)^2$$

$$= \gamma \left(\left(Cosh_\gamma(t,t_0) \right)^2 - \left(Sinh_\gamma(t,t_0) \right)^2 \right)$$

$$= \gamma E_{\gamma \oplus_c (-\gamma)}(t,t_0)$$

$$\neq 0, \quad t \in \mathbb{T}^{\kappa^2}.$$

Therefore, y_1 and y_2, defined by (7.3), form a fundamental set of solutions of equation (7.2) and

$$y(t) = c_1 y_1(t) + c_2 y_2(t)$$

$$= c_1 Cosh_\gamma(t,t_0) + c_2 Sinh_\gamma(t,t_0), \quad t \in \mathbb{T}^{\kappa^2},$$

where $c_1, c_2 \in \mathbb{R}$, is the general solution of equation (7.2). Now we consider the equation

$$(D^\alpha)^2 y - 2f D^\alpha y + (f^2 - g^2) y = 0, \quad t \in \mathbb{T}^{\kappa^2}, \tag{7.4}$$

where $f, g \in \mathbb{R}$, $g \neq 0$, $f \pm g \in \mathcal{R}_c$. Let

$$y_1(t) = Ch_{fg}(t,t_0), \quad y_2(t) = Sh_{fg}(t,t_0), \quad t \in \mathbb{T}^{\kappa^2}. \tag{7.5}$$

Note that

$$D^\alpha y_1(t) = D^\alpha Ch_{fg}(t,t_0)$$

$$= f Ch_{fg}(t,t_0) + g Sh_{fg}(t,t_0),$$

$$(D^\alpha)^2 y_1(t) = f D^\alpha Ch_{fg}(t,t_0) + g D^\alpha Sh_{fg}(t,t_0)$$

$$= f\left(f Ch_{fg}(t,t_0) + g Sh_{fg}(t,t_0)\right)$$

$$+ g\left(g Ch_{fg}(t,t_0) + f Sh_{fg}(t,t_0)\right)$$

$$= \left(f^2 + g^2\right) Ch_{fg}(t,t_0) + 2fg Sh_{fg}(t,t_0),$$

$$D^\alpha y_2(t) = D^\alpha Sh_{fg}(t,t_0)$$

$$= g Ch_{fg}(t,t_0) + f Sh_{fg}(t,t_0),$$

$$(D^\alpha)^2 y_2(t) = g D^\alpha Ch_{fg}(t,t_0) + f D^\alpha Sh_{fg}(t,t_0)$$

$$= g\left(f Ch_{fg}(t,t_0) + g Sh_{fg}(t,t_0)\right)$$

$$+ f\left(g Ch_{fg}(t,t_0) + f Sh_{fg}(t,t_0)\right)$$

$$= \left(f^2 + g^2\right) Sh_{fg}(t,t_0) + 2fg Ch_{fg}(t,t_0), \quad t \in \mathbb{T}^{\kappa^2}.$$

Hence,

$$(D^\alpha)^2 y_1(t) - 2f D^\alpha y_1(t) + \left(f^2 - g^2\right) y_1(t)$$

$$= \left(f^2 + g^2\right) Ch_{fg}(t,t_0) + 2fg Sh_{fg}(t,t_0)$$

$$- 2f\left(f Ch_{fg}(t,t_0) + g Sh_{fg}(t,t_0)\right)$$

$$+ \left(f^2 - g^2\right) Ch_{fg}(t,t_0)$$

$$= 0, \quad t \in \mathbb{T}^{\kappa^2},$$

and

$$(D^\alpha)^2 y_2(t) - 2f D^\alpha y_2(t) + \left(f^2 - g^2\right) y_2(t)$$

$$= \left(f^2 + g^2\right) Sh_{fg}(t,t_0) + 2fg Ch_{fg}(t,t_0)$$

$$- 2f\left(g Ch_{fg}(t,t_0) + f Sh_{fg}(t,t_0)\right)$$

$$+ \left(f^2 - g^2 \right) Sh_{fg}(t, t_0)$$

$$= 0, \quad t \in \mathbb{T}^{\kappa^2}.$$

Therefore y_1 and y_2, defined by (7.5), are solutions of equation (7.4). Also,

$$
\begin{aligned}
(f+g) \oplus_c (f-g) &= \frac{(2f - k_1)k_0 + \mu(f + g - k_1)(f - g - k_1)}{k_0} \\
&= \frac{(2f - k_1)k_0 + \mu \left(f^2 - 2fk_1 + k_1^2 - g^2 \right)}{k_0} \quad \text{on} \quad \mathbb{T}^{\kappa^2},
\end{aligned}
$$

and

$$
k_0 + \mu \left((f+g) \oplus_c (f-g) - k_1 \right) = k_0 + \mu \left(\frac{(2f - k_1)k_0 + \mu \left(f^2 - 2fk_1 + k_1^2 + g^2 \right)}{k_0} - k_1 \right)
$$

$$
= \frac{k_0^2 + 2\mu f k_0 - 2\mu k_0 k_1 + \mu^2 f^2 - 2\mu^2 f k_1 + \mu^2 k_1^2 - \mu^2 g^2}{k_0}
$$

$$
= \frac{(k_0 + \mu f)^2 - 2\mu k_0 k_1 + 2\mu^2 f k_1 + \mu^2 k_1^2 - \mu^2 g^2}{k_0}
$$

$$
= \frac{(k_0 + \mu f)^2 - 2\mu k_1 (k_0 + \mu f) + \mu^2 k_1^2 - \mu^2 g^2}{k_0}
$$

$$
= \frac{(k_0 + \mu(f - k_1))^2 - \mu^2 g^2}{k_0}
$$

$$
= \frac{(k_0 + \mu(f - g - k_1)) (k_0 + \mu(f + g - k_1))}{k_0}
$$

$$
\neq 0 \quad \text{on} \quad \mathbb{T}^{\kappa^2}.
$$

Next,

$$
\begin{aligned}
W(y_1, y_2)(t) &= \begin{pmatrix} y_1(t) & y_2(t) \\ D^{\alpha} y_1(t) & D^{\alpha} y_2(t) \end{pmatrix} \\
&= \begin{pmatrix} Ch_{fg}(t, t_0) & Sh_{fg}(t, t_0) \\ fCh_{fg}(t, t_0) + gSh_{fg}(t, t_0) & gCh_{fg}(t, t_0) + fSh_{fg}(t, t_0) \end{pmatrix} \\
&= g \left(Ch_{fg}(t, t_0) \right)^2 + fCh_{fg}(t, t_0) Sh_{fg}(t, t_0) \\
&\quad - fCh_{fg}(t, t_0) Sh_{fg}(t, t_0) - g \left(Sh_{fg}(t, t_0) \right)^2
\end{aligned}
$$

$$= g\left(\left(Ch_{fg}(t,t_0)\right)^2 - \left(Sh_{fg}(t,t_0)\right)^2\right)$$

$$= gE_{(f+g)\oplus_c(f-g)}(t,t_0)$$

$$\neq 0, \quad t \in \mathbb{T}^{\kappa^2}.$$

Consequently, y_1 and y_2, defined by (7.5), form a fundamental set of solutions of equation (7.4) and

$$y(t) = c_1 y_1(t) + c_2 y_2(t)$$

$$= c_1 Ch_{fg}(t,t_0) + c_2 Sh_{fg}(t,t_0),$$

where $c_1, c_2 \in \mathbb{R}$, is its general solution. Now we consider the equation

$$(D^\alpha)^2 y + f^2 y = 0, \quad t \in \mathbb{T}^{\kappa^2}, \tag{7.6}$$

where $f \in \mathbb{R}$, $f \neq 0$, $\pm if \in \mathscr{R}_c$. Let

$$y_1(t) = Cos_f(t,t_0), \quad y_2(t) = Sin_f(t,t_0), \quad t \in \mathbb{T}^{\kappa^2}. \tag{7.7}$$

Then

$$D^\alpha y_1(t) = D^\alpha Cos_f(t,t_0)$$

$$= -f Sin_f(t,t_0),$$

$$(D^\alpha)^2 y_1(t) = -f D^\alpha Sin_f(t,t_0)$$

$$= -f^2 Cos_f(t,t_0),$$

$$D^\alpha y_2(t) = D^\alpha Sin_f(t,t_0)$$

$$= f Cos_f(t,t_0),$$

$$(D^\alpha)^2 y_2(t) = f D^\alpha Cos_f(t,t_0)$$

$$= -f^2 Sin_f(t,t_0), \quad t \in \mathbb{T}^{\kappa^2}.$$

Hence,

$$(D^\alpha)^2 y_1(t) + f^2 y_1(t) = -f^2 Cos_f(t,t_0) + f^2 Cos_f(t,t_0)$$

$$= 0,$$

$$(D^\alpha)^2 y_2(t) + f^2 y_2(t) = -f^2 Sin_f(t,t_0) + f^2 Sin_f(t,t_0)$$

$$= 0, \quad t \in \mathbb{T}^{\kappa^2}.$$

Therefore, the functions y_1 and y_2, defined by (7.7), are solutions of equation (7.6). Next,

$$(if) \oplus_c (-if) = \frac{-k_1 k_0 - \mu(if - k_1)(if + k_1)}{k_0},$$

$$k_0 + \mu((if) \oplus_c (-if) - k_1) = k_0 + \mu\left(\frac{-k_1 k_0 + \mu\left(f^2 + k_1^2\right)}{k_0} - k_1\right)$$

$$= \frac{k_0^2 - 2\mu k_0 k_1 + \mu^2 k_1^2 + \mu^2 f^2}{k_0}$$

$$= \frac{(k_0 - \mu k_1)^2 - (i\mu f)^2}{k_0}$$

$$= \frac{(k_0 - \mu(k_1 + if))(k_0 - \mu(k_1 - if))}{k_0}$$

$$= \frac{(k_0 + \mu(-if - k_1))(k_0 + \mu(if - k_1))}{k_0}$$

$$\neq 0 \quad \text{on} \quad \mathbb{T}^{\kappa^2},$$

and

$$W(y_1,y_2) = \det\begin{pmatrix} y_1(t) & y_2(t) \\ D^\alpha y_1(t) & D^\alpha y_2(t) \end{pmatrix}$$

$$= \det\begin{pmatrix} Cos_f(t,t_0) & Sin_f(t,t_0) \\ -f Sin_f(t,t_0) & f Cos_f(t,t_0) \end{pmatrix}$$

$$= f\left((Cos_f(t,t_0))^2 + (Sin_f(t,t_0))^2\right)$$

$$= f E_{(if)\oplus_c(-if)}(t,t_0)$$

$$\neq 0 \quad \text{on} \quad \mathbb{T}^{\kappa^2}.$$

Consequently, the functions y_1 and y_2, defined by (7.7), form a fundamental set of solutions of equation (7.6) and

$$y(t) = c_1 y_1(t) + c_2 y_2(t)$$

$$= c_1 Cos_f(t,t_0) + c_2 Sin_f(t,t_0),$$

where $c_1, c_2 \in \mathbb{R}$, is its general solution. Now we consider the equation

$$(D^\alpha)^2 y - 2fD^\alpha y + (f^2 + g^2) y = 0, \quad t \in \mathbb{T}^{\kappa^2}, \tag{7.8}$$

where $f, g \in \mathbb{R}$, $f \pm ig \in \mathscr{R}_c$, $g \neq 0$. Let

$$y_1(t) = C_{fg}(t, t_0), \quad y_2(t) = S_{fg}(t, t_0), \quad t \in \mathbb{T}^{\kappa^2}. \tag{7.9}$$

We have

$$
\begin{aligned}
D^\alpha y_1(t) &= D^\alpha C_{fg}(t, t_0) \\[2mm]
&= fC_{fg}(t, t_0) - gS_{fg}(t, t_0), \\[2mm]
(D^\alpha)^2 y_1(t) &= fD^\alpha C_{fg}(t, t_0) - gD^\alpha S_{fg}(t, t_0) \\[2mm]
&= f\left(fC_{fg}(t, t_0) - gS_{fg}(t, t_0) \right) \\[2mm]
&\quad - g\left(gC_{fg}(t, t_0) + fS_{fg}(t, t_0) \right) \\[2mm]
&= \left(f^2 - g^2 \right) C_{fg}(t, t_0) - 2fgS_{fg}(t, t_0), \\[2mm]
D^\alpha y_2(t) &= D^\alpha S_{fg}(t, t_0) \\[2mm]
&= gC_{fg}(t, t_0) + fS_{fg}(t, t_0), \\[2mm]
(D^\alpha)^2 y_2(t) &= gD^\alpha C_{fg}(t, t_0) + fD^\alpha S_{fg}(t, t_0) \\[2mm]
&= g\left(fC_{fg}(t, t_0) - gS_{fg}(t, t_0) \right) \\[2mm]
&\quad + f\left(gC_{fg}(t, t_0) + fS_{fg}(t, t_0) \right) \\[2mm]
&= \left(f^2 - g^2 \right) S_{fg}(t, t_0) + 2fgC_{fg}(t, t_0), \quad t \in \mathbb{T}^{\kappa^2}.
\end{aligned}
$$

Hence,

$$(D^\alpha)^2 y_1(t) - 2fD^\alpha y_1(t) + (f^2 + g^2) y_1(t)$$

$$= \left(f^2 - g^2\right) C_{fg}(t, t_0) - 2fg S_{fg}(t, t_0)$$

$$-2f^2 C_{fg}(t, t_0) + 2fg S_{fg}(t, t_0)$$

$$+ \left(f^2 + g^2\right) C_{fg}(t, t_0)$$

$$= 0, \quad t \in \mathbb{T}^{\kappa^2},$$

and

$$(D^\alpha)^2 y_2(t) - 2f D^\alpha y_2(t) + \left(f^2 + g^2\right) y_2(t)$$

$$= \left(f^2 - g^2\right) S_{fg}(t, t_0) + 2fg C_{fg}(t, t_0)$$

$$-2fg C_{fg}(t, t_0) - 2f^2 S_{fg}(t, t_0)$$

$$+ \left(f^2 + g^2\right) S_{fg}(t, t_0)$$

$$= 0, \quad t \in \mathbb{T}^{\kappa^2}.$$

Therefore, the functions y_1 and y_2, defined by (7.9), are solutions of equation (7.8). Next,

$$(f + ig) \oplus_c (f - ig) = \frac{(2f - k_1)k_0 + \mu(f + ig - k_1)(f - ig - k_1)}{k_0}$$

$$= \frac{(2f - k_1)k_0 + \mu\left((f - k_1)^2 + g^2\right)}{k_0}$$

$$= \frac{(2f - k_1)k_0 + \mu\left(f^2 - 2fk_1 + k_1^2 + g^2\right)}{k_0} \quad \text{on} \quad \mathbb{T}^{\kappa^2},$$

and

$$k_0 + \mu\left((f + ig) \oplus_c (f - ig) - k_1\right) = k_0 + \mu\left(\frac{(2f - k_1)k_0 + \mu\left(f^2 - 2fk_1 + k_1^2 + g^2\right)}{k_0} - k_1\right)$$

$$= \frac{k_0^2 - 2\mu k_0 k_1 + 2\mu f k_0 + \mu^2 f^2 - 2\mu^2 f k_1 + \mu^2 k_1^2 + \mu^2 g^2}{k_0}$$

$$= \frac{(k_0 + \mu f)^2 - 2\mu k_1 (k_0 + \mu f) + \mu^2 k_1^2 + \mu^2 g^2}{k_0}$$

$$= \frac{(k_0 + \mu(f - k_1))^2 + \mu^2 g^2}{k_0}$$

$$= \frac{(k_0 + \mu(f + ig - k_1))(k_0 + \mu(f - ig - k_1))}{k_0}$$

$$\neq 0 \quad \text{on} \quad \mathbb{T}^{\kappa^2}.$$

Also,

$$W(y_1, y_2)(t) = \det \begin{pmatrix} y_1(t) & y_2(t) \\ D^\alpha y_1(t) & D^\alpha y_2(t) \end{pmatrix}$$

$$= \det \begin{pmatrix} C_{fg}(t, t_0) & S_{fg}(t, t_0) \\ fC_{fg}(t, t_0) - gS_{fg}(t, t_0) & gC_{fg}(t, t_0) + fS_{fg}(t, t_0) \end{pmatrix}$$

$$= g \left(C_{fg}(t, t_0) \right)^2 + fC_{fg}(t, t_0)S_{fg}(t, t_0)$$

$$- fC_{fg}(t, t_0)S_{fg}(t, t_0) + g \left(S_{fg}(t, t_0) \right)^2$$

$$= g \left(\left(C_{fg}(t, t_0) \right)^2 + \left(S_{fg}(t, t_0) \right)^2 \right)$$

$$= gE_{(f+ig)\oplus_c(f-ig)}(t, t_0)$$

$$\neq 0, \quad t \in \mathbb{T}^{\kappa^2}.$$

Consequently, the functions y_1 and y_2, defined by (7.9), form a fundamental set of solutions for equation (7.8) and

$$y(t) = c_1 y_1(t) + c_2 y_2(t)$$

$$= c_1 C_{fg}(t, t_0) + c_2 S_{fg}(t, t_0), \quad t \in \mathbb{T}^{\kappa^2},$$

where $c_1, c_2 \in \mathbb{R}$, is its general solution.

Exercise 7.1.16 *Find the general solution for each of the following equations.*

1. $(D^\alpha)^2 y - 9y = 0$, $t \in \mathbb{T}^{\kappa^2}$,

2. $(D^\alpha)^2 y - 36y = 0$, $t \in \mathbb{T}^{\kappa^2}$,

3. $(D^\alpha)^2 y + 4y = 0$, $t \in \mathbb{T}^{\kappa^2}$,

4. $9(D^\alpha)^2 y + y = 0$, $t \in \mathbb{T}^{\kappa^2}$,

5. $(D^\alpha)^2 y - 4D^\alpha y - 12y = 0$, $t \in \mathbb{T}^{\kappa^2}$,

6. $(D^\alpha)^2 y - D^\alpha y + y = 0$, $t \in \mathbb{T}^{\kappa^2}$.

7.2 REDUCTION OF ORDER

In this section we suppose that the graininess function μ is delta differentiable on $\mathbb{T}^{\kappa}$. Consider the equation

$$(D^{\alpha})^2 y - 2\gamma D^{\alpha} y + \gamma^2 y = 0, \quad t \in \mathbb{T}^{\kappa^2}, \tag{7.10}$$

where $\gamma \in \mathbb{R}$, $\gamma > 0$, $\gamma \in \mathscr{R}_c$. Note that

$$y_1(t) = E_{\gamma}(t, t_0), \quad t \in \mathbb{T}^{\kappa^2},$$

is a solution of equation (7.10). We will search for another solution of equation (7.10) in the form

$$y_2(t) = x(t) E_{\gamma}(t, t_0), \quad t \in \mathbb{T}^{\kappa^2},$$

where $x \in \mathscr{C}_{rd}^2(\mathbb{T})$ will be determined so that y_1 and y_2 are linearly independent solutions of equation (7.10). To this end, note that

$$
\begin{aligned}
D^{\alpha} y_2(t) &= D^{\alpha}\left(x(t) E_{\gamma}(t, t_0)\right) \\[2mm]
&= (D^{\alpha} x(t)) E_{\gamma}^{\sigma}(t, t_0) + x(t) D^{\alpha} E_{\gamma}(t, t_0) \\[2mm]
&\quad - k_1(\alpha, t) x(t) E_{\gamma}^{\sigma}(t, t_0) \\[2mm]
&= (D^{\alpha} x(t))\left(1 + \mu(t)\frac{\gamma - k_1(\alpha, t)}{k_0(\alpha, t)}\right) E_{\gamma}(t, t_0) \\[2mm]
&\quad + \gamma x(t) E_{\gamma}(t, t_0) \\[2mm]
&\quad - k_1(\alpha, t)\left(1 + \mu(t)\frac{\gamma - k_1(\alpha, t)}{k_0(\alpha, t)}\right) x(t) E_{\gamma}(t, t_0) \\[2mm]
&= (D^{\alpha} x(t) - k_1(\alpha, t) x(t))\left(1 + \mu(t)\frac{\gamma - k_1(\alpha, t)}{k_0(\alpha, t)}\right) E_{\gamma}(t, t_0) \\[2mm]
&\quad + \gamma x(t) E_{\gamma}(t, t_0), \quad t \in \mathbb{T}^{\kappa^2},
\end{aligned}
$$

and

$$
\begin{aligned}
(D^{\alpha})^2 y_2(t) &= D^{\alpha}\left((D^{\alpha} x(t) - k_1(\alpha, t) x(t))\left(1 + \mu(t)\frac{\gamma - k_1(\alpha, t)}{k_0(\alpha, t)}\right) E_{\gamma}(t, t_0)\right) \\[2mm]
&\quad + \gamma D^{\alpha}\left(x(t) E_{\gamma}(t, t_0)\right) \\[2mm]
&= D^{\alpha}\left((D^{\alpha} x(t) - k_1(\alpha, t) x(t))\left(1 + \mu(t)\frac{\gamma - k_1(\alpha, t)}{k_0(\alpha, t)}\right)\right) E_{\gamma}^{\sigma}(t, t_0)
\end{aligned}
$$

$$+ (D^\alpha x(t) - k_1(\alpha,t)x(t)) \left(1 + \mu(t)\frac{\gamma - k_1(\alpha,t)}{k_0(\alpha,t)} \right) D^\alpha E_\gamma(t,t_0)$$

$$- k_1(\alpha,t)(D^\alpha x(t) - k_1(\alpha,t)x(t)) \left(1 + \mu(t)\frac{\gamma - k_1(\alpha,t)}{k_0(\alpha,t)} \right) E_\gamma^\sigma(t,t_0)$$

$$+ \gamma(D^\alpha x(t)) E_\gamma^\sigma(t,t_0) + \gamma x(t) D^\alpha E_\gamma(t,t_0)$$

$$- \gamma x(t)k_1(\alpha,t)E_\gamma^\sigma(t,t_0)$$

$$= D^\alpha \left((D^\alpha x(t) - k_1(\alpha,t)x(t)) \left(1 + \mu(t)\frac{\gamma - k_1(\alpha,t)}{k_0(\alpha,t)} \right) \right)$$

$$\times \left(1 + \mu(t)\frac{\gamma - k_1(\alpha,t)}{k_0(\alpha,t)} \right) E_\gamma(t,t_0)$$

$$+ \gamma(D^\alpha x(t) - k_1(\alpha,t)x(t)) \left(1 + \mu(t)\frac{\gamma - k_1(\alpha,t)}{k_0(\alpha,t)} \right) E_\gamma(t,t_0)$$

$$- k_1(\alpha,t)(D^\alpha x(t) - k_1(\alpha,t)x(t)) \left(1 + \mu(t)\frac{\gamma - k_1(\alpha,t)}{k_0(\alpha,t)} \right)^2 E_\gamma(t,t_0)$$

$$+ \gamma(D^\alpha x(t)) \left(1 + \mu(t)\frac{\gamma - k_1(\alpha,t)}{k_0(\alpha,t)} \right) E_\gamma(t,t_0)$$

$$+ \gamma^2 x(t) E_\gamma(t,t_0)$$

$$- \gamma x(t)k_1(\alpha,t) \left(1 + \mu(t)\frac{\gamma - k_1(\alpha,t)}{k_0(\alpha,t)} \right) E_\gamma(t,t_0)$$

$$= D^\alpha \left((D^\alpha x(t) - k_1(\alpha,t)x(t)) \left(1 + \mu(t)\frac{\gamma - k_1(\alpha,t)}{k_0(\alpha,t)} \right) \right)$$

$$\times \left(1 + \mu(t)\frac{\gamma - k_1(\alpha,t)}{k_0(\alpha,t)} \right) E_\gamma(t,t_0)$$

$$+ 2\gamma(D^\alpha x(t) - k_1(\alpha,t)x(t)) \left(1 + \mu(t)\frac{\gamma - k_1(\alpha,t)}{k_0(\alpha,t)} \right) E_\gamma(t,t_0)$$

$$- k_1(\alpha,t)(D^\alpha x(t) - k_1(\alpha,t)x(t)) \left(1 + \mu(t)\frac{\gamma - k_1(\alpha,t)}{k_0(\alpha,t)} \right)^2 E_\gamma(t,t_0)$$

$$+ \gamma^2 x(t) E_\gamma(t,t_0), \quad t \in \mathbb{T}^{\kappa^2}.$$

Hence,

$$0 = (D^\alpha)^2 y_2(t) - 2\gamma D^\alpha y_2(t) + \gamma^2 y_2(t)$$

$$= D^\alpha \left((D^\alpha x(t) - k_1(\alpha,t)x(t)) \left(1 + \mu(t)\frac{\gamma - k_1(\alpha,t)}{k_0(\alpha,t)}\right) \right)$$

$$\times \left(1 + \mu(t)\frac{\gamma - k_1(\alpha,t)}{k_0(\alpha,t)}\right) E_\gamma(t,t_0)$$

$$+ 2\gamma(D^\alpha x(t) - k_1(\alpha,t)x(t)) \left(1 + \mu(t)\frac{\gamma - k_1(\alpha,t)}{k_0(\alpha,t)}\right) E_\gamma(t,t_0)$$

$$- k_1(\alpha,t)(D^\alpha x(t) - k_1(\alpha,t)x(t)) \left(1 + \mu(t)\frac{\gamma - k_1(\alpha,t)}{k_0(\alpha,t)}\right)^2 E_\gamma(t,t_0)$$

$$+ \gamma^2 x(t) E_\gamma(t,t_0)$$

$$- 2\gamma(D^\alpha x(t) - k_1(\alpha,t)x(t)) \left(1 + \mu(t)\frac{\gamma - k_1(\alpha,t)}{k_0(\alpha,t)}\right) E_\gamma(t,t_0)$$

$$- 2\gamma^2 x(t) E_\gamma(t,t_0) + \gamma^2 x(t) E_\gamma(t,t_0)$$

$$= D^\alpha \left((D^\alpha x(t) - k_1(\alpha,t)x(t)) \left(1 + \mu(t)\frac{\gamma - k_1(\alpha,t)}{k_0(\alpha,t)}\right) \right)$$

$$\times \left(1 + \mu(t)\frac{\gamma - k_1(\alpha,t)}{k_0(\alpha,t)}\right) E_\gamma(t,t_0)$$

$$- k_1(\alpha,t)(D^\alpha x(t) - k_1(\alpha,t)x(t)) \left(1 + \mu(t)\frac{\gamma - k_1(\alpha,t)}{k_0(\alpha,t)}\right)^2 E_\gamma(t,t_0),$$

$t \in \mathbb{T}^{\kappa^2}$. Set

$$v(t) = (D^\alpha x(t) - k_1(\alpha,t)x(t)) \left(1 + \mu(t)\frac{\gamma - k_1(\alpha,t)}{k_0(\alpha,t)}\right), \quad t \in \mathbb{T}^{\kappa^2}.$$

Then

$$0 = (D^\alpha v(t)) \left(1 + \mu(t)\frac{\gamma - k_1(\alpha,t)}{k_0(\alpha,t)}\right) E_\gamma(t,t_0)$$

$$- k_1(\alpha,t)v(t) \left(1 + \mu(t)\frac{\gamma - k_1(\alpha,t)}{k_0(\alpha,t)}\right) E_\gamma(t,t_0)$$

$$= (D^\alpha v(t) - k_1(\alpha,t)v(t)) \left(1 + \mu(t)\frac{\gamma - k_1(\alpha,t)}{k_0(\alpha,t)}\right) E_\gamma(t,t_0), \quad t \in \mathbb{T}^{\kappa^2}.$$

Hence,

$$D^\alpha v(t) = k_1(\alpha,t)v(t), \quad t \in \mathbb{T}^{\kappa^2},$$

and

$$v(t) = c_1 E_{k_1}(t,t_0)$$

$$= c_1, \quad t \in \mathbb{T}^{\kappa^2},$$

where c_1 is a constant. Take $c_1 = 1$. Then

$$v(t) = 1, \quad t \in \mathbb{T}^{\kappa^2},$$

and

$$(D^\alpha x(t) - k_1(\alpha,t)x(t))\left(1 + \mu(t)\frac{\gamma - k_1(\alpha,t)}{k_0(\alpha,t)}\right) = 1, \quad t \in \mathbb{T}^{\kappa^2},$$

or

$$D^\alpha x(t) = k_1(\alpha,t)x(t) + \frac{k_0(\alpha,t)}{k_0(\alpha,t) + \mu(t)(\gamma - k_1(\alpha,t))}, \quad t \in \mathbb{T}^{\kappa^2}.$$

Consequently,

$$\begin{aligned}
x(t) &= c_2 E_{k_1}(t,t_0) + \int_{t_0}^t \frac{k_0(\alpha,s)}{k_0(\alpha,s) + \mu(s)(\gamma - k_1(\alpha,s))} E_0(\sigma(s),t)\Delta_{\alpha,t}s \\
&= c_2 + \int_{t_0}^t \frac{k_0(\alpha,s)}{k_0(\alpha,s) + \mu(s)(\gamma - k_1(\alpha,s))} E_0(\sigma(s),t)\Delta_{\alpha,t}s,
\end{aligned}$$

$t \in \mathbb{T}^{\kappa^2}$, where c_2 is a constant. Take $c_2 = 0$. Then

$$x(t) = \int_{t_0}^t \frac{k_0(\alpha,s)}{k_0(\alpha,s) + \mu(s)(\gamma - k_1(\alpha,s))} E_0(\sigma(s),t)\Delta_{\alpha,t}s, \quad t \in \mathbb{T}^{\kappa^2},$$

and

$$y_2(t) = \left(\int_{t_0}^t \frac{k_0(\alpha,s)}{k_0(\alpha,s) + \mu(s)(\gamma - k_1(\alpha,s))} E_0(\sigma(s),t)\Delta_{\alpha,t}s\right) E_\gamma(t,t_0),$$

$t \in \mathbb{T}^{\kappa^2}$. By the above computations, we have

$$\begin{aligned}
D^\alpha y_2(t) &= E_\gamma(t,t_0) + \gamma x(t)E_\gamma(t,t_0) \\
\\
&= (1 + \gamma x(t))E_\gamma(t,t_0), \quad t \in \mathbb{T}^{\kappa^2}.
\end{aligned}$$

From here,

$$\begin{aligned}
W(y_1,y_2)(t) &= \det\begin{pmatrix} y_1(t) & y_2(t) \\ D^\alpha y_1(t) & D^\alpha y_2(t) \end{pmatrix} \\
\\
&= \det\begin{pmatrix} E_\gamma(t,t_0) & x(t)E_\gamma(t,t_0) \\ \gamma E_\gamma(t,t_0) & (1 + \gamma x(t))E_\gamma(t,t_0) \end{pmatrix} \\
\\
&= (1 + \gamma x(t))\left(E_\gamma(t,t_0)\right)^2 - \gamma x(t)\left(E_\gamma(t,t_0)\right)^2 \\
\\
&= \left(E_\gamma(t,t_0)\right)^2 \\
\\
&\neq 0, \quad t \in \mathbb{T}^{\kappa^2}.
\end{aligned}$$

Consequently, y_1 and y_2 form a fundamental set of solutions of equation (7.10). Therefore the general solution of equation (7.10) is given by the expression

$$y(t) = c_1 y_1(t) + c_2 y_2(t)$$

$$= c_1 E_\gamma(t, t_0)$$

$$+ c_2 \left(\int_{t_0}^t \frac{k_0(\alpha, s)}{k_0(\alpha, s) + \mu(s)(\gamma - k_1(\alpha, s))} E_0(\sigma(s), t) \Delta_{\alpha, t} s \right) E_\gamma(t, t_0)$$

$$= \left(c_1 + c_2 \left(\int_{t_0}^t \frac{k_0(\alpha, s)}{k_0(\alpha, s) + \mu(s)(\gamma - k_1(\alpha, s))} E_0(\sigma(s), t) \Delta_{\alpha, t} s \right) \right) E_\gamma(t, t_0),$$

$t \in \mathbb{T}^{\kappa^2}$, where c_1 and c_2 are constants.

Example 7.2.1 *Let* $\mathbb{T} = \mathbb{N}$ *and*

$$k_1(\alpha, t) = (1 - \alpha)t^{2\alpha}, \quad k_0(\alpha, t) = \alpha t^{2(1-\alpha)}, \quad t \in \mathbb{T}, \quad \alpha \in (0, 1].$$

Consider the equation

$$\left(D^{\frac{1}{2}} y \right)^2 y - 2 D^{\frac{1}{2}} y + y = 0, \quad t \in \mathbb{T}.$$

The characteristic equation is

$$\lambda^2 - 2\lambda + 1 = 0.$$

Then

$$\lambda_1 = \lambda_2 = 1.$$

Here

$$\mu(t) = 1,$$

$$k_1\left(\frac{1}{2}, t \right) = \frac{1}{2}t,$$

$$k_0\left(\frac{1}{2}, t \right) = \frac{1}{2}t, \quad t \in \mathbb{T}.$$

We have

$$y_1(t) = E_1(t, 1)$$

$$= e^{\int_1^t \frac{1}{\mu(\tau)} \log\left(1 + \mu(\tau) \frac{1 - k_1\left(\frac{1}{2}, \tau \right)}{k_0\left(\frac{1}{2}, \tau \right)} \right) \Delta \tau}$$

$$= e^{\int_1^t \log\left(1+\frac{1-\frac{1}{2}\tau}{\frac{1}{2}\tau}\right)\Delta\tau}$$

$$= e^{\int_1^t \log\left(1+\frac{2-\tau}{\tau}\right)\Delta\tau}$$

$$= e^{\int_1^t \log\frac{\tau+2-\tau}{\tau}\Delta\tau}$$

$$= e^{\sum_{\tau=1}^{t-1}\log\frac{2}{\tau}}$$

$$= \prod_{\tau=1}^{t-1}\frac{2}{\tau}, \quad t \in \mathbb{T}, \quad t \geq 2,$$

is a solution of the considered equation. Next,

$$
\begin{aligned}
y_2(t) &= \left(\int_1^t \frac{k_0\left(\frac{1}{2},s\right)}{k_0\left(\frac{1}{2},s\right)+1-k_1\left(\frac{1}{2},s\right)}E_0(\sigma(s),t)\Delta_{\alpha,t}s\right)E_1(t,1) \\
&= \left(\int_1^t \frac{k_0\left(\frac{1}{2},s\right)}{k_0\left(\frac{1}{2},s\right)+1-k_1\left(\frac{1}{2},s\right)}E_0(\sigma(s),t)\frac{E_0(t,\sigma(s))}{k_0\left(\frac{1}{2},s\right)}\Delta s\right)E_1(t,1) \\
&= \left(\int_1^t \frac{1}{k_0\left(\frac{1}{2},s\right)+1-k_1\left(\frac{1}{2},s\right)}\Delta s\right)E_1(t,1) \\
&= \left(\int_1^t \frac{1}{\frac{1}{2}s+1-\frac{1}{2}s}\Delta s\right)E_1(t,1) \\
&= \left(\int_1^t \Delta s\right)E_1(t,1) \\
&= (t-1)E_1(t,1) \\
&= (t-1)\prod_{\tau=1}^{t-1}\frac{2}{\tau}, \quad t \in \mathbb{T}, \quad t \geq 2.
\end{aligned}
$$

Hence, the general solution of the considered equation is

$$
\begin{aligned}
y(t) &= c_1 y_1(t) + c_2 y_2(t) \\
&= c_1\prod_{\tau=1}^{t-1}\frac{2}{\tau} + c_2(t-1)\prod_{\tau=1}^{t-1}\frac{2}{\tau} \\
&= (c_1 - c_2 + c_2 t)\prod_{|tau=1}^{t-1}\frac{2}{\tau}
\end{aligned}
$$

$$= (c_3 + c_2 t) \prod_{\tau=1}^{t-1} \frac{2}{\tau},$$

where c_1 and c_2 are constants, $c_3 = c_1 - c_2$. Here ends the example.

Exercise 7.2.2 *Let $\mathbb{T} = 2^{\mathbb{N}_0}$,*

$$k_1(\alpha,t) = (1-\alpha)t^{4\alpha}, \quad k_0(\alpha,t) = \alpha t^{4(1-\alpha)}, \quad \alpha \in (0,1], \quad t \in \mathbb{T}.$$

Find a fundamental set of solutions for each of the following equations.

1. $\left(D^{\frac{1}{3}}\right)^2 y - 2D^{\frac{1}{3}}y + y = 0$, $t \in \mathbb{T}$,

2. $\left(D^{\frac{1}{4}}\right)^2 y - 6D^{\frac{1}{4}}y + 9y = 0$, $t \in \mathbb{T}$,

3. $\left(D^{\frac{1}{2}}\right)^2 y - 4D^{\frac{1}{2}}y + 4y = 0$, $t \in \mathbb{T}$,

4. $\left(D^{\frac{1}{2}}\right)^2 y - D^{\frac{1}{2}}y + \frac{1}{4}y = 0$, $t \in \mathbb{T}$,

5. $\left(D^{\frac{1}{6}}\right)^2 y - 8D^{\frac{1}{6}}y + 16y = 0$, $t \in \mathbb{T}$.

7.3 METHOD OF FACTORING

Consider the equation

$$D^\alpha \left(D^\alpha y - py\right)(t) - q(t)\left(D^\alpha y(t) - p(t)y(t)\right) = 0, \quad t \in \mathbb{T}^{\kappa^2}, \tag{7.11}$$

where $p \in \mathcal{C}_{rd}^1(\mathbb{T})$, $q \in \mathcal{C}_{rd}(\mathbb{T})$, $p,q \in \mathcal{R}_c$. Set

$$v(t) = D^\alpha y(t) - p(t)y(t), \quad t \in \mathbb{T}^{\kappa^2}.$$

Then

$$D^\alpha v(t) = q(t)v(t), \quad t \in \mathbb{T}^{\kappa^2}.$$

Hence,

$$v(t) = c_1 E_q(t,t_0), \quad r \in \mathbb{T}^{\kappa^2},$$

where c_1 is a constant. From here,

$$D^\alpha y(t) = p(t)y(t) + c_1 E_q(t,t_0), \quad t \in \mathbb{T}^{\kappa^2}.$$

Then

$$y(t) = c_2 E_p(t,t_0) + c_1 \int_{t_0}^t E_q(s,t_0)E_g(\sigma(s),t)\Delta_{\alpha,t}s, \quad t \in \mathbb{T}^{\kappa^2},$$

where c_2 is a constant and

$$g(t) = \frac{(p(t) - k_1(\alpha,t))(\mu(t)k_1(\alpha,t) - k_0(\alpha,t))}{k_0(\alpha,t) + \mu(t)(p(t) - k_1(\alpha,t))}, \quad t \in \mathbb{T},$$

is the general solution of equation (7.11).

Example 7.3.1 *Let* $\mathbb{T} = \mathbb{N}_0$,

$$k_1(\alpha,t) = 1 - \alpha, \quad k_0(\alpha,t) = \alpha, \quad \alpha \in (0,1], \quad t \in \mathbb{T}.$$

Consider the equation

$$\left(D^{\frac{3}{4}}\right)^2 y - 2D^{\frac{3}{4}}y + y = 0, \quad t \in \mathbb{T}.$$

Here

$$\sigma(t) \;=\; t+1,$$

$$\mu(t) \;=\; 1,$$

$$\alpha \;=\; \frac{3}{4},$$

$$k_1\left(\frac{3}{4},t\right) \;=\; 1 - \frac{3}{4}$$

$$=\; \frac{1}{4},$$

$$k_0\left(\frac{3}{4},t\right) \;=\; \frac{3}{4}, \quad t \in \mathbb{T}.$$

We can rewrite the given equation in the form

$$D^{\frac{3}{4}}\left(D^{\frac{3}{4}}y - y\right) - \left(D^{\frac{3}{4}}y - y\right) = 0, \quad t \in \mathbb{T}.$$

Set

$$v(t) = D^{\frac{3}{4}}y(t) - y(t), \quad t \in \mathbb{T}.$$

Then

$$D^{\frac{3}{4}}v(t) = v(t), \quad t \in \mathbb{T},$$

and

$$v(t) \;=\; c_1 E_1(t,0)$$

$$=\; c_1 e^{\int_0^t \frac{1}{\mu(s)} \log\left(1 + \mu(s)\frac{1 - k_1\left(\frac{3}{4},t\right)}{k_0\left(\frac{3}{4},t\right)}\right)\Delta s}$$

$$=\; c_1 e^{\int_0^t \log\left(1 + \frac{1 - \frac{1}{4}}{\frac{3}{4}}\right)\Delta s}$$

$$= c_1 e^{t \log 2}$$
$$= c_1 2^t, \quad t \in \mathbb{T},$$

where c_1 is a constant. Hence,

$$D^{\frac{3}{4}} y(t) = y(t) + v(t), \quad t \in \mathbb{T},$$

and

$$
\begin{aligned}
y(t) &= c_2 E_1(t,0) + \int_0^t v(s) E_g(\sigma(s),t) \Delta_{\alpha,t} s \\[2mm]
&= c_2 E_1(t,0) + \int_0^t v(s) E_g(\sigma(s),t) \frac{E_0(t,\sigma(s))}{k_0\left(\frac{3}{4},s\right)} \Delta s \\[2mm]
&= c_2 E_1(t,0) + \frac{4}{3} \int_0^t v(s) E_g(\sigma(s),t) E_{\ominus_c 0}(\sigma(s),t) \Delta s \\[2mm]
&= c_2 E_1(t,0) + \frac{4}{3} \int_0^t v(s) E_{g \oplus_c (\ominus_c 0)}(\sigma(s),t) \Delta s \\[2mm]
&= c_2 E_1(t,0)
\end{aligned}
$$

$$+ \frac{4}{3} \int_0^t v(s) \left(1 + \mu(s) \frac{(g \oplus_c (\ominus_c 0))(s) - k_1\left(\frac{3}{4},s\right)}{k_0\left(\frac{3}{4},s\right)} \right) E_{g \oplus_c (\ominus_c 0)}(s,t) \Delta s,$$

where c_2 is a constant,

$$
\begin{aligned}
g(s) &= \frac{\left(1 - k_1\left(\frac{3}{4},s\right)\right)\left(\mu(s) k_1\left(\frac{3}{4},s\right) - k_0\left(\frac{3}{4},s\right)\right)}{k_0\left(\frac{3}{4},s\right) + \mu(s)\left(1 - k_1\left(\frac{3}{4},s\right)\right)} \\[3mm]
&= \frac{\left(1 - \frac{1}{4}\right)\left(\frac{1}{4} - \frac{3}{4}\right)}{\frac{3}{4} + 1 - \frac{1}{4}} \\[3mm]
&= \frac{\frac{3}{4}\left(-\frac{1}{2}\right)}{\frac{3}{2}} \\[3mm]
&= -\frac{1}{4},
\end{aligned}
$$

$$
\begin{aligned}
\ominus_c 0 &= \frac{k_0\left(\frac{3}{4},s\right) k_1\left(\frac{3}{4},s\right)}{k_0\left(\frac{3}{4},s\right) - k_1\left(\frac{3}{4},s\right)} \\[3mm]
&= \frac{\frac{3}{4}\left(\frac{1}{4}\right)}{\frac{3}{4} - \frac{1}{4}} \\[3mm]
&= \frac{\frac{3}{16}}{\frac{1}{2}} \\[3mm]
&= \frac{3}{8},
\end{aligned}
$$

$$
(g \oplus_c (\ominus_c 0))(s) = \frac{\left(-\frac{1}{4} + \frac{3}{8} - \frac{1}{4}\right)\frac{3}{4} + \left(-\frac{1}{4} - \frac{1}{4}\right)\left(\frac{3}{8} - \frac{1}{4}\right)}{\frac{3}{4}}
$$

$$
= \frac{-\frac{3}{32} - \frac{1}{16}}{\frac{3}{4}}
$$

$$
= \frac{-\frac{5}{32}}{\frac{3}{4}}
$$

$$
= -\frac{20}{96}
$$

$$
= -\frac{5}{24},
$$

$$
1 + \mu(s)\frac{(g \oplus_c (\ominus_c 0))(s) - k_1\left(\frac{3}{4},s\right)}{k_0\left(\frac{3}{4},s\right)} = 1 + \frac{-\frac{5}{24} - \frac{1}{4}}{\frac{3}{4}}
$$

$$
= 1 - \frac{\frac{11}{24}}{\frac{3}{4}}
$$

$$
= 1 - \frac{11}{18}
$$

$$
= \frac{7}{18}, \quad s \in \mathbb{T},
$$

and

$$
E_{g \oplus_c (\ominus_c 0)}(s,t) = e^{\int_t^s \frac{1}{\mu(\tau)}\log\left(1+\mu(\tau)\frac{(g \oplus_c (\ominus_c 0))(\tau) - k_1\left(\frac{3}{4},\tau\right)}{k_0\left(\frac{3}{4},\tau\right)}\right)\Delta\tau}
$$

$$
= e^{-\int_s^t \log\frac{7}{18}\Delta\tau}
$$

$$
= e^{-(t-s)\log\frac{7}{18}}
$$

$$
= \left(\frac{18}{7}\right)^{t-s}, \quad t,s \in \mathbb{T}, \quad s \leq t.
$$

Then

$$
y(t) = c_2 E_1(t,0) + \frac{4}{3}\left(\frac{7}{18}\right)\int_0^t v(s)\left(\frac{18}{7}\right)^{t-s}\Delta s
$$

$$
= c_2 E_1(t,0) + \frac{14}{27}\sum_{s=0}^{t-1} v(s)\left(\frac{18}{7}\right)^{t-s}
$$

$$= c_2 2^t + \frac{14}{27} c_1 \sum_{s=0}^{t-1} 2^s \frac{2^{t-s} s^{2t-2s}}{7^{t-s}}$$

$$= c_2 2^t + \frac{14}{27} c_1 2^t \sum_{s=0}^{t-1} \left(\frac{9}{7}\right)^{t-s}$$

$$= c_2 2^t + c_3 \left(\frac{18}{7}\right)^t \sum_{s=0}^{t-1} \left(\frac{7}{9}\right)^s,$$

where $c_3 = \dfrac{14}{27} c_1$. This ends the example.

Theorem 7.3.2 Let $f, g \in \mathscr{C}_{rd}(\mathbb{T})$. Consider the equation

$$(D^\alpha)^2 y + f(t) D^\alpha y + g(t) y = 0, \quad t \in \mathbb{T}^{\kappa^2}. \tag{7.12}$$

If either one of the two conditions

(i)

$$f(t) = -p^\sigma(t) - q(t),$$

$$g(t) = -D^\alpha p(t) + k_1(\alpha, t) p^\sigma(t) + q(t) p(t), \quad \alpha \in (0,1], \quad t \in \mathbb{T}^{\kappa^2},$$

(ii)

$$f(t) = -p - q(t),$$

$$g(t) = pq(t), \quad t \in \mathbb{T}^{\kappa^2},$$

where p is a constant,

is satisfied, then equation (7.12) can be written in the factored form (7.11).

Proof 7.3.3   1. Suppose (i). Then

$$0 = (D^\alpha)^2 y(t) + (-p^\sigma(t) - q(t)) D^\alpha y(t)$$

$$+ (-D^\alpha p(t) + k_1(\alpha, t) p^\sigma(t) + q(t) p(t)) y(t)$$

$$= (D^\alpha)^2 y(t)$$

$$- (p^\sigma(t) D^\alpha y(t) + D^\alpha p(t) y(t) - k_1(\alpha, t) p^\sigma(t) y(t))$$

$$- q(t) D^\alpha y(t) + q(t) p(t) y(t)$$

$$= (D^\alpha)^2 y(t) - D^\alpha(py)(t) - q(t)(D^\alpha y(t) - p(t)y(t))$$

$$= D^\alpha (D^\alpha y - py)(t) - q(t)(D^\alpha y(t) - p(t)y(t)), \quad t \in \mathbb{T}^{\kappa^2}.$$

2. *Suppose* (ii). *Then*

$$0 = (D^\alpha)^2 y(t) + (-p - q(t))D^\alpha y(t) + pq(t)y(t)$$

$$= (D^\alpha)^2 y(t) - pD^\alpha y(t) - q(t)D^\alpha y(t) + pq(t)y(t)$$

$$= D^\alpha (D^\alpha y(t) - py(t)) - q(t)(D^\alpha y(t) - py(t)), \quad t \in \mathbb{T}^{\kappa^2}.$$

This completes the proof. □

Example 7.3.4 *Let* $\mathbb{T} = 2^{\mathbb{N}_0}$,

$$k_1(\alpha,t) = (1-\alpha)t^{4\alpha}, \quad k_0(\alpha,t) = \alpha t^{4(1-\alpha)}, \quad \alpha \in (0,1], \quad t \in \mathbb{T}^{\kappa^2}.$$

Consider the equation

$$\left(D^{\frac{1}{4}}\right)^2 y - (t + 4t^2)D^{\frac{1}{4}}y + \left(\frac{13}{4}t^3 - \frac{3}{4}t^4\right)y = 0, \quad t \in \mathbb{T}.$$

Here

$$\alpha = \frac{1}{4},$$

$$k_1\left(\frac{1}{4}, t\right) = \frac{3}{4}t,$$

$$k_0\left(\frac{1}{4}, t\right) = \frac{1}{4}t^3,$$

$$\sigma(t) = 2t,$$

$$f(t) = -t - 4t^2,$$

$$g(t) = \frac{13}{4}t^3 - \frac{3}{4}t^4, \quad t \in \mathbb{T}.$$

Take

$$p(t) = t^2,$$

$$q(t) = t, \quad t \in \mathbb{T}.$$

Then

$$p^{\Delta}(t) = \sigma(t) + t$$

$$= 2t + t$$

$$= 3t,$$

$$p^{\sigma}(t) = (\sigma(t))^2$$

$$= (2t)^2$$

$$= 4t^2,$$

$$D^{\frac{1}{4}} p(t) = k_1 \left(\frac{1}{4}, t\right) p(t) + k_0 \left(\frac{1}{4}, t\right) p^{\Delta}(t)$$

$$= \frac{3}{4} t \left(t^2\right) + \frac{1}{4} t^3 (3t)$$

$$= \frac{3}{4} t^3 + \frac{3}{4} t^4,$$

$$-p^{\sigma}(t) - q(t) = -4t^2 - t$$

$$= f(t), \quad t \in \mathbb{T},$$

and

$$-D^{\frac{1}{4}} p(t) + k_1 \left(\frac{1}{4}, t\right) p^{\sigma}(t) + q(t) p(t) = -\frac{3}{4} t^3 - \frac{3}{4} t^4$$

$$+ \frac{3}{4} t \left(4t^2\right) + t^3$$

$$= \frac{13}{4} t^3 - \frac{3}{4} t^4$$

$$= g(t), \quad t \in \mathbb{T}.$$

Hence, and considering Theorem 7.3.2, we conclude that the considered equation can be written in the factored form

$$D^{\frac{1}{4}} \left(D^{\frac{1}{4}} y - t^2 y\right) - t \left(D^{\frac{1}{4}} y - t^2 y\right) = 0, \quad t \in \mathbb{T}.$$

This ends the example.

Exercise 7.3.5 *Let* $\mathbb{T} = 3^{\mathbb{N}_0}$ *and*

$$k_1(\alpha,t) = (1-\alpha)t^{3\alpha}, \quad k_0(\alpha,t) = \alpha t^{3(1-\alpha)}, \quad \alpha \in (0,1], \quad t \in \mathbb{T}.$$

Prove that the equation

$$\left(D^{\frac{1}{3}}\right)^2 y - \left(3t + t^2\right) D^{\frac{1}{3}} y + \left(t^2 + t^3\right) y = 0, \quad t \in \mathbb{T},$$

can be written in the factored form (7.11).

7.4 NONCONSTANT COEFFICIENTS

Consider the equation

$$(D^\alpha)^2 y - \left(q^{\circledR}(t) + k_1(\alpha,t)q\right) y^\sigma - k_1(\alpha,t)qy = 0, \tag{7.13}$$

$t \in \mathbb{T}^{\kappa^2}$, where $q \in \mathscr{R}_c$ is a constant, $q \neq 0$,

$$(k_0(\alpha,t) - \mu(t)k_1(\alpha,t))^2 - (\mu(t))^2 k_1(\alpha,t)q \neq 0, \quad \alpha \in (0,1], \quad t \in \mathbb{T}.$$

We will show that

$$y_1(t) = E_q(t,t_0), \quad t \in \mathbb{T},$$

is a solution of equation (7.13). We have

$$q^{\circledR}(t) = \frac{qk_0(\alpha,t)(q - k_1(\alpha,t))}{k_0(\alpha,t) + \mu(t)(q - k_1(\alpha,t))} - k_1(\alpha,t)q,$$

$$q^{\circledR}(t) + k_1(\alpha,t)q = \frac{qk_0(\alpha,t)(q - k_1(\alpha,t))}{k_0(\alpha,t) + \mu(t)(q - k_1(\alpha,t))}$$

$$\left(q^{\circledR}(t) + k_1(\alpha,t)q\right) E_q^\sigma(t,t_0) = q(q - k_1(\alpha,t))E_q(t,t_0),$$

$$D^\alpha E_q(t,t_0) = qE_q(t,t_0),$$

$$(D^\alpha)^2 E_q(t,t_0) = qD^\alpha E_q(t,t_0)$$

$$= q^2 E_q(t,t_0), \quad t \in \mathbb{T}^{\kappa^2}.$$

Hence,

$$(D^\alpha)^2 y - \left(q^{\circledR}(t) + k_1(\alpha,t)q\right) y^\sigma - k_1(\alpha,t)qy$$

$$= q^2 E_q(t,t_0) - (q^2 - k_1(\alpha,t)q) E_q(t,t_0)$$

$$-k_1(\alpha,t)qE_q(t,t_0)$$

$$= 0, \quad t \in \mathbb{T}^{\kappa^2}.$$

Note that

$$y^\sigma = y + \mu y^\Delta$$

$$= y + \frac{\mu}{k_0}\left(k_1 y + k_0 y^\Delta\right) - \frac{\mu k_1}{k_0} y$$

$$= \left(1 - \frac{\mu k_1}{k_0}\right) y + \frac{\mu}{k_0} D^\alpha y \quad \text{on} \quad \mathbb{T}^{\kappa^2}.$$

Then equation (7.13) takes the form

$$0 = (D^\alpha)^2 y(t) - \frac{q k_0(\alpha,t)(q - k_1(\alpha,t))}{k_0(\alpha,t) + \mu(t)(q - k_1(\alpha,t))}$$

$$\times \left(\frac{k_0(\alpha,t) - \mu(t)k_1(\alpha,t)}{k_0(\alpha,t)} y(t) + \frac{\mu(t)}{k_0(\alpha,t)} D^\alpha y(t)\right)$$

$$-k_1(\alpha,t)q y(t)$$

$$= (D^\alpha)^2 y(t) - \frac{q\mu(t)(q - k_1(\alpha,t))}{k_0(\alpha,t) + \mu(t)(q - k_1(\alpha,t))} D^\alpha y(t)$$

$$+ \left(-\frac{q(q - k_1(\alpha,t))(k_0(\alpha,t) - \mu(t)k_1(\alpha,t))}{k_0(\alpha,t) + \mu(t)(q - k_1(\alpha,t))} - k_1(\alpha,t)q\right) y(t),$$

$t \in \mathbb{T}^{\kappa^2}$. Let

$$a(t) = -\frac{q\mu(t)(q - k_1(\alpha,t))}{k_0(\alpha,t) + \mu(t)(q - k_1(\alpha,t))},$$

$$b(t) = -\frac{q(q - k_1(\alpha,t))(k_0(\alpha,t) - \mu(t)k_1(\alpha,t))}{k_0(\alpha,t) + \mu(t)(q - k_1(\alpha,t))} - k_1(\alpha,t)q,$$

$t \in \mathbb{T}$. Therefore, equation (7.13) can be rewritten in the form

$$(D^\alpha)^2 y + a(t)D^\alpha y + b(t)y = 0, \quad t \in \mathbb{T}^{\kappa^2}.$$

We will find another solution y_2 of equation (7.13) using Abel's formula, so that

$$y_2(t_0) = 0, \quad D^\alpha y_2(t_0) = 1.$$

We have

$$W(y_1, y_2)(t_0) = \det\begin{pmatrix} y_1(t_0) & y_2(t_0) \\ D^\alpha y_1(t_0) & D^\alpha y_2(t_0) \end{pmatrix}$$

$$= \det \begin{pmatrix} 1 & 0 \\ q & 1 \end{pmatrix}$$

$$= 1$$

$$\neq 0, \quad t \in \mathbb{T}^\kappa.$$

Let

$$p = -\frac{(a+k_1)(k_0 - \mu k_1)}{k_0} + \frac{b\mu}{k_0}, \quad \text{on} \quad \mathbb{T}.$$

Note that

$$p = -\frac{\left(-\frac{q\mu(q-k_1)}{k_0\mu(q-k_1)} + k_1\right)(k_0 - \mu k_1)}{k_0}$$

$$-\frac{\mu q(q-k_1)(k_0 - \mu k_1)}{k_0(k_0 + \mu(q-k_1))} - \frac{\mu q k_1}{k_0}$$

$$= -\frac{k_1(k_0 - \mu k_1)}{k_0} - \frac{\mu q k_1}{k_0}$$

$$= -\frac{k_1(k_0 + \mu(q-k_1))}{k_0},$$

$$\frac{p - k_1}{k_0} = -\frac{k_1(k_0 + \mu(q-k_1))}{k_0^2} - \frac{k_1}{k_0}$$

$$= -\frac{k_1(2k_0 + \mu(q-k_1))}{k_0^2},$$

$$1 + \mu \frac{p - k_1}{k_0} = 1 - \frac{\mu k_1(2k_0 + \mu(q-k_1))}{k_0^2}$$

$$= \frac{k_0^2 - 2\mu k_0 k_1 + \mu^2 k_1^2 - \mu^2 k_1 q}{k_0^2}$$

$$= \frac{(k_0 - \mu k_1)^2 - \mu^2 k_1 q}{k_0^2}$$

$$\neq 0 \quad \text{on} \quad \mathbb{T}.$$

From here, it follows that

$$E_p(t, t_0) \neq 0, \quad t \in \mathbb{T},$$

and

$$W(y_1, y_2)(t) \neq 0, \quad t \in \mathbb{T}^\kappa.$$

Then, using Abel's formula, we obtain

$$E_p(t,t_0)W(y_1,y_2)(t_0) = W(y_1,y_2)(t), \quad t \in \mathbb{T}^\kappa,$$

or

$$
\begin{aligned}
E_p(t,t_0) &= \det \begin{pmatrix} y_1(t) & y_2(t) \\ D^\alpha y_1(t) & D^\alpha y_2 3(t) \end{pmatrix} \\[2mm]
&= \det \begin{pmatrix} E_q(t,t_0) & y_2(t) \\ qE_q(t,t_0) & D^\alpha y_2(t) \end{pmatrix} \\[2mm]
&= E_q(t,t_0)\left(D^\alpha y_2(t) - qy_2(t)\right), \quad t \in \mathbb{T}^\kappa.
\end{aligned}
$$

Therefore

$$
\begin{aligned}
D^\alpha y_2(t) &= qy_2(t) + E_p(t,t_0)E_{\ominus_c q}(t,t_0) \\[2mm]
&= qy_2(t) + E_{p\ominus_c q}(t,t_0), \quad t \in \mathbb{T}^{\kappa^2}.
\end{aligned}
$$

Let

$$g(t) = \frac{(q - k_1(\alpha,t))(\mu(t)k_1(\alpha,t) - k_0(\alpha,t))}{k_0(\alpha,t) + \mu(t)(q - k_1(\alpha,t))}, \quad t \in \mathbb{T}.$$

Then

$$y_2(t) = \int_{t_0}^t E_{p\ominus_c q}(s,t_0)E_g(\sigma(s),t)\Delta_{\alpha,t}s, \quad t \in \mathbb{T}^{\kappa^2}.$$

Therefore a general solution of equation (7.13) is given by the expression

$$
\begin{aligned}
y(t) &= c_1 E_q(t,t_0) \\[2mm]
&\quad + c_2 \int_{t_0}^t E_{p\ominus_c q}(s,t_0)E_g(\sigma(s),t)\Delta_{\alpha,t}s, \quad t \in \mathbb{T}^{\kappa^2},
\end{aligned}
$$

where c_1 and c_2 are constants.

Exercise 7.4.1 *Let* $\mathbb{T} = \mathbb{Z}$,

$$k_1(\alpha,t) = 1 - \alpha, \quad k_0(\alpha,t) = \alpha, \quad \alpha \in (0,1], \quad t \in \mathbb{T}.$$

Find a general solution of the equation

$$\left(D^{\frac{1}{3}}\right)^2 y - \left(3^{\textcircled{2}}(t) + 2\right)y^\sigma - 2y = 0, \quad t \in \mathbb{T}^{\kappa^2}.$$

Now we consider the equation

$$(D^\alpha)^2 y + f(t)D^\alpha y + g(t)y = 0, \quad t \in \mathbb{T}^{\kappa^2}, \tag{7.14}$$

where $f,g \in \mathscr{C}_{rd}(\mathbb{T})$ and

$$(k_0(\alpha,t) - \mu(t)k_1(\alpha,t))^2 - f(t)\mu(t)(k_0(\alpha,t) - \mu(t)k_1(\alpha,t))$$

$$+ g(t)(\mu(t))^2 \neq 0, \quad t \in \mathbb{T}, \quad \alpha \in (0,1].$$

Suppose that $z \in \mathscr{C}_{rd}^1(\mathbb{T})$, $z \in \mathscr{R}_c$, satisfies the equation

$$D^\alpha z + z^\sigma z - k_1(\alpha, t) z^\sigma + a(t) z + b(t) = 0, \quad t \in \mathbb{T}^\kappa.$$

We will show that

$$y_1(t) = E_z(t, t_0), \quad t \in \mathbb{T},$$

is a solution of equation (7.14). We have

$$D^\alpha y_1(t) = z(t) E_z(t, t_0),$$

$$(D^\alpha)^2 y_1(t) = D^\alpha z(t) E_z(t, t_0) + z^\sigma(t) z(t) E_z(t, t_0)$$

$$-k_1(\alpha, t) z^\sigma(t) E_z(t, t_0)$$

$$= (D^\alpha z(t) + z^\sigma(t) z(t) - k_1(\alpha, t) z^\sigma(t)) E_z(t, t_0), \quad t \in \mathbb{T}^{\kappa^2}.$$

Then

$$(D^\alpha)^2 y_1(t) + f(t) D^\alpha y_1(t) + g(t) y_1(t)$$

$$= (D^\alpha z(t) + z^\sigma(t) z(t) - k_1(\alpha, t) z^\sigma(t)) E_z(t, t_0)$$

$$+ f(t) z(t) E_z(t, t_0) + g(t) E_z(t, t_0)$$

$$= (D^\alpha z(t) + z^\sigma(t) z(t) - k_1(\alpha, t) z^\sigma(t) + f(t) z(t) + g(t)) E_z(t, t_0)$$

$$= 0, \quad t \in \mathbb{T}^{\kappa^2}.$$

Let

$$h(t) = -\frac{(f(t) + k_1(\alpha, t))(k_0(\alpha, t) - \mu(t) k_1(\alpha, t))}{k_0(\alpha, t)}$$

$$+ \frac{g(t)\mu(t)}{k_0(\alpha, t)}, \quad t \in \mathbb{T}, \quad \alpha \in (0, 1].$$

Then

$$\frac{h - k_1}{k_0} = \frac{-(f + k_1)(k_0 - \mu k_1) + g\mu - k_0 k_1}{k_0^2}$$

$$= \frac{-f k_0 + f \mu k_1 + g\mu - 2 k_0 k_1 + \mu k_1^2}{k_0^2},$$

$$1 + \mu \frac{h - k_1}{k_0} = \frac{k_0^2 - 2\mu k_0 k_1 + \mu^2 k_1^2 - f\mu(k_0 - \mu k_1) + g\mu^2}{k_0^2}$$

$$= \frac{(k_0 - \mu k_1)^2 - f\mu(k_0 - \mu k_1) + g\mu}{k_0^2}$$

$$\neq 0, \quad \alpha \in (0, 1], \quad t \in \mathbb{T}.$$

Now we will find another solution y_2 of equation (7.14) so that

$$y_2(t_0) = 0, \quad D^\alpha y_2(t_0) = 1,$$

and y_1 and y_2 are linearly independent. By Abel's formula, we have

$$W(y_1, y_2)(t) = E_h(t, t_0) W(y_1, y_2)(t_0)$$

$$= E_h(t, t_0) \det \begin{pmatrix} y_1(t) & y_2(t) \\ D^\alpha y_1(t) & D^\alpha y_2(t) \end{pmatrix}$$

$$= E_h(t, t_0) \det \begin{pmatrix} 1 & 0 \\ z(t_0) & 1 \end{pmatrix}$$

$$= E_h(t, t_0)$$

$$\neq 0, \quad t \in \mathbb{T}^\kappa.$$

Hence,

$$E_h(t, t_0) = \det \begin{pmatrix} y_1(t) & y_2(t) \\ D^\alpha y_1(t) & D^\alpha y_2(t) \end{pmatrix}$$

$$= \det \begin{pmatrix} E_z(t, t_0) & y_2(t) \\ z(t) E_z(t, t_0) & D^\alpha y_2(t) \end{pmatrix}$$

$$= (D^\alpha y_2(t) - z(t) y_2(t)) E_z(t, t_0), \quad t \in \mathbb{T}^\kappa,$$

or

$$D^\alpha y_2(t) = z(t) y_2(t) + E_{h \ominus_c z}(t, t_0), \quad t \in \mathbb{T}^\kappa.$$

Then

$$y_2(t) = \int_{t_0}^t E_{h \ominus_c z}(s, t_0) E_{h_1}(\sigma(s), t) \Delta_{\alpha, t} s, \quad t \in \mathbb{T}^\kappa,$$

where

$$h_1 = \frac{(z - k_1)(\mu k_1 - k_0)}{k_0 + \mu(z - k_1)}, \quad \alpha \in (0, 1], \quad t \in \mathbb{T}.$$

We have

$$
\begin{aligned}
h_1 - k_1 &= \frac{(z-k_1)(\mu k_1 - k_0) - k_1 k_0 - \mu k_1(z-k_1)}{k_0 + \mu(z-k_1)} \\[2mm]
&= \frac{\mu z k_1 - z k_0 - \mu k_1^2 + k_1 k_0 - k_1 k_0 - \mu k_1 z + \mu k_1^2}{k_0 + \mu(z-k_1)} \\[2mm]
&= -\frac{z k_0}{k_0 + \mu(z-k_1)},
\end{aligned}
$$

$$
\begin{aligned}
1 + \mu \frac{h_1 - k_1}{k_0} &= 1 - \frac{\mu z}{k_0 + \mu(z-k_1)} \\[2mm]
&= \frac{k_0 + \mu(z-k_1) - \mu z}{k_0 + \mu(z-k_1)} \\[2mm]
&= \frac{k_0 - \mu k_1}{k_0 + \mu(z-k_1)} \\[2mm]
&\neq 0, \quad \alpha \in (0,1], \quad t \in \mathbb{T}.
\end{aligned}
$$

Consequently, $h_1 \in \mathscr{R}_c$ and a general solution of equation (7.14) is given by

$$
\begin{aligned}
y(t) &= c_1 y_1(t) + c_2 y_2(t) \\[2mm]
&= c_1 E_z(t, t_0) \\[2mm]
&\quad + c_2 \int_{t_0}^{t} E_{h\ominus_c z}(s, t_0) E_{h_1}(\sigma(s), t) \Delta_{\alpha,t} s, \quad t \in \mathbb{T}^{\kappa^2}.
\end{aligned}
$$

7.5 CONFORMABLE EULER-CAUCHY EQUATIONS

In this section we suppose that $\mathbb{T} \subseteq (0, \infty)$.

Definition 7.5.1 *Let $a, b \in \mathbb{R}$. The equations of the form*

$$
t\sigma(t)(D^\alpha)^2 y + (a + k_0(\alpha, t) - \mu(t)k_1(\alpha, t))tD^\alpha y + by = 0, \quad t \in \mathbb{T}^{\kappa^2}, \tag{7.15}
$$

will be called conformable Euler-Cauchy equations.

Theorem 7.5.2 *Let $a^2 - 4b \neq 0$,*

$$
\frac{-a \pm \sqrt{a^2 - 4b}}{2t} \in \mathscr{R}_c, \quad t \in \mathbb{T},
$$

and

$$
\lambda_{1,2} = \frac{-a \pm \sqrt{a^2 - 4b}}{2}.
$$

Then $E_{\frac{\lambda_1}{t}}(t, t_0)$ and $E_{\frac{\lambda_2}{t}}(t, t_0)$ are solutions of equation (7.15).

Proof 7.5.3 *Let*

$$y(t) = E_{\frac{\lambda}{t}}(t,t_0), \quad t \in \mathbb{T},$$

where $\lambda = \lambda_1$ or $\lambda = \lambda_2$. Note that

$$\lambda^2 + a\lambda + b = 0.$$

Then

$$
\begin{aligned}
D^\alpha y(t) &= \frac{\lambda}{t} E_{\frac{\lambda}{t}}(t,t_0) \\[2mm]
&= \frac{\lambda}{t} y(t), \quad t \in \mathbb{T}^\kappa, \\[2mm]
t D^\alpha y(t) &= \lambda y(t), \quad t \in \mathbb{T}^\kappa, \\[2mm]
(D^\alpha)^2 y(t) &= \lambda D^\alpha \left(\frac{y(t)}{t} \right) \\[2mm]
&= \lambda \frac{t D^\alpha y(t) - y(t) D^\alpha t}{t\sigma(t)} + \lambda k_1(\alpha,t) \frac{y(t)}{t} \\[2mm]
&= \lambda \frac{t D^\alpha y(t) - t y(t) k_1(\alpha,t) - k_0(\alpha,t) y(t)}{t\sigma(t)} \\[2mm]
&\quad + \lambda k_1(\alpha,t) \frac{y(t)}{t}, \quad t \in \mathbb{T}^{\kappa^2}, \\[2mm]
t\sigma(t)(D^\alpha)^2 y(t) &= \lambda \left(t D^\alpha y(t) - t y(t) k_1(\alpha,t) - k_0(\alpha,t) y(t) \right) \\[2mm]
&\quad + \lambda k_1(\alpha,t)\sigma(t) y(t) \\[2mm]
&= \lambda \Big(\lambda - t k_1(\alpha,t) - k_0(\alpha,t) \\[2mm]
&\quad + k_1(\alpha,t)\sigma(t) \Big) y(t), \quad t \in \mathbb{T}^{\kappa^2}.
\end{aligned}
$$

Therefore

$$t\sigma(t)(D^\alpha)^2 y(t) + (a + t k_1(\alpha,t) + k_0(\alpha,t) - \sigma(t)k_1(\alpha,t)) t D^\alpha y(t) + b y(t)$$

$$= \lambda^2 y(t) + \lambda \left(-t k_1(\alpha,t) - k_0(\alpha,t) + \sigma(t)k_1(\alpha,t) \right) y(t)$$

$$+ \lambda \left(a + t k_1(\alpha,t) + k_0(\alpha,t) - \sigma(t)k_1(\alpha,t) \right) y(t)$$

$$+by(t)$$

$$= \left(\lambda^2 + a\lambda + b\right) y(t)$$

$$= 0, \quad t \in \mathbb{T}^{\kappa^2}.$$

This completes the proof. □

Theorem 7.5.4 *Let a, b, λ_1 and λ_2 be as in Theorem 7.5.2. Then*

$$E_{\frac{\lambda_1}{t}}(t,t_0), \quad E_{\frac{\lambda_2}{t}}(t,t_0), \quad t \in \mathbb{T},$$

form a fundamental set of solutions for the Euler-Cauchy equation (7.15).

Proof 7.5.5 *Let*
$$y_1(t) = E_{\frac{\lambda_1}{t}}(t,t_0), \quad y_2(t) = E_{\frac{\lambda_2}{t}}(t,t_0), \quad t \in \mathbb{T}.$$

We have

$$
\begin{aligned}
W(y_1, y_2)(t) &= \det \begin{pmatrix} y_1(t) & y_2(t) \\ D^\alpha y_1(t) & D^\alpha y_2(t) \end{pmatrix} \\[2mm]
&= \det \begin{pmatrix} E_{\frac{\lambda_1}{t}}(t,t_0) & E_{\frac{\lambda_2}{t}}(t,t_0) \\ \dfrac{\lambda_1}{t} E_{\frac{\lambda_1}{t}}(t,t_0) & \dfrac{\lambda_2}{t} E_{\frac{\lambda_2}{t}}(t,t_0) \end{pmatrix} \\[2mm]
&= \frac{\lambda_2}{t} E_{\frac{\lambda_1}{t}}(t,t_0) E_{\frac{\lambda_2}{t}}(t,t_0) - \frac{\lambda_1}{t} E_{\frac{\lambda_1}{t}}(t,t_0) E_{\frac{\lambda_2}{t}}(t,t_0) \\[2mm]
&= \frac{\lambda_2 - \lambda_1}{t} E_{\frac{\lambda_1}{t}}(t,t_0) E_{\frac{\lambda_2}{t}}(t,t_0) \\[2mm]
&\neq 0, \quad t \in \mathbb{T}^\kappa.
\end{aligned}
$$

This completes the proof. □

Example 7.5.6 *Let $\mathbb{T} = \mathbb{N}$,*

$$k_1(\alpha,t) = (1 - \alpha)(1 + t^2)^\alpha, \quad k_0(\alpha,t) = \alpha(1 + t^2)^{1-\alpha}, \quad t \in \mathbb{T}, \quad \alpha \in (0,1].$$

Consider the equation

$$t(t+1) \left(D^{\frac{1}{2}}\right)^2 y + 3t D^{\frac{1}{2}} y + 2y = 0, \quad t \in \mathbb{T}.$$

Here

$$\alpha = \frac{1}{2},$$

$$a = 3,$$

$$b = 2,$$

$$k_1\left(\frac{1}{2},t\right) = \frac{1}{2}(1+t^2)^{\frac{1}{2}},$$

$$k_0\left(\frac{1}{2},t\right) = \frac{1}{2}(1+t^2)^{\frac{1}{2}}, \quad t \in \mathbb{T}.$$

Then

$$a + tk_1\left(\frac{1}{2},t\right) + k_0\left(\frac{1}{2},t\right) - \sigma(t)k_1\left(\frac{1}{2},t\right)$$

$$= 3 + \frac{t}{2}(1+t^2)^{\frac{1}{2}} + \frac{1}{2}(1+t^2)^{\frac{1}{2}} - (t+1)\frac{1}{2}(1+t^2)^{\frac{1}{2}}$$

$$= 3, \quad t \in \mathbb{T}.$$

Therefore

$$\lambda_1 = \frac{-3+\sqrt{9-8}}{2} = \frac{-3+1}{2} = -1,$$

$$\lambda_2 = \frac{-3-\sqrt{9-8}}{2} = \frac{-3-1}{2} = -2,$$

and

$$E_{-\frac{1}{t}}(t,t_0), \quad E_{-\frac{2}{t}}(t,t_0), \quad t \in \mathbb{T},$$

form a fundamental set of solutions of the considered equation.

Exercise 7.5.7 *Let* $\mathbb{T} = 2^{\mathbb{N}_0}$,

$$k_1(\alpha,t) = (1-\alpha)t^{\alpha}, \quad k_0(\alpha,t) = \alpha t^{1-\alpha}, \quad \alpha \in (0,1], \quad t \in \mathbb{T}.$$

Find a fundamental set of solutions for each of the following equations.

1.
$$2t^2\left(D^{\frac{1}{2}}\right)^2 y + \left(4 - \frac{1}{2}t^{\frac{3}{2}} + \frac{1}{2}t^{\frac{1}{2}}\right)tD^{\frac{1}{2}}y + 3y = 0, \quad t \in \mathbb{T},$$

2.
$$2t^2\left(D^{\frac{1}{3}}\right)^2 y + \left(5 - \frac{2}{3}t^{\frac{4}{3}} + \frac{1}{3}t^{\frac{2}{3}}\right)tD^{\frac{1}{3}}y + 4y = 0, \quad t \in \mathbb{T},$$

3.
$$2t^2\left(D^{\frac{1}{4}}\right)^2 y + \left(7 - \frac{3}{4}t^{\frac{5}{4}} + \frac{1}{4}t^{\frac{3}{4}}\right)tD^{\frac{1}{4}}y + 6y = 0, \quad t \in \mathbb{T}.$$

Now we suppose that $a \neq 0$ and

$$a^2 - 4b = 0, \quad \frac{\lambda}{t} = -\frac{a}{2t} \in \mathcal{R}_c.$$

Then

$$a = -2\lambda,$$

$$b = \frac{a^2}{4}$$

$$= \frac{4\lambda^2}{4}$$

$$= \lambda^2.$$

The equation (7.15) can be rewritten in the form

$$t\sigma(t)(D^\alpha)^2 y + (-2\lambda + k_0(\alpha,t) - \mu(t)k_1(\alpha,t))tD^\alpha y + \lambda^2 y = 0, \quad t \in \mathbb{T}^{\kappa^2}. \tag{7.16}$$

Let

$$g(t) = 1 + \int_{t_0}^t \frac{\lambda}{s + \mu(s)\frac{\lambda - sk_1(\alpha,s)}{k_0(\alpha,s)}} E_0(\sigma(s),t)\Delta_{\alpha,t}s, \quad t, t_0 \in \mathbb{T}.$$

Note that

$$D^\alpha g(t) = D^\alpha 1$$

$$+ D^\alpha \left(\int_{t_0}^t \frac{\lambda}{s + \mu(s)\frac{\lambda - sk_1(\alpha,s)}{k_0(\alpha,s)}} E_0(\sigma(s),t)\Delta_{\alpha,t}s \right)$$

$$= k_1(\alpha,t)$$

$$+ \int_{t_0}^t \frac{\lambda}{s + \mu(s)\frac{\lambda - sk_1(\alpha,s)}{k_0(\alpha,s)}} D_t^\alpha E_0(\sigma(s),t)\Delta_{\alpha,t}s|$$

$$- k_1(\alpha,t)\int_{t_0}^t \frac{\lambda}{s + \mu(s)\frac{\lambda - sk_1(\alpha,s)}{k_0(\alpha,s)}} E_0(\sigma(s),\sigma(t))\Delta_{\alpha,t}s$$

$$+ \frac{\lambda}{t + \mu(t)\frac{\lambda - tk_1(\alpha,t)}{k_0(\alpha,t)}} E_0(\sigma(t),\sigma(t))$$

$$= k_1(\alpha,t) + \frac{\lambda}{t + \mu(t)\frac{\lambda - tk_1(\alpha,t)}{k_0(\alpha,t)}}$$

$$+ \int_{t_0}^{t} \frac{\lambda}{s + \mu(s)^{\frac{\lambda - sk_1(\alpha,s)}{k_0(\alpha,s)}}} \left(\frac{k_1(\alpha,t)E_0(t,\sigma(s))}{E_0(t,\sigma(s))E_0(\sigma(t),\sigma(s))} \right.$$

$$\left. + k_1(\alpha,t)\frac{1}{E_0(t,\sigma(s))} \right) \Delta_{\alpha,t}s$$

$$- k_1(\alpha,t) \int_{t_0}^{t} \frac{\lambda}{s + \mu(s)^{\frac{\lambda - sk_1(\alpha,s)}{k_0(\alpha,s)}}} E_0(\sigma(s),\sigma(t))\Delta_{\alpha,t}s$$

$$= k_1(\alpha,t) + \frac{\lambda}{t + \mu(t)^{\frac{\lambda - tk_1(\alpha,t)}{k_0(\alpha,t)}}}$$

$$+ k_1(\alpha,t) \int_{t_0}^{t} \frac{\lambda}{s + \mu(s)^{\frac{\lambda - sk_1(\alpha,s)}{k_0(\alpha,s)}}} E_0(\sigma(s),\sigma(t))\Delta_{\alpha,t}s$$

$$+ k_1(\alpha,t) \int_{t_0}^{t} \frac{\lambda}{s + \mu(s)^{\frac{\lambda - sk_1(\alpha,s)}{k_0(\alpha,s)}}} E_0(\sigma(s),t)\Delta_{\alpha,t}s$$

$$- k_1(\alpha,t) \int_{t_0}^{t} \frac{\lambda}{s + \mu(s)^{\frac{\lambda - sk_1(\alpha,s)}{k_0(\alpha,s)}}} E_0(\sigma(s),\sigma(t))\Delta_{\alpha,t}s$$

$$= k_1(\alpha,t) \left(1 + \int_{t_0}^{t} \frac{\lambda}{s + \mu(s)^{\frac{\lambda - sk_1(\alpha,s)}{k_0(\alpha,s)}}} E_0(\sigma(s),t)\Delta_{\alpha,t}s \right)$$

$$+ \frac{\lambda}{t + \mu(t)^{\frac{\lambda - tk_1(\alpha,t)}{k_0(\alpha,t)}}}$$

$$= k_1(\alpha,t)g(t) + \frac{\lambda}{t + \mu(t)^{\frac{\lambda - tk_1(\alpha,t)}{k_0(\alpha,t)}}}, \quad t \in \mathbb{T}^{\kappa}.$$

Consequently, g satisfies the equation

$$D^{\alpha}g = k_1(\alpha,t)g + \frac{\lambda}{t + \mu(t)^{\frac{\lambda - tk_1(\alpha,t)}{k_0(\alpha,t)}}}, \quad t \in \mathbb{T}^{\kappa}.$$

Now we will prove that

$$y(t) = E_{\frac{\lambda}{t}}(t,t_0)g(t), \quad t \in \mathbb{T},$$

satisfies equation (7.16). Indeed, we have

$$D^{\alpha}y(t) = D^{\alpha}\left(E_{\frac{\lambda}{t}}(t,t_0)g(t) \right)$$

$$= \frac{\lambda}{t}E_{\frac{\lambda}{t}}(t,t_0)g(t) + E_{\frac{\lambda}{t}}(\sigma(t),t_0)D^{\alpha}g(t)$$

$$-k_1(\alpha,t)E_{\frac{\lambda}{t}}(\sigma(t),t_0)g(t)$$

$$= \frac{\lambda}{t}E_{\frac{\lambda}{t}}(t,t_0)g(t)+E_{\frac{\lambda}{t}}(\sigma(t),t_0)\left(D^{\alpha}g(t)-k_1(\alpha,t)g(t)\right)$$

$$= \frac{\lambda}{t}E_{\frac{\lambda}{t}}(t,t_0)g(t)$$

$$+\left(1+\mu(t)\frac{\frac{\lambda}{t}-k_1(\alpha,t)}{k_0(\alpha,t)}\right)E_{\frac{\lambda}{t}}(t,t_0)\left(D^{\alpha}g(t)-k_1(\alpha,t)g(t)\right)$$

$$= \frac{\lambda}{t}E_{\frac{\lambda}{t}}(t,t_0)g(t)+\frac{\lambda}{t}E_{\frac{\lambda}{t}}(t,t_0)$$

$$= \frac{\lambda}{t}y(t)+\frac{\lambda}{t}E_{\frac{\lambda}{t}}(t,t_0),$$

$$(D^{\alpha})^2 y(t) = D^{\alpha}\left(\frac{\lambda}{t}y(t)+\frac{\lambda}{t}E_{\frac{\lambda}{t}}(t,t_0)\right)$$

$$= \lambda D^{\alpha}\left(\frac{y(t)}{t}\right)+\lambda D^{\alpha}\left(\frac{E_{\frac{\lambda}{t}}(t,t_0)}{t}\right)$$

$$= \lambda\frac{tD^{\alpha}y(t)-y(t)D^{\alpha}t}{t\sigma(t)}+\lambda k_1(\alpha,t)\frac{y(t)}{t}$$

$$+\lambda\frac{tD^{\alpha}E_{\frac{\lambda}{t}}(t,t_0)-E_{\frac{\lambda}{t}}(t,t_0)D^{\alpha}t}{t\sigma(t)}$$

$$+\lambda k_1(\alpha,t)\frac{E_{\frac{\lambda}{t}}(t,t_0)}{t}$$

$$= \lambda\frac{t\left(\frac{\lambda}{t}y(t)+\frac{\lambda}{t}E_{\frac{\lambda}{t}}(t,t_0)\right)-k_1(\alpha,t)ty(t)-k_0(\alpha,t)y(t)}{t\sigma(t)}$$

$$+\lambda\frac{k_1(\alpha,t)y(t)}{t}$$

$$+\lambda\frac{t\frac{\lambda}{t}E_{\frac{\lambda}{t}}(t,t_0)-k_1(\alpha,t)tE_{\frac{\lambda}{t}}(t,t_0)-k_0(\alpha,t)E_{\frac{\lambda}{t}}(t,t_0)}{t\sigma(t)}$$

$$+\lambda k_1(\alpha,t)\frac{E_{\frac{\lambda}{t}}(t,t_0)}{t}$$

$$= \lambda\frac{\lambda y(t)+\lambda E_{\frac{\lambda}{t}}(t,t_0)-k_1(\alpha,t)ty(t)-k_0(\alpha,t)y(t)}{t\sigma(t)}$$

$$+\lambda \frac{k_1(\alpha,t)y(t)}{t}$$

$$+\lambda \frac{\lambda E_{\frac{\lambda}{t}}(t,t_0) - tk_1(\alpha,t)E_{\frac{\lambda}{t}}(t,t_0) - k_0(\alpha,t)E_{\frac{\lambda}{t}}(t,t_0)}{t\sigma(t)}$$

$$+\lambda k_1(\alpha,t)\frac{E_{\frac{\lambda}{t}}(t,t_0)}{t}, \quad t \in \mathbb{T}^{\kappa^2}.$$

Therefore

$$t\sigma(t)(D^\alpha)^2 y(t) + (-2\lambda + tk_1(\alpha,t) + k_0(\alpha,t) - \sigma(t)k_1(\alpha,t))tD^\alpha y(t) + \lambda^2 y(t)$$

$$= \lambda \left(\lambda y(t) + \lambda E_{\frac{\lambda}{t}}(t,t_0) - k_1(\alpha,t)ty(t) - k_0(\alpha,t)y(t)\right)$$

$$+\lambda \sigma(t)k_1(\alpha,t)y(t)$$

$$+\lambda \left(\lambda E_{\frac{\lambda}{t}}(t,t_0) - tk_1(\alpha,t)E_{\frac{\lambda}{t}}(t,t_0) - k_0(\alpha,t)E_{\frac{\lambda}{t}}(t,t_0)\right)$$

$$+\lambda \sigma(t)k_1(\alpha,t)E_{\frac{\lambda}{t}}(t,t_0)$$

$$+(-2\lambda + tk_1(\alpha,t) + k_0(\alpha,t) - \sigma(t)k_1(\alpha,t))\left(\lambda y(t) + \lambda E_{\frac{\lambda}{t}}(t,t_0)\right)$$

$$+\lambda^2 y(t)$$

$$= \lambda(\lambda - tk_1(\alpha,t) - k_0(\alpha,t))E_{\frac{\lambda}{t}}(t,t_0)$$

$$+\lambda \sigma(t)k_1(\alpha,t)E_{\frac{\lambda}{t}}(t,t_0) + \lambda^2 E_{\frac{\lambda}{t}}(t,t_0)$$

$$+\lambda(-2\lambda + tk_1(\alpha,t) + k_0(\alpha,t) - \sigma(t)k_1(\alpha,t))E_{\frac{\lambda}{t}}(t,t_0)$$

$$= 0, \quad t \in \mathbb{T}^{\kappa^2}.$$

Let

$$y_1(t) = E_{\frac{\lambda}{t}}(t,t_0), \quad y_2(t) = g(t)E_{\frac{\lambda}{t}}(t,t_0), \quad t \in \mathbb{T}.$$

Then

$$D^\alpha y_1(t) = \frac{\lambda}{t}E_{\frac{\lambda}{t}}(t,t_0)$$

$$= \frac{\lambda}{t}y_1(t),$$

$$D^\alpha y_2(t) = \frac{\lambda}{t} y_2(t) + \frac{\lambda}{t} E_{\frac{\lambda}{t}}(t,t_0)$$

$$= \frac{\lambda}{t} y_1(t) + \frac{\lambda}{t} y_2(t), \quad t, t_0 \in \mathbb{T},$$

and

$$W(y_1, y_2)(t) = \det \begin{pmatrix} y_1(t) & y_2(t) \\ D^\alpha y_1(t) & D^\alpha y_2(t) \end{pmatrix}$$

$$= \det \begin{pmatrix} y_1(t) & y_2(t) \\ \frac{\lambda}{t} y_1(t) & \frac{\lambda}{t} y_2(t) + \frac{\lambda}{t} y_1(t) \end{pmatrix}$$

$$= \det \begin{pmatrix} y_1(t) & y_2(t) \\ 0 & \frac{\lambda}{t} y_1(t) \end{pmatrix}$$

$$= \frac{\lambda}{t} (y_1(t))^2$$

$$= \frac{\lambda}{t} \left(E_{\frac{\lambda}{t}}(t,t_0) \right)^2$$

$$\neq 0,$$

i.e.,

$$y_1(t) = E_{\frac{\lambda}{t}}(t,t_0), \quad y_2(t) = g(t) E_{\frac{\lambda}{t}}(t,t_0), \quad t \in \mathbb{T},$$

form a fundamental set of solutions for equation (7.16). Then the general solution of equation (7.16) is given by

$$y(t) = (c_1 + c_2 g(t)) E_{\frac{\lambda}{t}}(t,t_0), \quad t \in \mathbb{T}.$$

Example 7.5.8 *Let* $\mathbb{T} = 2^{\mathbb{N}_0}$,

$$k_1(\alpha,t) = (1-\alpha)t^{4\alpha}, \quad k_0(\alpha,t) = \alpha t^{4(1-\alpha)}, \quad \alpha \in (0,1], \quad t \in \mathbb{T}.$$

Consider the equation

$$2t^2 \left(D^{\frac{1}{4}} \right)^2 y + \left(2 - \frac{1}{4}t^2 + \frac{1}{4}t^3 \right) t D^{\frac{1}{4}} y + y = 0, \quad t \geq 4.$$

Here

$$\sigma(t) = 2t,$$

$$k_1 \left(\frac{1}{4}, t \right) = \frac{3}{4} t,$$

$$k_0\left(\frac{1}{4},t\right) = \frac{1}{4}t^3,$$

$$a = 2,$$

$$b = 1.$$

Then

$$\lambda = -1,$$

and

$$g(t) = 1 - \int_4^t \frac{1}{s - s\frac{1+\frac{3}{4}s^2}{\frac{1}{4}s^3}} E_0(2s,t)\Delta_{\frac{1}{4},t}s$$

$$= 1 - \int_4^t \frac{1}{s - \frac{4+3s^2}{s^2}} E_0(2s,t)\Delta_{\frac{1}{4},t}s$$

$$= 1 - \int_4^t \frac{s^2}{s^3 - 3s^2 - 4} E_0(2s,t)\Delta_{\frac{1}{4},t}s$$

$$= 1 - \int_4^t \frac{s^2}{s^3 - 3s^2 - 4} E_0(2s,t)\frac{E_0(t,2s)}{\frac{1}{4}s^3}\Delta s$$

$$= 1 - \int_4^t \frac{4}{s(s^3 - 3s^2 - 4)}\Delta s$$

$$= 1 - \sum_{s=4}^{\frac{t}{2}} \frac{4}{s^3 - 3s^2 - 4},$$

$$E_{-\frac{1}{t}}(t,4) = e_{\frac{-\frac{1}{t}-\frac{3}{4}t}{\frac{1}{4}t^3}}(t,4)$$

$$= e_{-\frac{4+3t^2}{t^4}}(t,4)$$

$$= e^{\int_4^t \frac{1}{s}\log\left(1-\frac{4+3s^2}{s^3}\right)\Delta s}$$

$$= e^{\int_4^t \frac{1}{s}\log\left(\frac{s^3-3s^2-4}{s^3}\right)\Delta s}$$

$$= e^{\sum_{s=4}^{\frac{t}{2}}\log\left(\frac{s^3-3s^2-4}{s^3}\right)}$$

$$= \prod_{s=4}^{\frac{t}{2}}\left(\frac{s^3 - 3s^2 - 4}{s^3}\right).$$

Consequently,

$$y(t) = \left(c_1 + c_2 - c_2 \sum_{s=4}^{\frac{t}{2}} \frac{4}{s^3 - 3s^2 - 4}\right) \prod_{s=4}^{\frac{t}{2}} \frac{s^3 - 3s^2 - 4}{s^3}$$

$$= \left(c_3 + c_4 \sum_{s=4}^{\frac{t}{2}} \frac{4}{s^3 - 3s^2 - 4}\right) \prod_{s=4}^{\frac{t}{2}} \frac{s^3 - 3s^2 - 4}{s^3}$$

is the general solution, where c_1 and c_2 are constants, $c_3 = c_1 + c_2$, $c_4 = -c_2$. This ends the example.

Exercise 7.5.9 *Let* $\mathbb{T} = 3^{\mathbb{N}_0}$,

$$k_1(\alpha, t) = (1 - \alpha)(1 + t)^\alpha, \quad k_0(\alpha, t) = \alpha(1 + t)^{1-\alpha}, \quad \alpha \in (0, 1], \quad t \in \mathbb{T}.$$

Find the general solution of each of the following equations.

1.

$$3t^2 \left(D^{\frac{1}{2}}\right)^2 y + \left(6 - t(1 + t)^{\frac{1}{2}} + \frac{1}{2}(1 + t)^{\frac{1}{2}}\right) t D^{\frac{1}{2}} y + 9y = 0, \quad t \in \mathbb{T}^{\kappa^2},$$

2.

$$3t^2 \left(D^{\frac{1}{4}}\right)^2 y + \left(8 - \frac{3}{2}t(1 + t)^{\frac{1}{4}} + \frac{1}{4}(1 + t)^{\frac{3}{4}}\right) t D^{\frac{1}{4}} y + 16y = 0, \quad t \in \mathbb{T}^{\kappa^2},$$

3.

$$3t^2 \left(D^{\frac{1}{3}}\right)^2 y + \left(10 - \frac{4}{3}t(1 + t)^{\frac{1}{3}} + \frac{1}{3}(1 + t)^{\frac{2}{3}}\right) t D^{\frac{1}{3}} y + 25y = 0, \quad t \in \mathbb{T}^{\kappa^2}.$$

7.6 VARIATION OF PARAMETERS

Consider the equation

$$(D^\alpha)^2 y + a(t)D^\alpha y + b(t)y = f(t), \tag{7.17}$$

where $a, b, f \in \mathcal{C}_{rd}(\mathbb{T})$. Suppose that y_1 and y_2 form a fundamental set of solutions for the corresponding homogeneous equation (7.1). We will search for a solution of equation (7.17) of the form

$$y(t) = p(t)y_1(t) + q(t)y_2(t), \quad t \in \mathbb{T}^{\kappa^2},$$

where the functions p and q will be determined below. We have

$$D^\alpha y(t) = D^\alpha(p(t)y_1(t)) + D^\alpha(q(t)y_2(t))$$

$$= (D^\alpha p(t))y_1^\sigma(t) + p(t)D^\alpha y_1(t)$$

$$-k_1(\alpha, t)p(t)y_1^\sigma(t)$$

$$+ (D^\alpha q(t)) y_2^\sigma(t) + q(t) D^\alpha y_2(t)$$

$$- k_1(\alpha, t) q(t) y_2^\sigma(t)$$

$$= \quad p(t) D^\alpha y_1(t) + q(t) D^\alpha y_2(t), \quad t \in \mathbb{T}^{\kappa^2},$$

provided p and q satisfy

$$(D^\alpha p(t)) y_1^\sigma(t) + (D^\alpha q(t)) y_2^\sigma(t) \quad = \quad k_1(\alpha, t) p(t) y_1^\sigma(t)$$

$$+ k_1(\alpha, t) q(t) y_2^\sigma(t), \quad t \in \mathbb{T}^{\kappa^2}. \tag{7.18}$$

Next,

$$(D^\alpha)^2 y(t) \quad = \quad D^\alpha \left(p(t) D^\alpha y_1(t) \right) + D^\alpha \left(q(t) D^\alpha y_2(t) \right)$$

$$= \quad p(t) (D^\alpha)^2 y_1(t) + (D^\alpha p(t)) (D^\alpha y_1)^\sigma(t)$$

$$- k_1(\alpha, t) p(t) (D^\alpha y_1)^\sigma(t)$$

$$+ q(t) (D^\alpha)^2 y_2(t) + (D^\alpha q(t)) (D^\alpha y_2)^\sigma(t)$$

$$- k_1(\alpha, t) q(t) (D^\alpha y_2)^\sigma(t), \quad t \in \mathbb{T}^{\kappa^2}.$$

Then

$$(D^\alpha)^2 y(t) + a(t) D^\alpha y(t) + b(t) y(t)$$

$$= \quad p(t) \left((D^\alpha)^2 y_1(t) + a(t) D^\alpha y_1(t) + b(t) y_1(t) \right)$$

$$+ q(t) \left((D^\alpha)^2 y_2(t) + a(t) D^\alpha y_2(t) +_b (t) y_2(t) \right)$$

$$+ (D^\alpha p(t)) (D^\alpha y_1)^\sigma(t) - k_1(\alpha, t) p(t) (D^\alpha y_1)^\sigma(t)$$

$$+ (D^\alpha q(t)) (D^\alpha y_2)^\sigma(t) - k_1(\alpha, t) q(t) (D^\alpha y_2)^\sigma(t)$$

$$= \quad (D^\alpha p(t)) (D^\alpha y_1)^\sigma(t) - k_1(\alpha, t) p(t) (D^\alpha y_1)^\sigma(t)$$

$$+ (D^\alpha q(t)) (D^\alpha y_2)^\sigma(t) - k_1(\alpha, t) q(t) (D^\alpha y_2)^\sigma(t)$$

$$= \quad f(t), \quad t \in \mathbb{T}^{\kappa^2}.$$

Thus, using (7.18), for p and q we get the system

$$\begin{cases} (D^\alpha p(t))y_1^\sigma(t) + (D^\alpha q(t))y_2^\sigma(t) & = & k_1(\alpha,t)p(t)y_1^\sigma(t) \\ & & +k_1(\alpha,t)q(t)y_2^\sigma(t) \\ (D^\alpha p(t))(D^\alpha y_1)^\sigma(t) + (D^\alpha q(t))(D^\alpha y_2)^\sigma(t) & = & f(t)+k_1(\alpha,t)p(t)(D^\alpha y_1)^\sigma(t) \\ & & +k_1(\alpha,t)q(t)(D^\alpha y_2)^\sigma(t), \quad t \in \mathbb{T}^{\kappa^2}. \end{cases}$$

From the last system, we get

$$\begin{cases} D^\alpha p(t) & = & k_1(\alpha,t)p(t) - f(t)\dfrac{y_2^\sigma(t)}{(W(y_1,y_2))^\sigma(t)} \\ D^\alpha q(t) & = & k_1(\alpha,t)q(t) + f(t)\dfrac{y_1^\sigma(t)}{(W(y_1,y_2))^\sigma(t)}, \quad t \in \mathbb{T}^{\kappa^2}. \end{cases}$$

We take

$$\begin{cases} p(t) & = & E_{k_1}(t,t_0) - \displaystyle\int_{t_0}^t f(s)\dfrac{y_2^\sigma(s)}{(W(y_1,y_2))^\sigma(s)}E_0(\sigma(s),t)\Delta_{\alpha,t}s \\ q(t) & = & E_{k_1}(t,t_0) + \displaystyle\int_{t_0}^t f(s)\dfrac{y_1^\sigma(s)}{(W(y_1,y_2))^\sigma(s)}E_0(\sigma(s),t)\Delta_{\alpha,t}s, \quad t,t_0 \in \mathbb{T}^{\kappa^2}. \end{cases}$$

Consequently,

$$y(t) = c_1 y_1(t) + c_2 y_2(t)$$

$$+ \left(E_{k_1}(t,t_0) - \int_{t_0}^t f(s)\dfrac{y_2^\sigma(s)}{(W(y_1,y_2))^\sigma(s)}E_0(\sigma(s),t)\Delta_{\alpha,t}s \right) y_1(t)$$

$$+ \left(E_{k_1}(t,t_0) + \int_{t_0}^t f(s)\dfrac{y_1^\sigma(s)}{(W(y_1,y_2))^\sigma(s)}E_0(\sigma(s),t)\Delta_{\alpha,t}s \right) y_2(t), \quad t,t_0 \in \mathbb{T}^{\kappa^2},$$

where c_1 and c_2 are constants, is the general solution of equation (7.17).

Example 7.6.1 *Let* $\mathbb{T} = 2^{\mathbb{N}_0}$,

$$k_1(\alpha,t) = (1-\alpha)t^{2\alpha}, \quad k_0(\alpha,t) = \alpha t^{2(1-\alpha)}, \quad \alpha \in (0,1], \quad t \in \mathbb{T}.$$

Consider the equation

$$2t^2\left(D^{\frac{1}{2}}\right)^2 y + \left(-3 - \frac{1}{2}t^2 + \frac{1}{2}t\right)tD^{\frac{1}{2}}y + 2y = 2t^3, \quad t \in \mathbb{T}. \tag{7.19}$$

Here

$$\sigma(t) = 2t,$$

$$k_1\left(\frac{1}{2},t\right) = \frac{1}{2}t,$$

$$k_0\left(\frac{1}{2},t\right) = \frac{1}{2}t, \quad t \in \mathbb{T}.$$

Note that the equation

$$2t^2\left(D^{\frac{1}{2}}\right)^2 y + \left(-3 - \frac{1}{2}t^2 + \frac{1}{2}t\right)tD^{\frac{1}{2}}y + 2y = 0, \quad t \in \mathbb{T}, \qquad (7.20)$$

is an Euler-Cauchy equation. Then

$$a = -3, \quad b = 2.$$

The functions

$$y_1(t) = E_{\frac{1}{t}}(t,t_0) \quad and \quad y_2(t) = E_{\frac{2}{t}}(t,t_0), \quad t,t_0 \in \mathbb{T},$$

form a fundamental set of solutions for equation (7.20). We have

$$W(y_1,y_2)(t) = \det\begin{pmatrix} E_{\frac{1}{t}}(t,t_0) & E_{\frac{2}{t}}(t,t_0) \\ \frac{1}{t}E_{\frac{1}{t}}(t,t_0) & \frac{2}{t}E_{\frac{2}{t}}(t,t_0) \end{pmatrix}$$

$$= \frac{2}{t}E_{\frac{1}{t}}(t,t_0)E_{\frac{2}{t}}(t,t_0) - \frac{1}{t}E_{\frac{1}{t}}(t,t_0)E_{\frac{2}{t}}(t,t_0)$$

$$= \frac{1}{t}E_{\frac{1}{t}}(t,t_0)E_{\frac{2}{t}}(t,t_0).$$

Note that equation (7.19) can be rewritten in the form

$$\left(D^{\frac{1}{2}}\right)^2 y + \frac{-6 - t^2 + t}{4t^2}D^{\frac{1}{2}}y + \frac{1}{t^2}y = t, \quad t \in \mathbb{T}.$$

Here

$$p(t) = E_{k_1}(t,t_0) - \int_{t_0}^{t}\frac{s}{E_{\frac{1}{t}}(2s,t_0)}E_0(2s,t)\Delta_{\alpha,t}s,$$

$$q(t) = E_{k_1}(t,t_0) + \int_{t_0}^{t}\frac{s}{E_{\frac{2}{t}}(2s,t_0)}E_0(2s,t)\Delta_{\alpha,t}s$$

Hence,

$$y(t) = (c_1 + p(t))E_{\frac{1}{t}}(t,t_0) + (c_2 + q(t))E_{\frac{2}{t}}(2t,t_0), \quad t,t_0 \in \mathbb{T},$$

where c_1 and c_2 are constants, is a general solution of equation (7.19). This ends the example.

Exercise 7.6.2 *Let* $\mathbb{T} = 3^{\mathbb{N}_0}$,

$$k_1(\alpha,t) = (1-\alpha)(1+t)^{\alpha}, \quad k_0(\alpha,t) = \alpha(1+t)^{1-\alpha}, \quad \alpha \in (0,1], \quad t \in \mathbb{T}.$$

Find the general solution of each of the following equations.

1.

$$3t^2 \left(D^{\frac{1}{2}}\right)^2 y + \left(6 - t(1+t)^{\frac{1}{2}} + \frac{1}{2}(1+t)^{\frac{1}{2}}\right) t D^{\frac{1}{2}} y + 9y = t^2 - 1,$$

2.

$$3t^2 \left(D^{\frac{1}{4}}\right)^2 y + \left(8 - \frac{3}{2}t(1+t)^{\frac{1}{4}} + \frac{1}{4}(1+t)^{\frac{3}{4}}\right) t D^{\frac{1}{4}} y + 16y = t^2 + 2t + 3,$$

3.

$$3t^2 \left(D^{\frac{1}{3}}\right)^2 y + \left(10 - \frac{4}{3}t(1+t)^{\frac{1}{3}} + \frac{1}{3}(1+t)^{\frac{2}{3}}\right) t D^{\frac{1}{3}} y + 25y = t^3 + 3t^2 + t + 1.$$

7.7 ADVANCED PRACTICAL PROBLEMS

Problem 7.7.1 *Find a general solution for each of the following equations.*

1. $(D^{\alpha})^2 y - 121y = 0$, $t \in \mathbb{T}^{\kappa^2}$,

2. $(D^{\alpha})^2 y + 9y$, $t \in \mathbb{T}^{\kappa^2}$,

3. $(D^{\alpha})^2 y + 3D^{\alpha} y - 2y = 0$, $t \in \mathbb{T}^{\kappa^2}$,

4. $(D^{\alpha})^2 y + D^{\alpha} y + 5y$, $t \in \mathbb{T}^{\kappa^2}$.

Problem 7.7.2 *Let* $\mathbb{T} = \mathbb{N}_0^2$,

$$k_1(\alpha,t) = (1-\alpha)t^{3\alpha}, \quad k_0(\alpha,t) = \alpha t^{3(1-\alpha)}, \quad \alpha \in (0,1], \quad t \in \mathbb{T}.$$

Find a fundamental set of solutions for each of the following equations.

1. $\left(D^{\frac{1}{3}}\right)^2 y - 6D^{\frac{1}{2}} y + 9y = 0$, $t \in \mathbb{T}^{\kappa^2}$,

2. $\left(D^{\frac{1}{4}}\right)^2 y - 10D^{\frac{1}{4}} y + 25y = 0$, $t \in \mathbb{T}^{\kappa^2}$,

3. $\left(D^{\frac{1}{8}}\right)^2 - 2D^{\frac{1}{8}} y + y = 0$, $t \in \mathbb{T}^{\kappa^2}$,

4. $\left(D^{\frac{1}{3}}\right)^2 y - 8D^{\frac{1}{3}} y + 16y = 0$, $t \in \mathbb{T}^{\kappa^2}$,

5. $\left(D^{\frac{1}{2}}\right)^2 y - \frac{2}{3}D^{\frac{1}{2}} y + \frac{1}{9}y = 0$, $t \in \mathbb{T}^{\kappa^2}$.

Problem 7.7.3 *Let* $\mathbb{T} = 2\mathbb{Z}$,

$$k_1(\alpha,t) = (1-\alpha)t^{5\alpha}, \quad k_0(\alpha,t) = \alpha t^{5(1-\alpha)}, \quad \alpha \in (0,1], \quad t \in \mathbb{T}.$$

Prove that the equation

$$\left(D^{\frac{1}{5}}\right)^2 - (2t+2)D^{\frac{1}{5}}y + \left(-\frac{1}{5}t^4 + t^2 + \frac{8}{5}t\right)y = 0, \quad t \in \mathbb{T}^{\kappa^2},$$

can be written in the factored form (7.11).

Problem 7.7.4 *Let* $\mathbb{T} = \mathbb{Z}$,

$$k_1(\alpha,t) = 1-\alpha, \quad k_0(\alpha,t) = \alpha, \quad \alpha \in (0,1], \quad t \in \mathbb{T}.$$

Find a general solution of the equation

$$\left(D^{\frac{1}{5}}\right)^2 y - \left(5^{\textcircled{2}}(t) + 4\right)y^\sigma - 4y = 0, \quad t \in \mathbb{T}^{\kappa^2}.$$

Problem 7.7.5 *Let* $\mathbb{T} = 3^{\mathbb{N}_0}$,

$$k_1(\alpha,t) = (1-\alpha)t^\alpha, \quad k_0(\alpha,t) = \alpha t^{1-\alpha}, \quad \alpha \in (0,1], \quad t \in \mathbb{T}.$$

Find a fundamental set of solutions for each of the following equations.

1.

$$3t^2 \left(D^{\frac{1}{2}}\right)^2 y + \left(9 - t^{\frac{3}{2}} + \frac{1}{2}t^{\frac{1}{2}}\right) tD^{\frac{1}{2}}y + 8y = 0, \quad t \in \mathbb{T},$$

2.

$$3t^2 \left(D^{\frac{1}{3}}\right)^2 y + \left(12 - \frac{4}{3}t^{\frac{4}{3}} + \frac{1}{3}t^{\frac{2}{3}}\right) tD^{\frac{1}{3}}y + 11y = 0, \quad t \in \mathbb{T},$$

3.

$$3t^2 \left(D^{\frac{1}{4}}\right)^2 y + \left(15 - \frac{3}{2}t^{\frac{5}{4}} + \frac{1}{4}t^{\frac{3}{4}}\right) tD^{\frac{1}{4}}y + 14y = 0, \quad t \in \mathbb{T}.$$

Problem 7.7.6 *Let* $\mathbb{T} = 4^{\mathbb{N}_0}$,

$$k_1(\alpha,t) = (1-\alpha)(1+t)^\alpha, \quad k_0(\alpha,t) = \alpha(1+t)^{1-\alpha}, \quad \alpha \in (0,1], \quad t \in \mathbb{T}.$$ '

Find a general solution for each of the following equations.

1.

$$4t^2 \left(D^{\frac{1}{2}}\right)^2 y + \left(6 - \frac{3}{2}t(1+t)^{\frac{1}{2}} + \frac{1}{2}(1+t)^{\frac{1}{2}}\right) tD^{\frac{1}{2}}y + 9y = 0,$$

2.

$$4t^2 \left(D^{\frac{1}{4}}\right)^2 y + \left(8 - \frac{9}{4}t(1+t)^{\frac{1}{4}} + \frac{1}{4}(1+t)^{\frac{3}{4}}\right) tD^{\frac{1}{4}}y + 16y = 0,$$

3.

$$4t^2 \left(D^{\frac{1}{3}}\right)^2 y + \left(10 - 2t(1+t)^{\frac{1}{3}} + \frac{1}{3}(1+t)^{\frac{2}{3}}\right) tD^{\frac{1}{3}} y + 25y = 0.$$

Problem 7.7.7 *Let* $\mathbb{T} = 4^{\mathbb{N}_0}$,

$$k_1(\alpha,t) = (1-\alpha)(1+t)^\alpha, \quad k_0(\alpha,t) = \alpha(1+t)^{1-\alpha}, \quad \alpha \in (0,1], \quad t \in \mathbb{T}.$$

Find a general solution for each of the following equations.

1.

$$4t^2 \left(D^{\frac{1}{2}}\right)^2 y + \left(-6 - \frac{3}{2}t(1+t)^{\frac{1}{2}} + \frac{1}{2}(1+t)^{\frac{1}{2}}\right) tD^{\frac{1}{2}} y + 9y = t + 10,$$

2.

$$4t^2 \left(D^{\frac{1}{4}}\right)^2 y + \left(-8 - \frac{9}{4}t(1+t)^{\frac{1}{4}} + \frac{1}{4}(1+t)^{\frac{3}{4}}\right) tD^{\frac{1}{4}} y + 16y = \frac{t^2+1}{t^4+1},$$

3.

$$4t^2 \left(D^{\frac{1}{3}}\right)^2 y + \left(-10 - 2t(1+t)^{\frac{1}{3}} + \frac{1}{3}(1+t)^{\frac{2}{3}}\right) tD^{\frac{1}{3}} y + 25y = t^4 + t^3 + t^2 + t + 1.$$

CHAPTER 8

Second-Order Self-Adjoint Conformable Dynamic Equations

8.1 SELF-ADJOINT DYNAMIC EQUATIONS

Let p be continuous and q be ld-continuous functions on some time-scale interval $\mathscr{I} \subseteq \mathbb{T}_\kappa^\kappa$ with $p(t) > 0$ for all $t \in \mathscr{I}$. In this section we are concerned with the conformable second-order (formally) self-adjoint homogeneous dynamic equation $M_c x = 0$, or for a continuous function h, the associated non-homogeneous equation

$$M_c x = h, \quad M_c x(t) := \widehat{D}_\alpha \left[p D^\alpha x \right](t) + q(t)x(t), \tag{8.1}$$

for $t \in \mathscr{I}$, $\alpha \in (0,1]$. We interpret D^α using (1.1) and $\widehat{D}_\alpha$ using (1.9) for k_0 and k_1 satisfying (A1). Note that one could also study the equation

$$L_c = h, \quad L_c x(t) := D^\alpha \left[p\widehat{D}_\alpha x \right](t) + q(t)x(t)$$

for $t \in \mathscr{I}$, $\alpha \in (0,1]$, though here we will concentrate on (1.1).

Definition 8.1.1 *Let $\mathbb{D}$ denote the set of all functions $x : \mathbb{T} \to \mathbb{R}$ defined on $\mathscr{I}$ such that x^Δ is continuous and $\widehat{D}_\alpha \left[p D^\alpha x \right]$ is ld-continuous on $\mathscr{I}$. Then x is a solution of the homogeneous equation $M_c x = 0$ for M_c in (8.1) on $\mathscr{I}$ provided $x \in \mathbb{D}$ and*

$$M_c x(t) = 0 \quad for \ all \quad t \in \mathscr{I}.$$

We next state a theorem concerning the existence-uniqueness of solutions of initial value problems for the non-homogeneous self-adjoint equation $M_c x = h$.

Theorem 8.1.2 *Assume k_0, k_1 satisfy (A1). Let $\alpha \in (0,1]$, $t_0 \in \mathscr{I}$, and let D^α be as in (1.1) and $\widehat{D}_\alpha$ be as in (1.9). Assume p is continuous and q, h are ld-continuous on $\mathscr{I}$ with $p(t) \neq 0$, and suppose $x_0, x_1 \in \mathbb{R}$ are given constants. Then the initial value problem*

$$M_c x = h(t), \quad x(t_0) = x_0, \quad D^\alpha x(t_0) = x_1$$

has a unique solution that exists on all of $\mathscr{I}$.

Proof 8.1.3 *The idea of the proof is to use the induction principle for time scales. This is a straightforward modification of the proof in [4, Theorem 3.1].*

Definition 8.1.4 (Wronskian) *Let $k_0, k_1 : [0,1] \times \mathscr{I} \to [0,\infty)$ be continuous and satisfy (A1). If $x,y : \mathscr{I} \to \mathbb{R}$ are differentiable on $\mathscr{I}$, then the conformable Wronskian of x and y is given by*

$$W(x,y)(t) = \det \begin{pmatrix} x(t) & y(t) \\ D^\alpha x(t) & D^\alpha y(t) \end{pmatrix} \quad \text{for} \quad t \in \mathscr{I}, \tag{8.2}$$

for D^α given in (1.1).

Theorem 8.1.5 (Conformable Lagrange Identity) *Let M_c be given as in (8.1) and $\alpha \in (0,1]$. Assume*

$$k_0^\rho(\alpha,t) - v(t)k_1^\rho(\alpha,t) \neq 0, \quad t \in \mathscr{I}.$$

If $x,y \in \mathbb{D}$, then

$$x(M_c y) - y(M_c x) = \left(\frac{k_0^\rho - v k_1^\rho}{k_0^\rho} \right) \widehat{D}_\alpha[pW(x,y)] + \frac{k_1^\rho(k_0 + v k_1)}{k_0^\rho} pW(x,y), \quad t \in \mathbb{T}_\kappa^\kappa,$$

for M_c given in (8.1). Equivalently, for $t,b \in \mathbb{T}_\kappa^\kappa$ we have

$$E_0(t,b)\widehat{D}_\alpha \left[\frac{pW(x,y)(t)}{E_0(t,b)} \right] = x(M_c y) - y(M_c x). \tag{8.3}$$

Proof 8.1.6 *Let $x,y \in \mathbb{D}$. Then $x,y \in \mathbb{D}$ implies that $D^\alpha x$ and in particular x^Δ is continuous, so that $x^\nabla = x^{\Delta\rho}$; likewise $y^\nabla = y^{\Delta\rho}$. Using the product rule from Lemma 1.9.9 we have (suppressing all arguments)*

$$
\begin{aligned}
\widehat{D}_\alpha[pW(x,y)] &= \widehat{D}_\alpha[xpD^\alpha y - ypD^\alpha x] \\
&= x\widehat{D}_\alpha[pD^\alpha y] + \left(\widehat{D}_\alpha x\right)[pD^\alpha y]^\rho - k_1 x[pD^\alpha y]^\rho \\
&\quad - y\widehat{D}_\alpha[pD^\alpha x] - \left(\widehat{D}_\alpha y\right)[pD^\alpha x]^\rho + k_1 y[pD^\alpha x]^\rho \\
&= x(M_c y) - y(M_c x) + [pD^\alpha y]^\rho\left(k_0 x^\nabla\right) - [pD^\alpha x]^\rho\left(k_0 y^\nabla\right) \\
&= x(M_c y) - y(M_c x) + k_0\left[x^\Delta pD^\alpha y - y^\Delta pD^\alpha x\right]^\rho \\
&= x(M_c y) - y(M_c x) + k_0\left[x^\Delta pk_1 y - y^\Delta pk_1 x\right]^\rho \\
&= x(M_c y) - y(M_c x) - \left(\frac{k_1}{k_0}\right)^\rho k_0[pW(x,y)]^\rho
\end{aligned}
$$

on $\mathbb{T}_\kappa^\kappa$. Note that for any nabla differentiable f,

$$k_0 f^\rho = (k_0 + v k_1)f - v\widehat{D}_\alpha f.$$

It follows that

$$\widehat{D}_\alpha[pW] = x(M_c y) - y(M_c x) - \left(\frac{k_1}{k_0}\right)^\rho \left[(k_0 + v k_1)pW - v\widehat{D}_\alpha(pW)\right].$$

Putting the $\widehat{D}_\alpha[pW]$ terms on the left-hand side, we have

$$\left(1 - v\left(\frac{k_1}{k_0}\right)^\rho\right)\widehat{D}_\alpha[pW] = x(M_c y) - y(M_c x) - \left(\frac{k_1}{k_0}\right)^\rho (k_0 + vk_1)pW,$$

which is the first equation in the theorem; this can be rewritten as

$$\widehat{D}_\alpha[pW] = \left(\frac{k_0^\rho}{k_0^\rho - vk_1^\rho}\right)[x(M_c y) - y(M_c x)] - \frac{k_1^\rho(k_0 + vk_1)}{k_0^\rho - vk_1^\rho}pW. \tag{8.4}$$

Now let

$$\zeta(t) := k_1(\alpha,t) - \frac{k_0(\alpha,t)k_1^\rho(\alpha,t)}{k_0^\rho(\alpha,t) - v(t)k_1^\rho(\alpha,t)}. \tag{8.5}$$

Observe that $E_0 = \widehat{E}_\zeta$ by the equivalence of exponentials, and

$$\frac{\widehat{E}_\zeta}{\widehat{E}_\zeta^\rho} = \frac{k_0^\rho - vk_1^\rho}{k_0^\rho}.$$

Additionally, by the quotient rule we see that

$$\widehat{E}_\zeta(t,b)\widehat{D}_\alpha\left[\frac{pW(x,y)(t)}{\widehat{E}_\zeta(t,b)}\right] = \frac{\widehat{E}_\zeta}{\widehat{E}_\zeta^\rho}\left[\widehat{D}_\alpha(pW) - \zeta pW\right] + k_1 pW. \tag{8.6}$$

Substitution of (8.4) into (8.6) yields

$$\begin{aligned}
\widehat{E}_\zeta(t,b)\widehat{D}_\alpha\left[\frac{pW(x,y)(t)}{\widehat{E}_\zeta(t,b)}\right] &= \frac{\widehat{E}_\zeta}{\widehat{E}_\zeta^\rho}\left[\left(\frac{k_0^\rho}{k_0^\rho - vk_1^\rho}\right)[x(M_c y) - y(M_c x)] - \frac{k_1^\rho(k_0 + vk_1)}{k_0^\rho - vk_1^\rho}pW\right.\\
&\qquad\left. -\zeta pW\right] + k_1 pW\\
&= x(M_c y) - y(M_c x),
\end{aligned}$$

and thus (8.3) holds as well after recalling $\widehat{E}_\zeta = E_0$.

Remark 8.1.7 *Consider ζ defined in (8.5). If $\mathbb{T} = \mathbb{R}$, then $v \equiv 0$ and $\rho(t) = t$ for all $t \in \mathbb{T}$, hence $\zeta(t) = 0$ for all $t \in \mathscr{I}$. Similarly, for any time scale $\mathbb{T}$, if $\alpha = 1$, then $k_1 \equiv 0$ and $k_0 \equiv 1$, thus $\zeta(t) = 0$ for all $t \in \mathscr{I}$.*

Definition 8.1.8 (Inner Product) *Let $\alpha \in (0,1]$, let $\widehat{E}_0$ be as given in (1.10), and let ζ be as in (8.5), with*

$$k_0^\rho(\alpha,t) - v(t)k_1^\rho(\alpha,t) \neq 0, \qquad t \in \mathscr{I}.$$

Define the (weighted) inner product of rd-continuous functions $f,g \in C(\mathscr{I})$ on $[a,b]_\mathbb{T} \subseteq \mathscr{I}$ to be

$$\langle f,g\rangle = \int_a^b \frac{f(t)g(t)\widehat{E}_0(b,\rho(t))}{\widehat{E}_\zeta(t,b)k_0(\alpha,t)}\nabla t = \int_a^b \frac{f(t)g(t)}{\widehat{E}_\zeta(t,b)}\nabla_{\alpha,b}t, \quad \nabla_{\alpha,b}t := \frac{\widehat{E}_0(b,\rho(t))\nabla t}{k_0(\alpha,t)}. \tag{8.7}$$

As $\widehat{E}_\zeta = E_0$ by the equivalence of exponential functions, this inner product can also be written as

$$\langle f,g \rangle = \int_a^b \frac{f(t)g(t)\widehat{E}_0(b,\rho(t))}{E_0(t,b)k_0(\alpha,t)}\nabla t = \int_a^b \frac{f(t)g(t)}{E_0(t,b)}\nabla_{\alpha,b}t, \quad \nabla_{\alpha,b}t := \frac{\widehat{E}_0(b,\rho(t))\nabla t}{k_0(\alpha,t)}.$$

Corollary 8.1.9 (Green's Formula; Self-Adjoint Operator) *Let M_c be given as in (8.1) and $\alpha \in (0,1]$. Assume*

$$k_0^\rho(\alpha,t) - v(t)k_1^\rho(\alpha,t) \neq 0, \qquad t \in \mathscr{I}.$$

If $x,y \in \mathbb{D}$, then Green's formula

$$\langle x, M_c y \rangle - \langle M_c x, y \rangle = p(b)W(x,y)(b) - \frac{p(a)W(x,y)(a)}{E_0(a,b)\widehat{E}_0(a,b)} \tag{8.8}$$

holds. Moreover, the operator M_c is formally self-adjoint with respect to the inner product (8.7); that is, the identity

$$\langle x, M_c y \rangle = \langle M_c x, y \rangle$$

holds if and only if $x,y \in \mathbb{D}$ and x,y satisfy the self-adjoint boundary conditions

$$p(b)W(x,y)(b) = \frac{p(a)W(x,y)(a)}{E_0(a,b)\widehat{E}_0(a,b)}, \tag{8.9}$$

where we have used the conformable Wronskian matrix on time scales from Definition 8.1.4.

Proof 8.1.10 *From Theorem 8.1.5 we have the Lagrange identity (8.3) given by*

$$\widehat{E}_\zeta(t,b)\widehat{D}_\alpha\left[\frac{pW(x,y)(t)}{\widehat{E}_\zeta(t,b)}\right] = x(M_c y) - y(M_c x),$$

with ζ defined in (8.5). If we multiply both sides of this equation by

$$\frac{\widehat{E}_0(b,\rho(t))}{\widehat{E}_\zeta(t,b)k_0(\alpha,t)}$$

and integrate from a to b we obtain

$$\int_a^b \widehat{D}_\alpha\left[\frac{pW(x,y)(t)}{\widehat{E}_\zeta(t,b)}\right]\nabla_{\alpha,b}t = \langle x, M_c y \rangle - \langle M_c x, y \rangle.$$

By Theorem 1.9.7 we have

$$\int_a^b \widehat{D}_\alpha\left[\frac{pW(x,y)(t)}{\widehat{E}_\zeta(t,b)}\right]\nabla_{\alpha,b}t = \frac{pW(x,y)(b)}{\widehat{E}_\zeta(b,b)} - \frac{pW(x,y)(a)}{\widehat{E}_\zeta(a,b)}\widehat{E}_0(b,a),$$

so that Green's formula (8.8) holds. Thus $\langle x, M_c y \rangle = \langle M_c x, y \rangle$ if and only if $x,y \in \mathbb{D}$ satisfy the self-adjoint boundary conditions (8.9). This completes the proof. □

Corollary 8.1.11 (Abel's Formula) *Let M_c be given as in (8.1) and $\alpha \in (0,1]$. Assume*

$$k_0^\rho(\alpha,t) - v(t)k_1^\rho(\alpha,t) \neq 0, \qquad t \in \mathscr{I}.$$

If $x,y \in \mathbb{D}$ are solutions of $M_cx = 0$ on $\mathscr{I}$, then for fixed $a \in \mathscr{I}$ the Wronskian satisfies

$$W(x,y)(t) = \frac{p(b)W(x,y)(b)}{p(t)E_0(b,t)\widehat{E}_0(b,t)} \overset{(8.9)}{=} \frac{p(a)W(x,y)(a)}{p(t)E_0(a,t)\widehat{E}_0(a,t)}$$

for $t \in \mathscr{I}$.

Proof 8.1.12 *As in (8.3) and in the proof of Corollary 8.1.9, for $x,y \in \mathbb{D}$ we have*

$$E_0(t,b)\widehat{D}_\alpha \left[\frac{pW(x,y)(t)}{E_0(t,b)} \right] = x(M_cy) - y(M_cx).$$

If x,y are solutions of (8.1) on $\mathscr{I}$, then $M_cx = 0 = M_cy$ and

$$E_0(t,b)\widehat{D}_\alpha \left[\frac{pW(x,y)(t)}{E_0(t,b)} \right] = 0.$$

As a result,

$$\widehat{D}_\alpha \left[\frac{pW(x,y)(t)}{E_0(t,b)} \right] = 0,$$

so that

$$\frac{p(t)W(x,y)(t)}{E_0(t,b)} = c\widehat{E}_0(t,b),$$

where c is the constant $c = p(b)W(x,y)(b)$. Consequently,

$$W(x,y)(t) = \frac{p(b)W(x,y)(b)}{p(t)E_0(b,t)\widehat{E}_0(b,t)} \overset{(8.9)}{=} \frac{p(a)W(x,y)(a)}{p(t)E_0(a,t)\widehat{E}_0(a,t)},$$

completing the proof. □

Corollary 8.1.13 *Let M_c be given as in (8.1) and $\alpha \in (0,1]$. Assume*

$$k_0^\rho(\alpha,t) - v(t)k_1^\rho(\alpha,t) \neq 0, \qquad t \in \mathscr{I}.$$

If $x,y \in \mathbb{D}$ are solutions of (8.1) on $\mathscr{I}$, then either $W(x,y)(t) = 0$ for all $t \in \mathscr{I}$, or $W(x,y)(t) \neq 0$ for all $t \in \mathscr{I}$.

Theorem 8.1.14 *Equation (8.1) on $\mathscr{I}$ has two linearly independent solutions, and every solution of (8.1) on $\mathscr{I}$ is a linear combination of these two solutions.*

Theorem 8.1.15 (Converse of Abel's Formula) *Let M_c be given as in (8.1) and $\alpha \in (0,1]$. Assume*

$$k_0^\rho(\alpha,t) - v(t)k_1^\rho(\alpha,t) \neq 0, \qquad t \in \mathscr{I}.$$

Let x be a solution of $M_cx = 0$ on $\mathscr{I}$ such that $x \neq 0$ on $\mathscr{I}$. If $y \in \mathbb{D}$ satisfies

$$W(x,y)(t) = \frac{c}{p(t)E_0(b,t)\widehat{E}_0(b,t)}$$

for some constant $c \in \mathbb{R}$, then y is also a solution of $M_cx = 0$.

Proof 8.1.16 *Suppose that x is a solution of $M_c x = 0$ such that $x \neq 0$ on $\mathscr{I}$, and assume $y \in \mathbb{D}$ satisfies $p(t)W(x,y)(t) = cE_0(t,b)\widehat{E}_0(t,b)$ for some constant $c \in \mathbb{R}$. By the Lagrange identity (Theorem 8.1.5) and Lagrange's formula (8.3) we have for $t \in \mathscr{I}$ that*

$$x(M_c y)(t) - y(M_c x)(t) = E_0(t,b)\widehat{D}_\alpha \left[\frac{p(t)W(x,y)(t)}{E_0(t,b)} \right] = E_0(t,b)\widehat{D}_\alpha \left[c\widehat{E}_0(t,b) \right] \equiv 0;$$

since $M_c x = 0$, this yields $x(M_c y)(t) \equiv 0$ for $t \in \mathscr{I}$. As $x \neq 0$ on $\mathscr{I}$, $(M_c y)(t) \equiv 0$ on $\mathscr{I}$. Thus, y is also a solution of $M_c x = 0$. This completes the proof. □

Now, consider a few other second-order conformable dynamic equations, such as

$$M_1 x = 0, \quad M_1 x := \widehat{D}_\alpha [D^\alpha x] + p_1 [D^\alpha x]^\rho + p_2 x, \tag{8.10}$$

$$M_2 x = 0, \quad M_2 x := \widehat{D}_\alpha [D^\alpha x] + a_1 D^\alpha x + a_2 x, \tag{8.11}$$

where $p_j, a_j : \mathbb{T} \to \mathbb{R}$ are ld-continuous for $j = 1, 2$. Let $\mathbb{D}_M$ denote the set of all functions $x : \mathbb{T} \to \mathbb{R}$ such that x is D^α-differentiable on $\mathbb{T}^\kappa$, $(D^\alpha x)$ is $\widehat{D}_\alpha$-differentiable on $\mathbb{T}^\kappa_\kappa$, and $\widehat{D}_\alpha [D^\alpha x]$ is ld-continuous on $\mathbb{T}^\kappa_\kappa$. We say x is a solution of $M_j x = 0$ on $\mathbb{T}$ provided $x \in \mathbb{D}_M$ and $M_j x = 0$ on $\mathbb{T}^\kappa_\kappa$ for $j = 1, 2$.

Theorem 8.1.17 *Let $\alpha \in (0,1]$. If p_2 is ld-continuous and $(p_1 + k_1) \in \widehat{\mathscr{R}^+_c}$, then the dynamic equation (8.10) can be written in self-adjoint form (8.1), with*

$$p(t) = \widehat{E}_{p_1 + k_1}(t,t_0) \quad and \quad q(t) = p_2(t)\widehat{E}_{p_1 + k_1}(t,t_0).$$

Moreover, in this case, x is a solution of (8.10) if and only if x is a solution of $M_c x = 0$ for the self-adjoint M_c operator in (8.1).

Proof 8.1.18 *If $p = \widehat{E}_{p_1 + k_1}$ and $q = p_2 \widehat{E}_{p_1 + k_1}$, then (suppressing the arguments)*

$$
\begin{aligned}
M_c x &= \widehat{D}_\alpha [p D^\alpha x] + qx = \widehat{D}_\alpha \left[\widehat{E}_{p_1 + k_1} D^\alpha x \right] + p_2 \widehat{E}_{p_1 + k_1} x \\
&= \widehat{E}_{p_1 + k_1} \widehat{D}_\alpha [D^\alpha x] + (p_1 + k_1)\widehat{E}_{p_1 + k_1} [D^\alpha x]^\rho \\
&\quad - k_1 \widehat{E}_{p_1 + k_1} [D^\alpha x]^\rho + p_2 \widehat{E}_{p_1 + k_1} x \\
&= \widehat{E}_{p_1 + k_1} \left\{ \widehat{D}_\alpha [D^\alpha x] + (p_1 + k_1) [D^\alpha x]^\rho - k_1 [D^\alpha x]^\rho + p_2 x \right\} \\
&= \widehat{E}_{p_1 + k_1} M_1 x.
\end{aligned}
$$

As $(p_1 + k_1) \in \widehat{\mathscr{R}^+_c}$, we know that $\widehat{E}_{p_1 + k_1} > 0$, and thus $M_c x = 0$ if and only if $M_1 x = 0$. This completes the proof. □

Theorem 8.1.19 *Let $\alpha \in (0,1]$. If a_2 is ld-continuous and $(-a_1) \in \widehat{\mathscr{R}^+_c}$, then the dynamic equation (8.11) can be written in self-adjoint form (8.1), with*

$$p(t) = \widehat{E}_\gamma(t,t_0) \quad and \quad q(t) = a_2(t)\widehat{E}^\rho_\gamma(t,t_0),$$

where

$$\gamma := \frac{(k_0 + \nu k_1)(a_1 + k_1)}{k_0 + \nu(a_1 + k_1)}.$$

Moreover, in this case, x is a solution of (8.11) if and only if x is a solution of $M_c x = 0$ for the self-adjoint M_c operator in (8.1).

Proof 8.1.20 *Since* $(-a_1) \in \widehat{\mathscr{R}_c^+}$, *we have*

$$k_0 + v(a_1 + k_1) = k_0 - v(-a_1 - k_1) > 0,$$

so that γ *is a well-defined function. Also,*

$$1 - v\left(\frac{\gamma - k_1}{k_0}\right) = \frac{k_0 + vk_1}{k_0 + v(a_1 + k_1)} > 0,$$

putting $\gamma \in \widehat{\mathscr{R}_c^+}$, *making* $\widehat{E}_\gamma$ *and thus p and q well defined as well. If* $p = \widehat{E}_\gamma$ *and* $q = a_2\widehat{E}_\gamma^\rho$, *then*

$$
\begin{aligned}
M_c x &= \widehat{D}_\alpha\left[pD^\alpha x\right] + qx = \widehat{D}_\alpha\left[\widehat{E}_\gamma D^\alpha x\right] + a_2\widehat{E}_\gamma^\rho x \\
&= \widehat{E}_\gamma^\rho \widehat{D}_\alpha\left[D^\alpha x\right] + \gamma\widehat{E}_\gamma D^\alpha x - k_1\widehat{E}_\gamma^\rho D^\alpha x + a_2\widehat{E}_\gamma^\rho x \\
&= \widehat{E}_\gamma^\rho\left\{\widehat{D}_\alpha\left[D^\alpha x\right] + \frac{\gamma}{1 - v\left(\frac{\gamma - k_1}{k_0}\right)}D^\alpha x - k_1 D^\alpha x + a_2 x\right\} \\
&= \widehat{E}_\gamma^\rho M_2 x.
\end{aligned}
$$

As $\gamma \in \widehat{\mathscr{R}_c^+}$ *we know that* $\widehat{E}_\gamma^\rho > 0$, *and thus* $M_c x = 0$ *if and only if* $M_2 x = 0$. *This completes the proof.* $\square$

Remark 8.1.21 *In the next theorem, we will assume for constants* $a, b \in \mathbb{R}$ *that*

$$(k_0^\rho(\alpha,t) - v(t)k_1^\rho(\alpha,t) - av(t))(k_0^\rho(\alpha,t) - v(t)k_1^\rho(\alpha,t)) + bv^2(t) \neq 0, \qquad t \in \mathbb{T}_\kappa.$$

If $\alpha = 1$, *then* $k_0 \equiv 1$ *and* $k_1 \equiv 0$, *and this assumption becomes*

$$1 - av(t) + bv^2(t) \neq 0, \qquad t \in \mathbb{T}_\kappa,$$

which is the standard assumption for second-order nabla equations on time scales. If $\mathbb{T} = \mathbb{R}$, *the assumption is merely* $k_0(\alpha,t) \neq 0$, *which is always true for* $\alpha \in (0,1]$.

Remark 8.1.22 *Consider the assumption*

$$k_0^\rho(\alpha,t) - v(t)k_1^\rho(\alpha,t) \neq 0, \qquad t \in \mathbb{T}_\kappa.$$

If t is a left-scattered or a left-dense-right-dense point, then this is clearly equivalent to the regressivity assumption

$$k_0(\alpha,t) - \mu(t)k_1(\alpha,t) \neq 0, \qquad t \in \mathbb{T}^\kappa.$$

Suppose t is a left-dense right-scattered point. Then $\sigma(t)$ *is a left-scattered point, and*

$$k_0^\rho(\alpha,\sigma(t)) - v(\sigma(t))k_1^\rho(\alpha,\sigma(t)) \neq 0$$

implies

$$k_0(\alpha,t) - \mu(t)k_1(\alpha,t) \neq 0.$$

Theorem 8.1.23 *Let $a, b \in \mathbb{R}$ be given constants such that $\lambda_j \in \mathbb{R}$ solves $\lambda^2 + a\lambda + b = 0$ for $j = 1, 2$, with $\lambda_1 \neq \lambda_2$. Assume*

$$k_0^\rho(\alpha, t) - v(t) k_1^\rho(\alpha, t) \neq 0, \qquad t \in \mathbb{T}_\kappa$$

and

$$(k_0^\rho(\alpha, t) - v(t) k_1^\rho(\alpha, t) - av(t))(k_0^\rho(\alpha, t) - v(t) k_1^\rho(\alpha, t)) + bv^2(t) \neq 0, \qquad t \in \mathbb{T}_\kappa.$$

Let

$$p(t) := \widehat{E}_\beta(t, t_0), \qquad q(t) := \frac{b(k_0(\alpha, t) + v(t) k_1(\alpha, t))}{k_0^\rho(\alpha, t) - v(t) k_1^\rho(\alpha, t)} \widehat{E}_\beta(t, t_0),$$

where

$$\beta := \frac{(k_0 + vk_1) \left[(a + k_1^\rho)(k_0^\rho - vk_1^\rho) - bv \right]}{(k_0^\rho - vk_1^\rho) k_0^\rho}.$$

Then the general solution of $M_c x = 0$ for M_c in (8.1) with p, q as given here, is

$$x(t) = c_1 E_{\lambda_1} + c_2 E_{\lambda_2}.$$

Proof 8.1.24 *Let $a, b \in \mathbb{R}$ be given constants such that $\lambda_j \in \mathbb{R}$ solves $\lambda^2 + a\lambda + b = 0$ for $j = 1, 2$, with $\lambda_1 \neq \lambda_2$. Then $a = -(\lambda_1 + \lambda_2)$ and $b = \lambda_1 \lambda_2$. The regressivity conditions we need to check are*

$$k_0 - \mu k_1 \neq 0, \qquad k_0 + \mu(\lambda - k_1) \neq 0.$$

By Remark 8.1.22, the assumption

$$k_0^\rho - vk_1^\rho \neq 0 \quad \text{implies} \quad k_0 - \mu k_1 \neq 0$$

on $\mathbb{T}^\kappa$. Moreover,

$$\left[k_0^\rho + v(\lambda_1 - k_1^\rho) \right] \left[k_0^\rho + v(\lambda_2 - k_1^\rho) \right] = \left(k_0^\rho - vk_1^\rho - av \right) \left(k_0^\rho - vk_1^\rho \right) + bv^2 \neq 0$$

on $\mathbb{T}_\kappa$ by assumption. It follows that

$$\left[k_0^\rho + v(\lambda_1 - k_1^\rho) \right] \neq 0, \qquad \left[k_0^\rho + v(\lambda_2 - k_1^\rho) \right] \neq 0,$$

and thus

$$\left[k_0 + \mu(\lambda_1 - k_1) \right] \neq 0, \qquad \left[k_0 + \mu(\lambda_2 - k_1) \right] \neq 0,$$

putting $\lambda_1, \lambda_2 \in \mathcal{R}_c$. Hence, E_{λ_1} and E_{λ_2} are well defined. We will show that $x = E_\lambda$ is a solution of $M_c x = 0$ for p and q as in the statement of the theorem, where λ represents, without loss of generality, either λ_1 or λ_2. To this end, note that $D^\alpha x = \lambda E_\lambda$, so that

$$\widehat{D}_\alpha[pD^\alpha x] = \lambda \widehat{D}_\alpha \left[\widehat{E}_\beta E_\lambda \right]$$

$$= \lambda \widehat{D}_\alpha \left[\widehat{E}_\beta \widehat{E}_{k_1 + \frac{k_0(\lambda - k_1)^\rho}{k_0^\rho + v(\lambda - k_1)^\rho}} \right]$$

$$= \lambda \left(\beta \widehat{\oplus}_c \left(k_1 + \frac{k_0(\lambda - k_1)^\rho}{k_0^\rho + v(\lambda - k_1)^\rho} \right) \right) \widehat{E}_\beta E_\lambda.$$

Consequently,

$$\widehat{D}_\alpha[pD^\alpha x] + qx = \lambda \left(\beta \widehat{\oplus}_c \left(k_1 + \frac{k_0(\lambda - k_1)^\rho}{k_0^\rho + v(\lambda - k_1)^\rho} \right) \right) \widehat{E}_\beta E_\lambda$$

$$+ \left(\frac{b(k_0 + vk_1)}{k_0^\rho - vk_1^\rho} \right) \widehat{E}_\beta E_\lambda$$

$$= \left(\frac{(k_0 + vk_1)(\lambda^2 + a\lambda + b)}{k_0^\rho + v(\lambda - k_1)^\rho} \right) \widehat{E}_\beta E_\lambda$$

$$= 0$$

as $\lambda^2 + a\lambda + b = 0$. Hence, $x_j = E_{\lambda_j}$ is a solution for $j = 1, 2$. Since $\lambda_1 \neq \lambda_2$, the Wronskian satisfies $W\left(E_{\lambda_1}, E_{\lambda_2}\right) = (\lambda_2 - \lambda_1)E_{\lambda_1 \oplus_c \lambda_2} \neq 0$, and the two solutions are linearly independent. This completes the proof. ☐

Example 8.1.25 *Let $a = -8$ and $b = 15$, $\mathbb{T} = h\mathbb{Z}$ for $h > 0$, $k_0 \equiv \alpha$, $k_1 \equiv 1 - \alpha$, and let $\alpha \in (0, 1]$ such that $\alpha \neq \dfrac{h}{1 + h}$. Clearly $\lambda_1 = 3$ and $\lambda_2 = 5$ solve $\lambda^2 + a\lambda + b = 0$, with $\lambda_1 \neq \lambda_2$. Checking the regressivity conditions,*

$$k_0^\rho(\alpha, t) - v(t)k_1^\rho(\alpha, t) = \alpha - h(1 - \alpha) \neq 0$$

as $\alpha \neq \dfrac{h}{1 + h}$, and

$$(k_0^\rho(\alpha, t) - v(t)k_1^\rho(\alpha, t) - av(t))(k_0^\rho(\alpha, t) - v(t)k_1^\rho(\alpha, t)) + bv^2(t)$$

$$= (\alpha - h(1 - \alpha) + 8h)(\alpha - h(1 - \alpha)) + 15h^2 > 0.$$

Let

$$p(t) := \widehat{E}_\beta(t, t_0), \qquad q(t) := \frac{15(\alpha + h(1 - \alpha))}{\alpha - h(1 - \alpha)} \widehat{E}_\beta(t, t_0),$$

where

$$\beta := \frac{(\alpha + h(1 - \alpha))\left[(-7 - \alpha)(\alpha - h(1 - \alpha)) - 15h\right]}{(\alpha - h(1 - \alpha))\alpha}.$$

Then the general solution of $M_c x = 0$ for M_c in (8.1) with p, q as given here, is

$$x(t) = c_1 E_3(t, t_0) + c_2 E_5(t, t_0).$$

In particular, if $\alpha = 1/5$ and $h = 1$ for example, then $\beta = 89$ and $q = -25\widehat{E}_{89}$, and $x = c_1 E_3 + c_2 E_5$ solves the self-adjoint equation $\widehat{D}_{\frac{1}{5}}\left[\widehat{E}_{89} D^{\frac{1}{5}} x\right] - 25\widehat{E}_{89} x = 0$.

Lemma 8.1.26 *Assume*

$$k_0(\alpha, t) - \mu(t)k_1(\alpha, t) \neq 0, \qquad t \in \mathbb{T}^\kappa.$$

If $f : \mathbb{T} \to \mathbb{R}$ is Δ-differentiable, then

$$\frac{1}{E_0^\sigma} D^\alpha [E_0 f] = D^\alpha f - k_1 f = k_0 f^\Delta. \tag{8.12}$$

Similarly, if $f : \mathbb{T} \to \mathbb{R}$ is ∇-differentiable, then

$$\frac{1}{\widehat{E}_0^\rho} \widehat{D}_\alpha [\widehat{E}_0 f] = \widehat{D}_\alpha f - k_1 f = k_0 f^\nabla. \tag{8.13}$$

Theorem 8.1.27 *Let $a, b \in \mathbb{R}$ with $1 - a\nu(t) + b\nu^2(t) \neq 0$, $t_0 \in \mathbb{T}$, and assume*

$$k_0(\alpha, t) - \mu(t)k_1(\alpha, t) \neq 0, \qquad t \in \mathbb{T}^\kappa.$$

Let

$$p(t) := \frac{\widehat{E}_{(a-b\nu)(k_0+\nu k_1)}(t, t_0)}{k_0(\alpha, t)E_0^\sigma(t, t_0)}, \qquad q(t) := \left(1 + \nu(t)\frac{k_1(\alpha, t)}{k_0(\alpha, t)}\right)\frac{k_0(\alpha, t)b\widehat{E}_{(a-b\nu)(k_0+\nu k_1)}(t, t_0)}{E_0(t, t_0)}.$$

If λ is a solution of the characteristic equation

$$\lambda^2 + a\lambda + b = 0,$$

then a solution of $M_c x = 0$ for M_c in (8.1) with p, q as given here, is

$$x(t) = E_{\lambda(k_0 - \mu k_1)}(t, t_0).$$

Proof 8.1.28 *Let p, q be as given in the statement of the theorem, and let*

$$x(t) = E_{\lambda(k_0 - \mu k_1)}(t, t_0) = E_0(t, t_0)E_{(k_0\lambda + k_1)}(t, t_0).$$

Then

$$D^\alpha x(t) = \lambda(k_0 - \mu k_1)(t)x(t)$$

implies

$$\begin{aligned}
p(t)D^\alpha x(t) &= \frac{\widehat{E}_{(a-b\nu)(k_0+\nu k_1)}(t, t_0)}{k_0(\alpha, t)E_0^\sigma(t, t_0)}\lambda(k_0 - \mu k_1)(t)E_{\lambda(k_0 - \mu k_1)}(t, t_0) \\
&= \frac{\lambda\widehat{E}_{(a-b\nu)(k_0+\nu k_1)}(t, t_0)}{E_0(t, t_0)}E_{\lambda(k_0 - \mu k_1)}(t, t_0) \\
&= \lambda\widehat{E}_{(a-b\nu)(k_0+\nu k_1)}(t, t_0)E_{(k_0\lambda + k_1)}(t, t_0) \\
&= \lambda\widehat{E}_0(t, t_0)\widehat{E}_{(k_0(a-b\nu)+k_1)}(t, t_0)E_{(k_0\lambda + k_1)}(t, t_0).
\end{aligned}$$

Let e_p and $\widehat{e}_p$ be the delta and nabla exponential functions on time scales, respectively. Then

$$p(t)D^\alpha x(t) = \lambda\widehat{E}_0(t, t_0)\widehat{e}_{(a-b\nu)}(t, t_0)e_\lambda(t, t_0)$$

$$= \lambda \widehat{E}_0(t,t_0)\widehat{e}_{(a-bv)}(t,t_0)\widehat{e}_{\frac{\lambda}{1+\lambda v}}(t,t_0)$$

$$= \lambda \widehat{E}_0(t,t_0)\widehat{e}_{\left(\frac{a-bv+\lambda}{1+\lambda v}\right)}(t,t_0).$$

As a result,

$$\widehat{D}_\alpha[p(t)D^\alpha x](t) = \lambda \widehat{D}_\alpha\left[\widehat{E}_0(t,t_0)\widehat{e}_{\left(\frac{a-bv+\lambda}{1+\lambda v}\right)}(t,t_0)\right]$$

$$\stackrel{(8.13)}{=} \lambda \widehat{E}_0^\rho(t,t_0)k_0(\alpha,t)\left(\frac{a-bv+\lambda}{1+\lambda v}\right)\widehat{e}_{\left(\frac{a-bv+\lambda}{1+\lambda v}\right)}(t,t_0).$$

Moreover,

$$q(t)x(t) = \left(1+v(t)\frac{k_1(\alpha,t)}{k_0(\alpha,t)}\right)\frac{k_0(\alpha,t)b\widehat{E}_{(a-bv)(k_0+vk_1)}(t,t_0)}{E_0(t,t_0)}E_{\lambda(k_0-\mu k_1)}(t,t_0)$$

$$= \left(1+v(t)\frac{k_1(\alpha,t)}{k_0(\alpha,t)}\right)k_0(\alpha,t)b\widehat{E}_0(t,t_0)\widehat{E}_{(k_0(a-bv)+k_1)}(t,t_0)E_{(k_0\lambda+k_1)}(t,t_0)$$

$$= b\widehat{E}_0^\sigma(t,t_0)k_0(\alpha,t)\widehat{E}_{(k_0(a-bv)+k_1)}(t,t_0)E_{(k_0\lambda+k_1)}(t,t_0)$$

$$= b\widehat{E}_0^\sigma(t,t_0)k_0(\alpha,t)\widehat{e}_{\left(\frac{a-bv+\lambda}{1+\lambda v}\right)}(t,t_0).$$

It follows that

$$\widehat{D}_\alpha[p(t)D^\alpha x](t)+q(t)x(t) = \left(\lambda\left(\frac{a-bv+\lambda}{1+\lambda v}\right)+b\right)\widehat{E}_0^\rho(t,t_0)k_0(\alpha,t)\widehat{e}_{\left(\frac{a-bv+\lambda}{1+\lambda v}\right)}(t,t_0)$$

$$= \left(\frac{\lambda^2+a\lambda+b}{1+\lambda v}\right)\widehat{E}_0^\rho(t,t_0)k_0(\alpha,t)\widehat{e}_{\left(\frac{a-bv+\lambda}{1+\lambda v}\right)}(t,t_0)$$

$$= 0,$$

as λ is a solution of the characteristic equation. This completes the proof. □

Example 8.1.29 *Assume*

$$k_0(\alpha,t)-\mu(t)k_1(\alpha,t)\neq 0, \qquad t\in\mathbb{T}^\kappa.$$

Let $\omega>0$ and $t_0\in\mathbb{T}$. In the previous theorem, Theorem 8.1.27, let $a=0$ and $b=\omega^2$; clearly $1+\omega^2 v^2(t)\neq 0$. Let

$$p(t):=\frac{\widehat{E}_{(-\omega^2 v)(k_0+vk_1)}(t,t_0)}{k_0(\alpha,t)E_0^\sigma(t,t_0)}, \quad q(t):=\left(1+v(t)\frac{k_1(\alpha,t)}{k_0(\alpha,t)}\right)\frac{k_0(\alpha,t)b\widehat{E}_{(-\omega^2 v)(k_0+vk_1)}(t,t_0)}{E_0(t,t_0)}.$$

$$\tag{8.14}$$

If λ is a solution of the characteristic equation

$$\lambda^2+\omega^2=0,$$

then a solution of $M_c x=0$ for M_c in (8.1) with p,q as given here, is

$$x(t) = E_{\lambda(k_0-\mu k_1)}(t,t_0)$$

$$
\begin{aligned}
&= E_0(t,t_0)E_{(k_0\lambda+k_1)}(t,t_0) \\
&= E_0(t,t_0)e_\lambda(t,t_0) \\
&= E_0(t,t_0)e_{\pm i\omega}(t,t_0).
\end{aligned}
$$

Recall the trigonometric functions on time scales defined by

$$
\cos_\omega = \frac{e_{i\omega}+e_{-i\omega}}{2}, \quad \sin_\omega = \frac{e_{i\omega}-e_{-i\omega}}{2i},
$$

where e_p is the exponential function on time scales. It follows that $M_c x = 0$ has the general solution

$$
x(t) = c_1 E_0(t,t_0)\cos_\omega(t,t_0)+c_2 E_0(t,t_0)\sin_\omega(t,t_0)
$$

for p,q given in (8.14). This ends the example.

8.2 REDUCTION-OF-ORDER THEOREMS

In this section we establish two related reduction-of-order theorems for the conformable self-adjoint dynamic equation $M_c x = 0$ for M_c given in (8.1).

Theorem 8.2.1 (Reduction of Order I) *Let M_c be given as in (8.1) and $\alpha \in (0,1]$. Assume*

$$
k_0^p(\alpha,t)-v(t)k_1^p(\alpha,t)\neq 0, \quad t\in\mathscr{I}.
$$

Let $t_0 \in \mathscr{I}$, and assume x is a solution of $M_c x = 0$ with $x \neq 0$ on $\mathscr{I}$. Then for $t \in \mathscr{I}$,

$$
y(t)=x(t)\int_{t_0}^t \frac{E_0(s,t_0)\widehat{E}_0(s,t_0)}{p(s)x(s)x^\sigma(s)k_0(\alpha,s)}\Delta s = x(t)\int_{t_0}^t \frac{E_0(s,t_0)E_0(\sigma(s),t)\widehat{E}_0(s,t_0)}{p(s)x(s)x^\sigma(s)}\Delta_{\alpha,t}s
$$

defines a second linearly independent solution y of $M_c x = 0$ on $\mathscr{I}$.

Proof 8.2.2 *By the converse of Abel's formula, Theorem 8.1.15, if $y \in \mathbb{D}_R$ satisfies*

$$
W(x,y)(t) = \frac{c}{p(t)E_0(b,t)\widehat{E}_0(b,t)}
$$

for some constant $c \in \mathbb{R}$, then y is also a solution of $M_c x = 0$. Checking the Wronskian, we have

$$
\begin{aligned}
W(x,y) &= xD^\alpha y - yD^\alpha x \\
&= x[k_1 y + k_0 y^\Delta] - y[k_1 x + k_0 x^\Delta] \\
&= k_0 xy^\Delta - k_0 yx^\Delta \\
&= k_0 x\left[x^\sigma\frac{E_0(t,t_0)\widehat{E}_0(t,t_0)}{pxx^\sigma k_0}+x^\Delta\int_{t_0}^t\frac{E_0(s,t_0)\widehat{E}_0(s,t_0)}{p(s)x(s)x^\sigma(s)k_0(\alpha,s)}\Delta s\right]-k_0 yx^\Delta \\
&= \frac{E_0(t,t_0)\widehat{E}_0(t,t_0)}{p(t)}+k_0 x^\Delta x\int_{t_0}^t\frac{E_0(s,t_0)\widehat{E}_0(s,t_0)}{p(s)x(s)x^\sigma(s)k_0(\alpha,s)}\Delta s-k_0 x^\Delta y
\end{aligned}
$$

$$= \frac{E_0(t,t_0)\widehat{E}_0(t,t_0)}{p(t)},$$

and the condition holds with $c = E_0(t_0,b)\widehat{E}_0(t_0,b)$. To show that $y \in \mathbb{D}_R$, one can modify the argument made by Messer in Chapter 4 (page 98) of Bohner and Peterson's 2003 monograph to the conformable setting. Thus, by Theorem 8.1.15, y is also a linearly independent solution of $M_c x = 0$. This completes the proof. □

Theorem 8.2.3 (Reduction of Order II) *Let M_c be given as in (8.1) and $\alpha \in (0,1]$. Assume*

$$k_0^{\rho}(\alpha,t) - \nu(t)k_1^{\rho}(\alpha,t) \neq 0, \qquad t \in \mathscr{I}.$$

Let $t_0 \in \mathscr{I}$, and assume x is a solution of $M_c x = 0$ with $x \neq 0$ on $\mathscr{I}$. Then y is a second linearly independent solution of $M_c x = 0$ if and only if y satisfies the first-order equation

$$D^{\alpha}[y/x](t) - k_1(t)[y/x](t) = \frac{cE_0(t,t_0)\widehat{E}_0(t,t_0)}{p(t)x(t)x^{\sigma}(t)}, \qquad t \in \mathscr{I}, \quad t \geq t_0, \qquad (8.15)$$

for some constant $c \in \mathbb{R}$ if and only if y is of the form

$$y(t) = c_1 x(t) + c_2 x(t) \int_{t_0}^{t} \frac{E_0(s,t_0)\widehat{E}_0(s,t_0)}{p(s)x(s)x^{\sigma}(s)k_0(\alpha,s)} \Delta s \qquad (8.16)$$

for $t \in \mathscr{I}$ with $t \geq t_0$, where $c_1, c_2 \in \mathbb{R}$ are constants. In the latter case,

$$c_1 = y(t_0)/x(t_0), \qquad c_2 = p(t_0)W(x,y)(t_0). \qquad (8.17)$$

Proof 8.2.4 *Assume x is a solution of $M_c x = 0$ with $x \neq 0$ on $\mathscr{I}$. Let y be any solution of $M_c x = 0$; we must show y is of the form (8.16). Using the Wronskian, set*

$$c_2 := p(t_0)W(x,y)(t_0).$$

By Abel's formula, Corollary 8.1.11 with $a = t_0$, we must have

$$x(t)D^{\alpha}y(t) - y(t)D^{\alpha}x(t) = W(x,y)(t) = \frac{c_2}{p(t)E_0(t_0,t)\widehat{E}_0(t_0,t)}, \qquad t \in \mathscr{I},$$

so that

$$\frac{xD^{\alpha}y - yD^{\alpha}x}{xx^{\sigma}} + k_1\frac{y}{x} = \frac{c_2 E_0(t,t_0)\widehat{E}_0(t,t_0)}{pxx^{\sigma}} + k_1\frac{y}{x}.$$

Thus, we see that

$$D^{\alpha}[y/x] = \frac{c_2 E_0(t,t_0)\widehat{E}_0(t,t_0)}{pxx^{\sigma}} + k_1\frac{y}{x} \qquad (8.18)$$

and y satisfies (8.15). Note that the left-hand side of (8.15) simplifies to $k_0[y/x]^{\Delta}$; divide it by k_0 and use the straight Δ-integral on time scales on both sides from t_0 to t to get

$$y(t)/x(t) - y(t_0)/x(t_0) = c_2 \int_{t_0}^{t} \frac{E_0(s,t_0)\widehat{E}_0(s,t_0)}{p(s)x(s)x^{\sigma}(s)k_0(\alpha,s)} \Delta s.$$

Recovering y from this yields

$$y(t) = c_1 x(t) + c_2 x(t) \int_{t_0}^{t} \frac{E_0(s,t_0)\widehat{E}_0(s,t_0)}{p(s)x(s)x^{\sigma}(s)k_0(\alpha,s)} \Delta s$$

provided $c_1 = y(t_0)/x(t_0)$.

Conversely, assume y is given by (8.16). By Theorem 8.2.1 and linearity, y is a solution of $M_c x = 0$ on $\mathscr{I}$ for $t \geq t_0$. Setting $t = t_0$ in (8.16) leads to c_1 in (8.17). By Abel's formula, Corollary 8.1.11 with $a = t_0$, the Wronskian satisfies

$$W(x,y)(t) = \frac{p(t_0)W(x,y)(t_0)}{p(t)E_0(t_0,t)\widehat{E}_0(t_0,t)};$$

to calculate $W(x,y)(t_0)$, we use (8.16) to obtain

$$W(x,y)(t_0) = (xD^{\alpha}y - yD^{\alpha}x)(t_0) = \frac{c_2}{p(t_0)}.$$

This ends the proof. □

8.3 DOMINANT AND RECESSIVE SOLUTIONS

In this section we lay the groundwork for further exploration of the non-homogeneous equation (8.1) by introducing the Pólya factorization for the conformable self-adjoint dynamic equation $M_c x = 0$, which in turn leads to a variation of parameters result for $M_c x = h$ and to the notion of dominant and recessive solutions. Again we assume throughout that the coefficient function p satisfies $p > 0$. We begin, though, with a discussion of oscillation and disconjugacy for the homogeneous equation.

Definition 8.3.1 *A function $x \in \mathbb{D}_R$ has a generalized zero at $t \in \mathscr{I}$ if $x(t) = 0$ or, if t is a left-scattered point and $x^{\rho}(t)x(t) < 0$.*

Definition 8.3.2 *The homogeneous self-adjoint equation (8.1), namely*

$$M_c x(t) = \widehat{D}_{\alpha}[pD^{\alpha}](t) + q(t)x(t) = 0, \qquad (8.19)$$

is disconjugate on a time-scale interval $[a,b]_{\mathbb{T}}$ if and only if the following hold.

1. *If x is a nontrivial solution of (8.19) with $x(a) = 0$, then x has no generalized zeros in $(a,b]_{\mathbb{T}}$.*

2. *If x is a nontrivial solution of (8.19) with $x(a) \neq 0$, then x has at most one generalized zero in $(a,b]_{\mathbb{T}}$.*

Definition 8.3.3 *Let $\omega = \sup \mathbb{T}$; if $\omega < \infty$, assume $\rho(\omega) = \omega$. Let $a \in \mathbb{T}$. Then (8.19) is oscillatory if and only if every nontrivial real-valued solution has infinitely many generalized zeros in $[a,\omega)_{\mathbb{T}}$. Equation (8.19) is non-oscillatory on $[a,\omega)_{\mathbb{T}}$ if and only if it is not oscillatory on $[a,\omega)_{\mathbb{T}}$.*

Lemma 8.3.4 *Let $\omega = \sup \mathbb{T}$; if $\omega < \infty$, assume $\rho(\omega) = \omega$. Let $a \in \mathbb{T}$. Then if (8.19) is non-oscillatory on $[a, \omega)_\mathbb{T}$, there exists $t_0 \in \mathbb{T}$ with $t_0 \geq a$ such that (8.19) has a positive solution on $[t_0, \omega)_\mathbb{T}$.*

Proof 8.3.5 *If (8.19) is non-oscillatory on $[a, \omega)_\mathbb{T}$, then there exists a nontrivial solution y of (8.19) that has only finitely many generalized zeros in $[a, \omega)_\mathbb{T}$. Set $\tau := \max\{t \in \mathbb{T} : y$ has a generalized zero at $t\}$. Then for any $t_0 \in \mathbb{T}$ with $t_0 > \tau$, either $y > 0$ on $[t_0, \omega)_\mathbb{T}$, or $-y > 0$ on $[t_0, \omega)_\mathbb{T}$. This completes the proof.* $\square$

Theorem 8.3.6 (Sturm Separation) *Assume $k_0(\alpha, t) - \mu(t)k_1(\alpha, t) > 0$ for $t \in \mathbb{T}^\kappa$. Let x and y be linearly independent solutions of (8.19) on $\mathbb{T}$. Then x and y have no common zeros in $\mathbb{T}^\kappa$. If x has a zero at $t_1 \in \mathbb{T}$ and a generalized zero at $t_2 \in \mathbb{T}$ with $t_2 > t_1$, then y has a generalized zero in $(t_1, t_2]_\mathbb{T}$. If x has generalized zeros at $t_1, t_2 \in \mathbb{T}$ with $t_2 > t_1$, then y has a generalized zero in $[t_1, t_2]_\mathbb{T}$.*

Proof 8.3.7 *If x and y share a common zero at $t_0 \in \mathbb{T}^\kappa$, then their Wronskian satisfies $W(x, y)(t_0) = 0$, making x and y linearly dependent, a contradiction. If x has a zero at $t_1 \in \mathbb{T}$ and a generalized zero at $t_2 \in \mathbb{T}$ with $t_2 > t_1$, then without loss of generality, assume $t_2 > \sigma(t_1)$ is the first generalized zero to the right of t_1, and that $x(t) > 0$ on $(t_1, t_2)_\mathbb{T}$ with $x(t_2) \leq 0$. Assume y is a linearly independent solution of (8.19) on $\mathbb{T}$ with no generalized zeros in $(t_1, t_2]_\mathbb{T}$; without loss of generality, $y(t) > 0$ for $t \in [t_1, t_2]_\mathbb{T}$. Note that on $\mathbb{T}^\kappa$,*

$$
\begin{aligned}
W(y, x) &= y D^\alpha x - x D^\alpha y \\
&= y(k_1 x + k_0 x^\Delta) - x(k_1 y + k_0 y^\Delta) \\
&= k_0 (y x^\Delta - x y^\Delta).
\end{aligned}
$$

It follows that for $t \in [t_1, t_2]_\mathbb{T}$,

$$
\begin{aligned}
\left(\frac{x}{y}\right)^\Delta (t) &= \frac{y x^\Delta - x y^\Delta}{y y^\sigma}(t) \\
&= \frac{W(y, x)}{y y^\sigma k_0}(t) \\
&= \frac{c E_0(t, a)\widehat{E}_0(t, a)}{p(t)y(t)y^\sigma(t)k_0(\alpha, t)}
\end{aligned}
$$

is of one sign on $[t_1, t_2)_\mathbb{T}$, since $y(t) > 0$ for $t \in [t_1, t_2]_\mathbb{T}$, $\widehat{E}_0(t, a), p(t), k_0(\alpha, t)$ are all positive on $\mathbb{T}$, and the assumption $k_0(\alpha, t) - \mu(t)k_1(\alpha, t) > 0$ for $t \in \mathbb{T}^\kappa$ guarantees $E_0(t, a) > 0$ on $\mathbb{T}$ as well. Consequently, (x/y) is monotone on $[t_1, t_2]_\mathbb{T}$. For any $\tau \in (t_1, t_2)_\mathbb{T}$,

$$
\frac{x(t_1)}{y(t_1)} = 0 \quad and \quad \frac{x(\tau)}{y(\tau)} > 0, \quad while \quad \frac{x(t_2)}{y(t_2)} \leq 0,
$$

a contradiction of monotonicity. Thus y must have a generalized zero in $(t_1, t_2]_\mathbb{T}$. If x has generalized zeros at $t_1, t_2 \in \mathbb{T}$ with $t_2 > t_1$, then without loss of generality, assume $t_2 > \sigma(t_1)$ is the first generalized zero to the right of t_1. If $x(t_1) = 0$, this has been dealt with in the previous case, so $x(t_1) \neq 0$, but t_1 must be a left-scattered point; assume without loss of *generality that $x(t) > 0$ on $[t_1, t_2)_{\mathbb{T}}$, with $x^\rho(t_1) < 0$ and $x(t_2) \leq 0$. Assume y is a linearly independent solution of (8.19) on $\mathbb{T}$ with no generalized zeros in $[t_1, t_2)_{\mathbb{T}}$; without loss of generality, $y(t) > 0$ for $t \in [t_1, t_2]_{\mathbb{T}}$ and $y^\rho(t_1) > 0$. As in the previous case, (x/y) is monotone on $[\rho(t_1), t_2]_{\mathbb{T}}$. However,*

$$\frac{x^\rho(t_1)}{y^\rho(t_1)} < 0 \quad and \quad \frac{x(t_1)}{y(t_1)} > 0, \quad while \quad \frac{x(t_2)}{y(t_2)} \leq 0,$$

a contradiction of monotonicity. Thus y must have a generalized zero in $[t_1, t_2]_{\mathbb{T}}$. This completes the proof. □

Theorem 8.3.8 (Disconjugacy) *Assume $k_0(\alpha, t) - \mu(t)k_1(\alpha, t) > 0$ for $t \in \mathbb{T}^\kappa$. If (8.19) has a positive solution on a time-scale interval $\mathcal{I} \subset \mathbb{T}$, then (8.19) is disconjugate on $\mathcal{I}$. Conversely, if (8.19) is disconjugate on $[\rho(a), \sigma(b)]_{\mathbb{T}}$ for some $a, b \in \mathbb{T}^\kappa_\kappa$ with $a < b$, then (8.19) has a positive solution on $[\rho(a), \sigma(b)]_{\mathbb{T}}$.*

Proof 8.3.9 *Assume (8.19) has a positive solution x on the interval $\mathcal{I} \subset \mathbb{T}$. If (8.19) is not disconjugate on $\mathcal{I}$, then it has a nontrivial solution y with at least two generalized zeros in $\mathcal{I}$. So, without loss of generality, there exist $t_1, t_2 \in \mathcal{I}$ such that $(t_1, t_2)_{\mathbb{T}} \neq \emptyset$ with*

$$y(t_1) \leq 0, \quad y(t_2) \leq 0, \quad y(t) > 0, \quad t \in (t_1, t_2)_{\mathbb{T}}.$$

It follows that for $t \in [t_1, t_2]_{\mathbb{T}}$,

$$\begin{aligned}
\left(\frac{y}{x}\right)^\Delta(t) &= \frac{xy^\Delta - yx^\Delta}{xx^\sigma}(t) \\
&= \frac{cE_0(t,a)\widehat{E}_0(t,a)}{p(t)x(t)x^\sigma(t)k_0(\alpha, t)}
\end{aligned}$$

is of one sign on $[t_1, t_2)_{\mathbb{T}}$, since $x(t) > 0$ for $t \in [t_1, t_2]_{\mathbb{T}}$. Consequently, (y/x) is monotone on $[t_1, t_2]_{\mathbb{T}}$. For any $\tau \in (t_1, t_2)_{\mathbb{T}}$,

$$\frac{y(t_1)}{x(t_1)} \leq 0 \quad and \quad \frac{y(\tau)}{x(\tau)} > 0, \quad while \quad \frac{y(t_2)}{x(t_2)} \leq 0,$$

a contradiction of monotonicity. Thus (8.19) is disconjugate on $\mathcal{I}$. Conversely, assume (8.19) is disconjugate on the compact interval $[\rho(a), \sigma(b)]_{\mathbb{T}}$ for some $a, b \in \mathbb{T}^\kappa_\kappa$ with $a < b$. Let u, y be the unique solutions of (8.19) satisfying

$$u^\rho(a) = 0, \ D^\alpha u^\rho(a) = k_0(\alpha, \rho(a)), \quad y^\sigma(b) = 0, \ D^\alpha y(b) = -k_0(\alpha, b),$$

respectively. Note that u then satisfies $u^{\Delta\rho}(a) = 1$, while y satisfies $y(b) = 0$ and $y^\Delta(b) = -1$ if b is a left-dense point, and $y(b) = \dfrac{\mu(b)k_0(\alpha, b)}{k_0(\alpha, b) - \mu(b)k_1(\alpha, b)} > 0$ with $y^\sigma(b) = 0$ if b is a left-scattered point. Since (8.19) is disconjugate on $[\rho(a), \sigma(b)]_{\mathbb{T}}$, we have $u(t) > 0$ for $t \in (\rho(a), \sigma(b)]_{\mathbb{T}}$ and $y(t) > 0$ for $t \in [\rho(a), \sigma(b))_{\mathbb{T}}$. If we take $x = u + y$, then x is a positive solution of (8.19) on $[\rho(a), \sigma(b)]_{\mathbb{T}}$. This completes the proof. □

Theorem 8.3.10 (Pólya Factorization) *Let M_c be as in (8.1), and assume $M_c x = 0$ has a positive solution $x > 0$ on an interval $[a,b)_\mathbb{T}$ with $b \leq \infty$. If*

$$k_0(\alpha,t) - \mu(t)k_1(\alpha,t) > 0, \quad t \in [a,b)_\mathbb{T}$$

then for any function $y \in \mathbb{D}$ we have a Pólya factorization of $M_c y$ on $[a,b)_\mathbb{T}$ given by

$$M_c y = \varphi_1 \widehat{D}_\alpha \{\varphi_2 D^\alpha [\varphi_1 y]\} \quad on \quad [a,b)_\mathbb{T},$$

where

$$\varphi_1(t) := \frac{E_0(t,a)}{x(t)} > 0, \quad \varphi_2(t) := \frac{p(t)x(t)x^\sigma(t)}{E_0(t,a)E_0^\sigma(t,a)} = \frac{p(t)}{\varphi_1(t)\varphi_1^\sigma(t)} > 0.$$

Proof 8.3.11 *Recall from (8.12) that*

$$\frac{1}{E_0^\sigma}D^\alpha[E_0 f] = D^\alpha f - k_1 f,$$

a fact that we will use in the calculations below. Assume $x > 0$ is a positive solution of $M_c x = 0$ on $[a,b)_\mathbb{T}$, and let $y \in \mathbb{D}$. Then

$$
\begin{aligned}
M_c y \overset{Thm\ 8.1.5}{=} & \frac{E_0(\cdot,a)}{x}\widehat{D}_\alpha \left\{ \frac{p}{E_0(\cdot,a)}[xD^\alpha y - yD^\alpha x] \right\} \\
= & \varphi_1 \widehat{D}_\alpha \left\{ \frac{pxx^\sigma}{E_0(\cdot,a)} \left[\frac{1}{x^\sigma}D^\alpha y - \frac{y}{xx^\sigma}D^\alpha x \right] \right\} \\
= & \varphi_1 \widehat{D}_\alpha \left\{ \frac{pxx^\sigma}{E_0(\cdot,a)} \left(D^\alpha \left[\frac{y}{x}\right] - k_1 \frac{y}{x} \right) \right\} \\
\overset{(8.12)}{=} & \varphi_1 \widehat{D}_\alpha \left\{ \frac{pxx^\sigma}{E_0(\cdot,a)} \left(\frac{1}{E_0^\sigma(\cdot,a)}D^\alpha[E_0(\cdot,a)y/x] \right) \right\} \\
= & \varphi_1 \widehat{D}_\alpha \{\varphi_2 D^\alpha [\varphi_1 y]\},
\end{aligned}
$$

for φ_1 and φ_2 as defined in the statement of the theorem. This completes the proof. □

Theorem 8.3.12 (Variation of Parameters) *Assume*

$$k_0(\alpha,t) - \mu(t)k_1(\alpha,t) > 0, \quad t \in [a,\infty)_\mathbb{T}.$$

Let h be a continuous function defined on $[a,\infty)_\mathbb{T}$, and let M_c be given as in (8.1). If the homogeneous equation $M_c x = 0$ has a positive solution $x > 0$ on $[a,\infty)_\mathbb{T}$, then the non-homogeneous equation $M_c y = h$ has a solution y given by

$$
\begin{aligned}
y(t) = & \frac{y(a)x(t)}{x(a)} + p(a)W(x,y)(a)x(t) \int_a^t \frac{E_0(s,a)\widehat{E}_0(s,a)}{(pxx^\sigma)(s)k_0(\alpha,s)}\Delta s \\
& + x(t) \int_a^t \frac{E_0(s,a)}{(pxx^\sigma)(s)k_0(\alpha,s)} \left(\int_a^s \frac{x(\omega)h(\omega)\widehat{E}_0(s,\rho(\omega))}{E_0(\omega,a)k_0(\alpha,\omega)}\nabla\omega \right) \Delta s.
\end{aligned}
$$

Proof 8.3.13 *Let $y \in \mathbb{D}$ be defined on $[a,\infty)_{\mathbb{T}}$, and assume $x > 0$ is a positive solution of $M_c x = 0$ on $[a,\infty)_{\mathbb{T}}$. As in Theorem 8.3.10, we use the Pólya factorization of $M_c y$ to get*

$$h(\omega) = M_c y(\omega) = \frac{E_0(\omega,a)}{x(\omega)} \widehat{D}_\alpha \left\{ \frac{p(\omega)x(\omega)x^\sigma(\omega)}{E_0(\omega,a)E_0^\sigma(\omega,a)} D^\alpha \left[\frac{E_0(\omega,a)}{x(\omega)} y(\omega) \right] \right\}.$$

Multiplying by $\widehat{E}_0(t,\rho(\omega))E_0(a,\omega)x(\omega)/k_0(\alpha,\omega)$ and integrating from a to t we arrive, via the fundamental theorem of conformable ∇ calculus, at

$$\int_a^t \frac{x(\omega)h(\omega)\widehat{E}_0(t,\rho(\omega))}{E_0(\omega,a)k_0(\alpha,\omega)} \nabla \omega$$

$$= \int_a^t \widehat{D}_\alpha \left\{ \frac{p(\omega)x(\omega)x^\sigma(\omega)}{E_0(\omega,a)E_0^\sigma(\omega,a)} D^\alpha \left[\frac{E_0(\omega,a)}{x(\omega)} y(\omega) \right] \right\} \frac{\widehat{E}_0(t,\rho(\omega))}{k_0(\alpha,\omega)} \nabla \omega$$

$$= \frac{p(t)x(t)x^\sigma(t)}{E_0(t,a)E_0^\sigma(t,a)} D^\alpha \left[\frac{E_0(t,a)}{x(t)} y(t) \right] - \frac{p(a)x(a)x^\sigma(a)}{x(a)x^\sigma(a)} W(x,y)(a)\widehat{E}_0(t,a)$$

$$= \frac{p(t)x(t)x^\sigma(t)}{E_0(t,a)E_0^\sigma(t,a)} D^\alpha \left[\frac{E_0(t,a)}{x(t)} y(t) \right] - p(a)W(x,y)(a)\widehat{E}_0(t,a)$$

using (8.12), since $x > 0$ is a solution. This leads to

$$D^\alpha \left[\frac{E_0(\cdot,a)}{x} y \right](s) \frac{E_0(t,\sigma(s))}{k_0(\alpha,s)} \Delta s = \frac{E_0(t,a)p(a)}{(pxx^\sigma)(s)k_0(\alpha,s)} W(x,y)(a)E_0(s,a)\widehat{E}_0(s,a)\Delta s$$

$$+ \frac{E_0(t,a)E_0(s,a)}{(pxx^\sigma)(s)k_0(\alpha,s)} \left(\int_a^s \frac{x(\omega)h(\omega)\widehat{E}_0(s,\rho(\omega))}{E_0(\omega,a)k_0(\alpha,\omega)} \nabla \omega \right) \Delta s.$$

Integrating this from a to t and using the fundamental theorem of conformable Δ calculus yields

$$\frac{y(t)}{x(t)} - \frac{y(a)}{x(a)} = p(a)W(x,y)(a) \int_a^t \frac{E_0(s,a)\widehat{E}_0(s,a)}{(pxx^\sigma)(s)k_0(\alpha,s)} \Delta s$$

$$+ \int_a^t \frac{E_0(s,a)}{(pxx^\sigma)(s)k_0(\alpha,s)} \left(\int_a^s \frac{x(\omega)h(\omega)\widehat{E}_0(s,\rho(\omega))}{E_0(\omega,a)k_0(\alpha,\omega)} \nabla \omega \right) \Delta s,$$

which can be rewritten in the form for y given in the statement of the theorem. Clearly the right-hand side of the form of y above reduces to $y(a)$ at a, and since $x > 0$ is a solution the conformable derivative reduces to $D^\alpha y(a)$ at a. This completes the proof. □

Corollary 8.3.14 *Let h be a continuous function defined on $[a,\infty)_{\mathbb{T}}$, and let M_c be given as in (8.1). Assume*

$$k_0(\alpha,t) - \mu(t)k_1(\alpha,t) > 0, \quad t \in [a,b)_{\mathbb{T}}.$$

If the homogeneous matrix equation (8.1) has a positive solution $x > 0$ on $[a,\infty)_{\mathbb{T}}$, then the non-homogeneous initial value problem

$$M_c y = h, \quad y(a) = y_a, \quad D^\alpha y(a) = y_a' \tag{8.20}$$

has a unique solution.

Proof 8.3.15 *By Theorem 8.3.12, the non-homogeneous initial value problem (8.20) has a solution. Suppose y_1 and y_2 both solve (8.20). Then $x = y_1 - y_2$ solves the homogeneous initial value problem*

$$M_c x = 0, \quad x(a) = 0, \quad D^\alpha x(a) = 0;$$

by Theorem 8.1.2, this has only the trivial solution $x \equiv 0$, and thus $y_1 = y_2$ is unique. This completes the proof. □

Theorem 8.3.16 (Trench Divergence) *Let M_c be as in (8.1). Let $a \in \mathbb{T}$, and $b := \sup \mathbb{T}$; if $b < \infty$, assume $\rho(b) = b$. Also assume*

$$k_0(\alpha, t) - \mu(t)k_1(\alpha, t) > 0, \quad t \in [a, b)_{\mathbb{T}},$$

and let

$$\delta_2(t) = \frac{E_0(t,a)E_0^\sigma(t,a)}{p(t)x(t)x^\sigma(t)}.$$

If $M_c x = 0$ has a positive solution x on an interval $[a,b)_{\mathbb{T}}$ with $b \le \infty$, then either

$$\int_a^b \frac{E_0(t,a)\widehat{E}_0(t,a)}{p(t)x(t)x^\sigma(t)k_0(\alpha,t)}\Delta t = \int_a^b \delta_2(t)\widehat{E}_0(t,a)\Delta_{\alpha,a}t = \infty,$$

or

$$\int_a^b \frac{\delta_2(t)\widehat{E}_0(t,a)}{\left(\int_t^b \delta_2(s)\widehat{E}_0(s,a)\Delta_{\alpha,a}s\right)\left(\int_{\sigma(t)}^b \delta_2(s)\widehat{E}_0(s,a)\Delta_{\alpha,a}s\right)}\Delta_{\alpha,a}t = \infty.$$

Proof 8.3.17 *Since $M_c x = 0$ has a positive solution x on $[a,b)_{\mathbb{T}}$, for any $y \in \mathbb{D}$ we have a Pólya factorization of $M_c y$ on $[a,b)_{\mathbb{T}}$ given by*

$$M_c y = \varphi_1 \widehat{D}_\alpha \{\varphi_2 D^\alpha[\varphi_1 y]\} \quad on \quad [a,b)_{\mathbb{T}},$$

where

$$\varphi_1(t) := \frac{E_0(t,a)}{x(t)} > 0, \quad \varphi_2(t) := \frac{p(t)x(t)x^\sigma(t)}{E_0(t,a)E_0^\sigma(t,a)} = \frac{1}{\delta_2(t)} > 0.$$

Either

$$\int_a^b \frac{\delta_2(s)\widehat{E}_0(s,a)E_0(a,\sigma(s))}{k_0(\alpha,s)}\Delta s = \int_a^b \delta_2(s)\widehat{E}_0(s,a)\Delta_{\alpha,a}s = \infty,$$

or

$$\int_a^b \frac{\delta_2(s)\widehat{E}_0(s,a)E_0(a,\sigma(s))}{k_0(\alpha,s)}\Delta s = \int_a^b \delta_2(s)\widehat{E}_0(s,a)\Delta_{\alpha,a}s < \infty. \qquad (8.21)$$

If (8.21) holds, set

$$\beta_2(t) = \frac{\delta_2(t)\widehat{E}_0(t,a)}{\left(\int_t^b \delta_2(s)\widehat{E}_0(s,a)\Delta_{\alpha,a}s\right)\left(\int_{\sigma(t)}^b \delta_2(s)\widehat{E}_0(s,a)\Delta_{\alpha,a}s\right)}$$

for $t \in [a,b)_{\mathbb{T}}$. Then by (8.21) we have

$$\int_a^b \beta_2(t)\Delta_{\alpha,a}t = \lim_{c \to b^-} \int_a^c \frac{\delta_2(t)\widehat{E}_0(t,a)}{\left(\int_t^b \delta_2(s)\widehat{E}_0(s,a)\Delta_{\alpha,a}s\right)\left(\int_{\sigma(t)}^b \delta_2(s)\widehat{E}_0(s,a)\Delta_{\alpha,a}s\right)}\Delta_{\alpha,a}t$$

$$= \lim_{c \to b^-} \int_a^c \frac{\delta_2(t)\widehat{E}_0(t,a)E_0(a,\sigma(t))}{\left(\int_b^t \frac{\delta_2(s)\widehat{E}_0(s,a)E_0(a,\sigma(s))}{k_0(\alpha,s)}\Delta s\right)\left(\int_b^{\sigma(t)} \frac{\delta_2(s)\widehat{E}_0(s,a)E_0(a,\sigma(s))}{k_0(\alpha,s)}\Delta s\right)} \frac{\Delta t}{k_0(\alpha,t)}$$

$$= \lim_{c \to b^-} \int_a^c \frac{E_0(t,a)\delta_2(t)\widehat{E}_0(t,a)}{\left(\int_b^t \frac{\delta_2(s)\widehat{E}_0(s,a)E_0(t,\sigma(s))}{k_0(\alpha,s)}\Delta s\right)\left(\int_b^{\sigma(t)} \frac{\delta_2(s)\widehat{E}_0(s,a)E_0(\sigma(t),\sigma(s))}{k_0(\alpha,s)}\Delta s\right)} \frac{\Delta t}{k_0(\alpha,t)}$$

$$= \lim_{c \to b^-} \int_a^c \left\{ D^\alpha \left[\frac{-E_0(t,a)}{\int_b^t \delta_2(s)\widehat{E}_0(s,a)\Delta_{\alpha,t}s} \right] \right.$$

$$\left. -k_1(\alpha,t)\frac{1}{\int_t^b \frac{\delta_2(s)\widehat{E}_0(s,a)E_0(a,\sigma(s))}{k_0(\alpha,s)}\Delta s} \right\} \frac{\Delta t}{k_0(\alpha,t)}$$

$$= \lim_{c \to b^-} \int_a^c \frac{1}{E_0^\sigma(t,a)} \left\{ E_0^\sigma(t,a)D^\alpha \left[\frac{-E_0(t,a)}{\int_b^t \frac{\delta_2(s)\widehat{E}_0(s,a)E_0(t,\sigma(s))}{k_0(\alpha,s)}\Delta s} \right] \right.$$

$$\left. -k_1(\alpha,t)E_0^\sigma(t,a)\frac{1}{\int_t^b \frac{\delta_2(s)\widehat{E}_0(s,a)E_0(a,\sigma(s))}{k_0(\alpha,s)}\Delta s} \right\} \frac{\Delta t}{k_0(\alpha,t)}$$

$$= \lim_{c \to b^-} \int_a^c \frac{1}{E_0^\sigma(t,a)} D^\alpha \left\{ E_0(t,a)\frac{1}{\int_t^b \delta_2(s)\widehat{E}_0(s,a)\Delta_{\alpha,a}s} \right\} \frac{\Delta t}{k_0(\alpha,t)}$$

$$= \lim_{c \to b^-} \int_a^c \left\{ D^\alpha \left(\frac{1}{\int_t^b \delta_2(s)\widehat{E}_0(s,a)\Delta_{\alpha,a}s} \right) \right.$$

$$\left. -k_1(\alpha,t)\left(\frac{1}{\int_t^b \delta_2(s)\widehat{E}_0(s,a)\Delta_{\alpha,a}s} \right) \right\} \frac{\Delta t}{k_0(\alpha,t)}$$

$$= \lim_{c \to b^-} \int_a^c \left\{ k_0(\alpha,t)\left(\frac{1}{\int_t^b \delta_2(s)\widehat{E}_0(s,a)\Delta_{\alpha,a}s} \right)^\Delta \right\} \frac{\Delta t}{k_0(\alpha,t)}$$

$$= \lim_{c \to b^-} \left(\frac{1}{\int_t^b \delta_2(s)\widehat{E}_0(s,a)\Delta_{\alpha,a}s} \right) \Big|_{t=a}^{c}$$

$$= \infty.$$

This completes the proof. □

Theorem 8.3.18 (Dominant and Recessive Solutions) *Let M_c be as in (8.1). Also assume*

$$k_0(\alpha,t) - \mu(t)k_1(\alpha,t) > 0, \quad t \in [a,b)_{\mathbb{T}}.$$

If $M_c x = 0$ is non-oscillatory on an interval $[a,b)_{\mathbb{T}} \subset (0,\infty)_{\mathbb{T}}$ with $b \leq \infty$, then there is a positive solution u, called a recessive solution at b, such that for any second linearly independent solution v, called a dominant solution at b,

$$\lim_{t \to b^-} \frac{u(t)}{v(t)} = 0, \quad \int_a^b \frac{E_0(t,a)\widehat{E}_0(t,a)}{p(t)u(t)u^\sigma(t)k_0(\alpha,t)}\Delta t = \infty, \quad and \quad \int_{t_0}^b \frac{E_0(s,a)\widehat{E}_0(s,a)}{p(s)v(s)v^\sigma(s)k_0(\alpha,s)}\Delta s < \infty,$$

where $t_0 < b$ is sufficiently close. Furthermore

$$\frac{p(t)D^\alpha v(t)}{v(t)} > \frac{p(t)D^\alpha u(t)}{u(t)} \tag{8.22}$$

for $t < b$ sufficiently close. Moreover, the recessive solution is unique up to multiplication by a nonzero constant.

Proof 8.3.19 *Since we are assuming that $M_c x = 0$ is non-oscillatory on $[a,b)_{\mathbb{T}}$, without loss of generality, it has a positive solution x on $[a,b)_{\mathbb{T}}$.*

Case I. Suppose for this solution $x > 0$ that

$$\int_a^b \frac{E_0(s,a)\widehat{E}_0(s,a)}{p(s)x(s)x^\sigma(s)k_0(\alpha,s)}\Delta s = \int_a^b \frac{E_0(s,a)E_0^\sigma(s,a)}{p(s)x(s)x^\sigma(s)}\widehat{E}_0(s,a)\Delta_{\alpha,a}s = \infty.$$

Let $u = x$, and let

$$v(t) = u(t)\int_a^t \frac{E_0(s,a)E_0^\sigma(s,a)}{p(s)u(s)u^\sigma(s)}\widehat{E}_0(s,a)\Delta_{\alpha,a}s > 0.$$

Then v is a second linearly independent solution of $M_c x = 0$ on $[a,b)_{\mathbb{T}}$ by reduction of order, with $v(a) = 0$. Note that

$$\lim_{t \to b^-}\frac{u(t)}{v(t)} = \lim_{t \to b^-}\frac{1}{\int_a^t \frac{E_0(s,a)E_0^\sigma(s,a)}{p(s)u(s)u^\sigma(s)}\widehat{E}_0(s,a)\Delta_{\alpha,a}s} = 0$$

and

$$\int_a^b \frac{E_0(t,a)\widehat{E}_0(t,a)}{p(t)u(t)u^\sigma(t)k_0(\alpha,t)}\Delta t = \infty$$

by the assumption on the integral in this case. Pick $t_0 \in [a,b)_{\mathbb{T}}$ so that $v(t) \neq 0$ on $[t_0,b)_{\mathbb{T}}$, which is possible by the positivity of the solution $u = x$. For $t \in [t_0,b)_{\mathbb{T}}$, by reduction of order

$$D^\alpha\left[\frac{u}{v}\right](t) - k_1(t)\frac{u(t)}{v(t)} = \frac{-E_0(t,a)\widehat{E}_0(t,a)}{p(t)v(t)v^\sigma(t)},$$

but we also have

$$\frac{1}{E_0^\sigma(t,a)}D^\alpha\left[E_0(\cdot,a)\frac{u}{v}\right](t) = D^\alpha\left[\frac{u}{v}\right](t) - k_1(t)\frac{u(t)}{v(t)},$$

so that a little manipulation yields

$$D^\alpha\left[E_0(\cdot,a)\frac{u}{v}\right](s)\frac{E_0(t,\sigma(s))}{k_0(\alpha,s)} = \frac{-E_0(t,a)E_0(s,a)\widehat{E}_0(s,a)}{p(s)v(s)v^\sigma(s)k_0(\alpha,s)}$$

for $t \in [t_0,b)_{\mathbb{T}}$. Integrating both sides of this last equality from t_0 to t we obtain

$$\frac{u(t)}{v(t)} - \frac{u(t_0)}{v(t_0)} = \int_{t_0}^t \frac{-E_0(s,a)\widehat{E}_0(s,a)}{p(s)v(s)v^\sigma(s)k_0(\alpha,s)}\Delta s.$$

Letting $t \to b^-$ we get

$$\int_{t_0}^b \frac{E_0(s,a)\widehat{E}_0(s,a)}{p(s)v(s)v^\sigma(s)k_0(\alpha,s)}\Delta s = \frac{u(t_0)}{v(t_0)} < \infty.$$

Finally, for $t \in [t_0, b)_{\mathbb{T}}$,

$$\frac{p(t)D^{\alpha}v(t)}{v(t)} - \frac{p(t)D^{\alpha}u(t)}{u(t)} = \frac{p(t)W(u,v)(t)}{u(t)v(t)} = \frac{E_0(t,a)\widehat{E}_0(t,a)}{u(t)v(t)} > 0.$$

Case II. Suppose for the solution $x > 0$ that

$$\int_a^b \frac{E_0(s,a)\widehat{E}_0(s,a)}{p(s)x(s)x^{\sigma}(s)k_0(\alpha,s)} \Delta s = \int_a^b \frac{E_0(s,a)E_0^{\sigma}(s,a)}{p(s)x(s)x^{\sigma}(s)} \widehat{E}_0(s,a)\Delta_{\alpha,a}s < \infty.$$

Then for

$$\delta_2(t) = \frac{E_0(t,a)E_0^{\sigma}(t,a)}{p(t)x(t)x^{\sigma}(t)},$$

$$\int_a^b \frac{\delta_2(t)\widehat{E}_0(t,a)}{\left(\int_t^b \delta_2(s)\widehat{E}_0(s,a)\Delta_{\alpha,a}s\right)\left(\int_{\sigma(t)}^b \delta_2(s)\widehat{E}_0(s,a)\Delta_{\alpha,a}s\right)} \Delta_{\alpha,a}t = \infty$$

by the Trench divergence theorem. If we let

$$u(t) = x(t)\int_t^b \delta_2(s)\widehat{E}_0(s,a)\Delta_{\alpha,a}s > 0,$$

then $u \in \mathbb{D}$ and we have a Pólya factorization of $M_c u$ on $[a,b)_{\mathbb{T}}$ given by

$$M_c u = \varphi_1 \widehat{D}_{\alpha}\{\varphi_2 D^{\alpha}[\varphi_1 u]\} \quad on \quad [a,b)_{\mathbb{T}},$$

where

$$\varphi_1(t) := \frac{E_0(t,a)}{x(t)} > 0, \quad \varphi_2(t) := \frac{p(t)x(t)x^{\sigma}(t)}{E_0(t,a)E_0^{\sigma}(t,a)} = \frac{1}{\delta_2(t)} > 0.$$

Thus,

$$
\begin{aligned}
D^{\alpha}[\varphi_1 u] &= D^{\alpha}\left[\frac{E_0(t,a)}{x(t)}x(t)\int_t^b \delta_2(s)\widehat{E}_0(s,a)\Delta_{\alpha,a}s\right] \\
&= D^{\alpha}\left[E_0(t,a)\int_t^b \delta_2(s)\widehat{E}_0(s,a)\Delta_{\alpha,a}s\right] \\
&\stackrel{(8.12)}{=} E_0^{\sigma}(t,a)\left\{D^{\alpha}\left[\int_t^b \delta_2(s)\widehat{E}_0(s,a)\Delta_{\alpha,a}s\right] - k_1(\alpha,t)\int_t^b \delta_2(s)\widehat{E}_0(s,a)\Delta_{\alpha,a}s\right\} \\
&= E_0^{\sigma}(t,a)k_0(\alpha,t)\left\{\int_t^b \delta_2(s)\widehat{E}_0(s,a)\Delta_{\alpha,a}s\right\}^{\Delta} \\
&= -\delta_2(t)\widehat{E}_0(t,a).
\end{aligned}
$$

This leads to

$$M_c u = \varphi_1 \widehat{D}_{\alpha}\{\varphi_2 D^{\alpha}[\varphi_1 u]\} = \varphi_1 \widehat{D}_{\alpha}\{-\widehat{E}_0(t,a)\} = 0,$$

so that u is a positive solution of $M_c x = 0$ on $[a,b)_{\mathbb{T}}$. Let

$$v_0(t) = u(t) \int_a^t \frac{\delta_2(s)\widehat{E}_0(s,a)}{\left(\int_s^b \delta_2(\eta)\widehat{E}_0(\eta,a)\Delta_{\alpha,a}\eta\right)\left(\int_{\sigma(s)}^b \delta_2(\eta)\widehat{E}_0(\eta,a)\Delta_{\alpha,a}\eta\right)} \Delta_{\alpha,a}s > 0.$$

Then

$$
\begin{aligned}
D^{\alpha}\left[\frac{v_0}{u}\right](t) - k_1(\alpha,t)\left(\frac{v_0}{u}\right)(t) &= k_0(\alpha,t)\left(\frac{v_0}{u}\right)^{\Delta}(t) \\
&= \frac{\delta_2(t)\widehat{E}_0(t,a)E_0(a,\sigma(t))}{\left(\int_t^b \delta_2(\eta)\widehat{E}_0(\eta,a)\Delta_{\alpha,a}\eta\right)\left(\int_{\sigma(t)}^b \delta_2(\eta)\widehat{E}_0(\eta,a)\Delta_{\alpha,a}\eta\right)} \\
&= \frac{E_0(t,a)\widehat{E}_0(t,a)}{p(t)u(t)u^{\sigma}(t)},
\end{aligned}
$$

which by reduction of order and (8.15) makes v_0 a solution of $M_c x = 0$ on $[a,b)_{\mathbb{T}}$. Note that

$$\lim_{t \to b^-} \frac{u(t)}{v_0(t)} = \lim_{t \to b^-} \frac{1}{\int_a^t \frac{\delta_2(s)\widehat{E}_0(s,a)}{\left(\int_s^b \delta_2(\eta)\widehat{E}_0(\eta,a)\Delta_{\alpha,a}\eta\right)\left(\int_{\sigma(s)}^b \delta_2(\eta)\widehat{E}_0(\eta,a)\Delta_{\alpha,a}\eta\right)}\Delta_{\alpha,a}s} = 0$$

since we are assuming the integral diverges to positive infinity in this case. Moreover, using the quotient rule from Lemma 1.9.9 we have

$$
\begin{aligned}
\frac{E_0(t,a)\widehat{E}_0(t,a)}{p(t)u(t)u^{\sigma}(t)} &= D^{\alpha}\left[\frac{v_0}{u}\right](t) - k_1(\alpha,t)\left(\frac{v_0}{u}\right)(t) \\
&= \frac{W(u,v_0)(t)}{u(t)u^{\sigma}(t)}.
\end{aligned}
\tag{8.23}
$$

Using (8.12) we have

$$D^{\alpha}\left[\frac{v_0}{u}\right](t) - k_1(\alpha,t)\frac{v_0(t)}{u(t)} = \frac{1}{E_0^{\sigma}(t,a)}D^{\alpha}\left[E_0(\cdot,a)\frac{v_0}{u}\right](t),$$

so that

$$\frac{1}{E_0^{\sigma}(t,a)}D^{\alpha}\left[E_0(\cdot,a)\frac{v_0}{u}\right](t) = \frac{W(u,v_0)(t)}{u(t)u^{\sigma}(t)} = \frac{E_0(t,a)\widehat{E}_0(t,a)}{p(t)u(t)u^{\sigma}(t)}.$$

Integrating both sides of this last equality from a to t via Theorem 1.9.7, we obtain

$$0 < \frac{v_0(t)}{u(t)} = \int_a^t \frac{E_0(s,a)\widehat{E}_0(s,a)}{p(s)u(s)u^{\sigma}(s)k_0(\alpha,s)}\Delta s.$$

Letting $t \to b^-$ we get one of the desired results, namely

$$\int_a^b \frac{E_0(s,a)\widehat{E}_0(s,a)}{p(s)u(s)u^{\sigma}(s)k_0(\alpha,s)}\Delta s = \infty.$$

Suppose v is any solution of $M_c x = 0$ such that u and v are linearly independent. Then

$$v(t) = c_1 u(t) + c_2 v_0(t),$$

where $c_2 \neq 0$. It follows that

$$\lim_{t \to b^-} \frac{u(t)}{v(t)} = \lim_{t \to b^-} \frac{u(t)}{c_1 u(t) + c_2 v_0(t)} = \lim_{t \to b^-} \frac{\frac{u(t)}{v_0(t)}}{c_1 \frac{u(t)}{v_0(t)} + c_2} = 0.$$

Again let v be a fixed solution of (8.1) such that u and v are linearly independent, but this time, pick $t_0 \in [a,b)_\mathbb{T}$ so that $v(t) \neq 0$ on $[t_0,b)_\mathbb{T}$, which is possible by the non-oscillatory nature of $M_c x = 0$. Then for $t \in [t_0,b)_\mathbb{T}$, similar to the calculations above in (8.23), we have

$$
\begin{aligned}
D^\alpha \left[\frac{u}{v} \right](t) - k_1(\alpha,t) \frac{u(t)}{v(t)} &= \frac{1}{E_0^\sigma(t,a)} D^\alpha \left[E_0(\cdot,a) \frac{u}{v} \right](t) \\
&= \frac{W(v,u)(t)}{v(t)v^\sigma(t)} \\
&= \frac{c_2 E_0(t,a)\widehat{E}_0(t,a)}{p(t)v(t)v^\sigma(t)},
\end{aligned}
$$

where $c_2 \neq 0$. Integrating both sides of this last equality from t_0 to t we obtain

$$\frac{u(t)}{v(t)} - \frac{u(t_0)}{v(t_0)} = c_2 \int_{t_0}^t \frac{E_0(s,a)\widehat{E}_0(s,a)}{p(s)v(s)v^\sigma(s)k_0(\alpha,s)} \Delta s.$$

Letting $t \to b^-$ we get another one of the desired results, that is,

$$\int_{t_0}^b \frac{E_0(s,a)\widehat{E}_0(s,a)}{p(s)v(s)v^\sigma(s)k_0(\alpha,s)} \Delta s < \infty.$$

We now show that (8.22) holds for $t < b$ sufficiently close. Above we saw that $v(t) \neq 0$ on $[t_0,b)_\mathbb{T}$. Note that the expression

$$\frac{p(t)D^\alpha v(t)}{v(t)}$$

is the same if $v(t)$ is replaced by $-v(t)$. Hence without loss of generality we can assume $v > 0$ on $[t_0,b)_\mathbb{T}$. For $t \in [t_0,b)_\mathbb{T}$, consider

$$\frac{p(t)D^\alpha v(t)}{v(t)} - \frac{p(t)D^\alpha u(t)}{u(t)} = \frac{p(t)W(u,v)(t)}{v(t)u(t)} = \frac{c_3 E_0(t,t_0)\widehat{E}_0(t,t_0)}{v(t)u(t)},$$

where

$$c_3 = \frac{p(t)W(u,v)(t)}{E_0(t,t_0)\widehat{E}_0(t,t_0)}$$

is a (nonzero) constant by Abel's formula. It remains to show that $c_3 > 0$. Since

$$\lim_{t \to b^-} \frac{v(t)}{u(t)} = \infty,$$

the ordinary Δ-derivative $(v/u)^\Delta > 0$ near b. Consequently, we see that

$$0 < k_0(\alpha,t) \left(\frac{v}{u} \right)^\Delta (t)$$

$$= D^\alpha \left[\frac{v}{u} \right] (t) - k_1(\alpha, t) \frac{v(t)}{u(t)}$$

$$= \frac{W(u,v)(t)}{u(t)u^\sigma(t)} = \frac{c_3 E_0(t, t_0) \widehat{E}_0(t, t_0)}{p(t)u(t)u^\sigma(t)},$$

and we get the desired result that $c_3 > 0$. This completes the proof. □

Remark 8.3.20 *The proof of the previous theorem is non-standard and new on all time scales. Usually, one uses the Pólya factorization of $M_c x$ to derive a Trench factorization, and then proceeds from there to prove the theorem on dominant and recessive solutions. As shown above, however, the proof is possible even if a Trench factorization cannot be found; all one needs is the Pólya factorization and what we have called the Trench divergence, proved earlier. Thus we get the same result with a weaker assumption.*

8.4 RICCATI EQUATION

Now we will consider the associated conformable nabla Riccati dynamic equation

$$Rz = 0, \quad \text{where} \quad Rz := \widehat{D}_\alpha z + q + \frac{k_0 z^\rho (z - k_1 p)^\rho}{k_0^\rho p^\rho + v(z - k_1 p)^\rho}, \tag{8.24}$$

where p is continuous and q is ld-continuous on $\mathscr{I}$ such that $p(t) > 0$ for all $t \in \mathscr{I}$. Also, assume throughout that $\alpha \in (0, 1]$.

Definition 8.4.1 *Denote by $\mathbb{D}_R$ the set of all differentiable functions z such that $\widehat{D}_\alpha z$ is ld-continuous and $k_0^\rho p^\rho + v(z - k_1 p)^\rho > 0$ on $\mathscr{I}$. A function $z \in \mathbb{D}_R$ is a solution of (8.24) on $\mathscr{I}$ if and only if $Rz(t) = 0$ for all $t \in \mathscr{I}$.*

Remark 8.4.2 *If $\alpha = 1$, then $k_0 \equiv 1$, $k_1 \equiv 0$, and (8.24) simplifies to*

$$Rz := \widehat{D}_\alpha z + q + \frac{(z^\rho)^2}{p^\rho + vz^\rho}$$

for any time scale $\mathbb{T}$, while the positivity condition is

$$k_0^\rho p^\rho + v(z - k_1 p)^\rho = p^\rho + vz^\rho > 0,$$

both of which agree with the classic time scales form. If $\mathbb{T} = \mathbb{R}$, then $v \equiv 0$, $\rho(t) = t$, and (8.24) reduces to

$$Rz := D^\alpha z + q + \frac{z(z - k_1 p)}{p},$$

while the positivity condition is

$$k_0^\rho p^\rho + v(z - k_1 p)^\rho = k_0 p > 0,$$

both of which agree with [2, 3].

Example 8.4.3 *Let* $\mathbb{T} = \mathbb{R}$, $\alpha \in (0,1]$, *and let* k_0, k_1 *satisfy* (A1). *Furthermore, let* $\theta_-, \theta_+ \in \mathbb{R}$ *with* $\theta_- < 0 < \theta_+$ *such that*

$$h_1(\theta_-, 0) := \int_0^{\theta_-} \frac{1}{k_0(\alpha, s)} ds = -\frac{\pi}{2}, \quad h_1(\theta_+, 0) := \int_0^{\theta_+} \frac{1}{k_0(\alpha, s)} ds = \frac{\pi}{2}.$$

Then the solution of the initial value problem

$$Rz = 0, \quad Rz := D^\alpha z + 1 + z^2 - k_1 z, \quad z(0) = 0$$

is given by

$$z(t) = -\tan(h_1(t, 0)), \quad t \in \mathscr{I} := (\theta_-, \theta_+).$$

Here we have solved (8.24) *for* $\mathbb{T} = \mathbb{R}$ *with* $p \equiv q \equiv 1$. $\triangle$

Theorem 8.4.4 (Factorization Theorem) *Let* M_c *be as in* (8.1), R *as in* (8.24), *and let* $x \in \mathbb{D}_R$ *such that* x *has no generalized zeros in* $\mathscr{I}$. *If* z *is defined by*

$$z = \frac{pD^\alpha x}{x} \tag{8.25}$$

on $\mathscr{I}$, *then* $z \in \mathbb{D}_R$ *and* $M_c x = xRz$ *on* $\mathscr{I}$.

Proof 8.4.5 *Assume* $x \in \mathbb{D}_R$ *such that* x *has no generalized zeros in* $\mathscr{I}$, *and let* z *have the form* (8.25). *Then*

$$z - k_1 p \stackrel{(8.25)}{=} \frac{pD^\alpha x}{x} - k_1 p = \frac{p}{x}(D^\alpha x - k_1 x) = \frac{k_0 p}{x} x^\Delta.$$

Taking ρ *of both sides and using the fact that* $x \in \mathbb{D}_R$ *implies* $x^{\Delta\rho} - x^\nabla$,

$$(z - k_1 p)^\rho = \left(\frac{k_0 p}{x}\right)^\rho x^\nabla.$$

Multiplying by ν *and using the nabla formula* $x - x^\rho = \nu x^\nabla$, *we have*

$$\nu(z - k_1 p)^\rho = \left(\frac{k_0 p}{x}\right)^\rho \nu x^\nabla = \left(\frac{k_0 p}{x}\right)^\rho (x - x^\rho) = \left(\frac{k_0 p}{x}\right)^\rho x - k_0^\rho p^\rho.$$

Consequently,

$$k_0^\rho p^\rho + \nu(z - k_1 p)^\rho = \left(\frac{k_0 p}{x}\right)^\rho x > 0$$

on $\mathscr{I}$ *since* x *has no generalized zeros in* $\mathscr{I}$. *Thus,* $z \in \mathbb{D}_R$. *Note that* (8.25) *also implies*

$$\frac{zx}{p} = k_1 x + k_0 x^\Delta.$$

If we solve for x^Δ *and take* ρ *of both sides, we have*

$$x^\nabla = \left(\frac{z - k_1 p}{k_0 p}\right)^\rho x^\rho.$$

Using the nabla formula $x^\rho = x - \nu x^\nabla$, we have

$$x^\nabla = \left(\frac{z - k_1 p}{k_0 p}\right)^\rho (x - \nu x^\nabla);$$

rearranging and resolving for x^∇ yields

$$x^\nabla = \frac{x(z - k_1 p)^\rho}{(k_0 p)^\rho + \nu(z - k_1 p)^\rho}.$$

Then

$$
\begin{aligned}
M_c x &= \widehat{D}_\alpha[p D^\alpha x] + qx \\
&\overset{(8.25)}{=} \widehat{D}_\alpha[zx] + qx \\
&= z^\rho \widehat{D}_\alpha x + x \widehat{D}_\alpha z - k_1 z^\rho x + qx \\
&= z^\rho \left(k_1 x + k_0 x^\nabla - k_1 x\right) + \left(\widehat{D}_\alpha z + q\right) x \\
&= k_0 z^\rho \left(\frac{x(z - k_1 p)^\rho}{(k_0 p)^\rho + \nu(z - k_1 p)^\rho}\right) + \left(\widehat{D}_\alpha z + q\right) x \\
&\overset{(8.24)}{=} xRz
\end{aligned}
$$

on $\mathscr{I}$. This completes the proof. □

Theorem 8.4.6 *The self-adjoint equation $M_c x = 0$ for M_c in (8.1) has a positive solution on $\mathbb{T}$ if and only if the Riccati equation (8.24) has a solution z on $\mathbb{T}^\kappa$.*

Proof 8.4.7 *First assume $M_c x = 0$ has a solution $x \in \mathbb{D}_R$ with $x > 0$ on $\mathbb{T}$, and let z have the Riccati substitution form (8.25). By Theorem 8.4.4 we have $z \in \mathbb{D}_R$ and $xRz = M_c x = 0$ on $\mathbb{T}$, making z a solution of (8.24) on $\mathbb{T}^\kappa$, since x positive means x has no generalized zeros.*

Conversely, let z be a solution of the Riccati equation (8.24) on $\mathbb{T}^\kappa$. Then $z \in \mathbb{D}_R$ and hence

$$0 < k_0^\rho p^\rho + \nu(z - k_1 p)^\rho = (k_0 p)^\rho \left[1 + \nu \left(\frac{z/p - k_1}{k_0}\right)^\rho\right]$$

on $\mathbb{T}_\kappa^\kappa$, and z is continuous on $\mathbb{T}^\kappa$. In particular, (z/p) is continuous, and the conformable exponential

$$E_{\frac{z}{p}} = e_{\frac{z/p - k_1}{k_0}} = \widehat{e}_{\frac{\left(\frac{z/p - k_1}{k_0}\right)^\rho}{1 + \nu \left(\frac{z/p - k_1}{k_0}\right)^\rho}} = \widehat{E}_{\frac{k_0 \left(\frac{z/p - k_1}{k_0}\right)^\rho}{1 + \nu \left(\frac{z/p - k_1}{k_0}\right)^\rho} + k_1}$$

exists and is well defined. Let $t_0 \in \mathbb{T}$, and let

$$x(t) = E_{\frac{z}{p}}(t, t_0), \quad t \in \mathbb{T}.$$

Then x is continuous and positive on $\mathbb{T}$. Note also that $D^\alpha x = zx/p$ is continuous, and if we solve this equation for z we get (8.25). Furthermore,

$$\widehat{D}_\alpha[p D^\alpha x] \overset{(8.25)}{=} \widehat{D}_\alpha[zx] = z^\rho \widehat{D}_\alpha x + x \widehat{D}_\alpha z - k_1 z^\rho x$$

is ld-continuous on $\mathbb{T}_\kappa^\kappa$, putting $x \in \mathbb{D}_R$. Moreover,

$$\widehat{D}_\alpha[pD^\alpha x] = z^\rho k_0 x^\nabla + x\widehat{D}_\alpha z = k_0 z^\rho \left(\frac{\left(\frac{z/p-k_1}{k_0}\right)^\rho}{1+v\left(\frac{z/p-k_1}{k_0}\right)^\rho} \right) x + x\widehat{D}_\alpha z = -qx$$

using (8.24), since $Rz = 0$. This shows that x is a solution of $M_c x = 0$ on $\mathbb{T}$, completing the proof. □

8.5 CAUCHY FUNCTION AND VARIATION OF CONSTANTS FORMULA

In this section we discuss the Cauchy function and derive a variation of constants formula for the conformable non-homogeneous self-adjoint dynamic equation $M_c x = h$ for M_c given in (8.1), where we assume h is a continuous function on some interval $\mathscr{I} \subseteq \mathbb{T}_\kappa^\kappa$. The following theorem follows from a standard argument, using the linearity of D^α in (1.1) and $\widehat{D}_\alpha$ in (1.9), and Theorem 8.1.2.

Theorem 8.5.1 *If x_1 and x_2 are linearly independent solutions of the conformable homogeneous equation $M_c x = 0$ on $\mathscr{I}$, and y is a particular solution of the conformable non-homogeneous equation $M_c x = h$ on $\mathscr{I}$, then*

$$x = c_1 x_1 + c_2 x_2 + y$$

is a general solution of $M_c x = h$ for constants $c_1, c_2 \in \mathbb{R}$.

Definition 8.5.2 (Cauchy Function) *Let M_c be as in (8.1). A function $x : \mathscr{I} \times \mathscr{I} \to \mathbb{R}$ is the Cauchy function for $M_c x = 0$ provided for each fixed $s \in \mathscr{I}$, $x(\cdot, s)$ is the solution of the initial value problem*

$$M_c x(\cdot, s) = 0, \quad x(s, s) = 0, \quad D^\alpha x(\rho(s), s) = \frac{1}{p^\rho(s)}.$$

It is easy to verify the following example.

Example 8.5.3 *If $q, h \equiv 0$ in (8.1), then the Cauchy function for*

$$\widehat{D}_\alpha[pD^\alpha x](t) = 0$$

is given by

$$x(t, s) = \int_s^t \frac{\widehat{E}_0(\tau, \rho(s))}{p(\tau)} \Delta_{\alpha,t} \tau$$

for all $t, s \in \mathscr{I}$. △

For M_c in (8.1), a formula for the Cauchy function for $M_c x = 0$ is given in the next theorem.

Theorem 8.5.4 *If u and v are linearly independent solutions of $M_c x = 0$ for M_c in (8.1), then the Cauchy function $x(t,s)$ for $M_c x = 0$ is given by*

$$x(t,s) = \frac{u(s)v(t) - v(s)u(t)}{p^\rho(s)[u(s)D^\alpha v^\rho(s) - v(s)D^\alpha u^\rho(s)]} \quad \text{for} \quad t,s \in \mathscr{I}. \tag{8.26}$$

Proof 8.5.5 *Let $y(t,s)$ be defined by the right-hand side of equation (8.26). Then note that for each fixed s, $y(\cdot,s)$ is a linear combination of the solutions u and v and as such is a solution of (8.1). Clearly $y(s,s) = 0$. Also note that*

$$D^\alpha y(t,s) = \frac{u(s)D^\alpha v(t) - v(s)D^\alpha u(t)}{p^\rho(s)[u(s)D^\alpha v^\rho(s) - v(s)D^\alpha u^\rho(s)]},$$

from which we have

$$D^\alpha y(\rho(s),s) = \frac{u(s)D^\alpha v^\rho(s) - v(s)D^\alpha u^\rho(s)}{p^\rho(s)[u(s)D^\alpha v^\rho(s) - v(s)D^\alpha u^\rho(s)]} = \frac{1}{p^\rho(s)}.$$

From the uniqueness of solutions of initial value problems (Theorem 8.1.2) we have that for each fixed s,

$$x(t,s) = y(t,s),$$

which gives us the desired result. □

Theorem 8.5.6 (Variation of Constants Formula) *Assume h is continuous on $\mathscr{I}$ and $a \in \mathscr{I}$. Let $x(t,s)$ be the Cauchy function for $M_c x = 0$ for M_c in (8.1). Then*

$$x(t) = \int_a^t \frac{x(t,s)h(s)}{k_0(\alpha,s)} \nabla s, \quad t \in \mathscr{I} \tag{8.27}$$

is the solution of the initial value problem

$$M_c x = h(t), \quad x(a) = 0, \quad D^\alpha x(a) = 0.$$

Proof 8.5.7 *In the proof we will use the fact from time scales calculus that if f, f^Δ, and f^∇ are continuous, then*

$$\left[\int_a^t f(t,s)\nabla s \right]^\Delta = \int_a^t f^\Delta(t,s)\nabla s + f(\sigma(t),\sigma(t))$$

and

$$\left[\int_a^t f(t,s)\nabla s \right]^\nabla = \int_a^t f^\nabla(t,s)\nabla s + f(\rho(t),t).$$

Let $x(t,s)$ be the Cauchy function for $M_c x = 0$, and set

$$x(t) = \int_a^t \frac{x(t,s)h(s)}{k_0(\alpha,s)} \nabla s.$$

Note that $x(a) = 0$. Taking the conformable derivative D^α of x,

$$D^\alpha x(t) \quad = \quad k_1 x + k_0 x^\Delta$$

$$
\begin{aligned}
&= k_1(t)x(t) + k_0\left[\int_a^t \frac{x^\Delta(t,s)h(s)}{k_0(\alpha,s)}\nabla s + \frac{x(\sigma(t),\sigma(t))h^\sigma(t)}{k_0(\alpha,\sigma(t))}\right]\\
&= \int_a^t \frac{D^\alpha[x(t,s)]h(s)}{k_0(\alpha,s)}\nabla s
\end{aligned}
$$

since the Cauchy function satisfies $x(\sigma(t),\sigma(t)) = 0$. Note that in the integral, D^α denotes the derivative with respect to the first variable t; thus $D^\alpha x(a) = 0$. Multiply by p on both sides to get

$$
(pD^\alpha x)(t) = \int_a^t \frac{p(t)D^\alpha[x(t,s)]h(s)}{k_0(\alpha,s)}\nabla s. \tag{8.28}
$$

Then

$$
\begin{aligned}
\widehat{D}_\alpha[pD^\alpha x] &= k_1(pD^\alpha x) + k_0(pD^\alpha x)^\nabla\\
&\overset{(8.28)}{=} \int_a^t \frac{k_1(t)p(t)D^\alpha[x(t,s)]h(s)}{k_0(\alpha,s)}\nabla s + \int_a^t \frac{k_0(t)\,(pD^\alpha[x(\cdot,s)])^\nabla(t)h(s)}{k_0(\alpha,s)}\nabla s\\
&\quad + p^\rho(t)D^\alpha x(\rho(t),t)h(t)\\
&= \int_a^t \widehat{D}_\alpha[pD^\alpha x(\cdot,s)](t)\frac{h(s)}{k_0(\alpha,s)}\nabla s + h(t)\\
&= h(t) - \int_a^t q(t)x(t,s)\frac{h(s)}{k_0(\alpha,s)}\nabla s\\
&= h(t) - q(t)x(t),
\end{aligned}
$$

by all of the properties of the Cauchy function. Consequently, $M_c x(t) = h(t)$. This completes the proof. □

8.6 BOUNDARY VALUE PROBLEMS AND GREEN FUNCTIONS

In this section we are concerned with Green functions for a general two-point boundary value problem (abbreviated by BVP) for $M_c x = 0$, M_c given in (8.1). Also assume

$$
k_0(\alpha,t) - \mu(t)k_1(\alpha,t) > 0, \quad t \in [a,b)_\mathbb{T}
$$

throughout this section.

Theorem 8.6.1 (Existence and Uniqueness of Solutions for General Two-Point BVPs)
Assume that the homogeneous boundary value problem

$$
M_c x = 0, \quad \xi x(a) - \beta D^\alpha x(a) = \gamma x(b) + \delta D^\alpha x(b) = 0 \tag{8.29}
$$

has only the trivial solution. Then the nonhomogeneous BVP

$$
M_c x = h(t), \quad \xi x(a) - \beta D^\alpha x(a) = A, \quad \gamma x(b) + \delta D^\alpha x(b) = B, \tag{8.30}
$$

where A and B are given constants and h is continuous, has a unique solution.

Proof 8.6.2 *The proof is similar to the classical ($\alpha = 1$) case and thus is omitted.* □

In the next example we give a BVP of the type (8.29) which does not have just the trivial solution. In Example 8.6.4 we give a necessary and sufficient condition for some boundary value problems of the form (8.29) to have only the trivial solution.

Example 8.6.3 *Find all solutions of the BVP*

$$\widehat{D}_\alpha[pD^\alpha x](t) = 0, \quad D^\alpha x(a) = D^\alpha x(b) = 0, \quad t \in (a,b)_\mathbb{T}, \tag{8.31}$$

where $a < b$ for $a \in \mathbb{T}_\kappa$ and $b \in \mathbb{T}^\kappa$. The BVP (8.31) is equivalent to a BVP of the form (8.29) if we take $q(t) \equiv 0$, $\xi = \gamma = 0$, and $\beta = \delta = 1$. A general solution of the boundary value problem in (8.31) is

$$x(t) = c_1 E_0(t,a) + c_2 \int_a^t \frac{\widehat{E}_0(s,a)}{p(s)} \Delta_{\alpha,t} s, \quad t \in [a,b]_\mathbb{T}, \tag{8.32}$$

and the boundary conditions lead to the equations

$$D^\alpha x(a) = \frac{c_2 \widehat{E}_0(a,a)}{p(a)} = 0 \quad \text{and} \quad D^\alpha x(b) = \frac{c_2 \widehat{E}_0(b,a)}{p(b)} = 0.$$

Thus $c_2 = 0$, and $x(t) = c_1 E_0(t,a)$ solves (8.31) for any constant $c_1 \in \mathbb{R}$. △

Example 8.6.4 *Let*

$$D = \xi\gamma \int_a^b \frac{\widehat{E}_0(s,a)}{p(s)} \Delta_{\alpha,b} s + \frac{\beta\gamma E_0(b,a)}{p(a)} + \frac{\xi\delta \widehat{E}_0(b,a)}{p(b)}.$$

Using (8.32), it is straightforward to show that the BVP (8.29) with $q \equiv 0$ has only the trivial solution if and only if $D \neq 0$. In particular, $D = 0$ for the BVP in Example 8.6.3 and thus (8.31) has nontrivial solutions, as seen above. △

Theorem 8.6.5 (Green Function for General Two-Point BVPs) *Assume that the BVP (8.29) has only the trivial solution. For each fixed $s \in [a,b]_\mathbb{T}$, let $u(\cdot,s)$ be the unique solution of the BVP*

$$M_c u(\cdot,s) = 0, \quad \xi u(a,s) - \beta D^\alpha u(a,s) = 0,$$
$$\gamma u(b,s) + \delta D^\alpha u(b,s) = -\gamma x(b,s) - \delta D^\alpha x(b,s), \tag{8.33}$$

where $x(t,s)$ is the Cauchy function (8.26) for $M_c x = 0$, M_c in (8.1). Define the Green function $G : [a,b]_\mathbb{T} \times [a,b]_\mathbb{T} \to \mathbb{R}$ for the BVP (8.29) by

$$G(t,s) = \begin{cases} u(t,s) & : a \leq t \leq s \leq b, \\ v(t,s) & : a \leq s \leq t \leq b, \end{cases}$$

where $v(t,s) := u(t,s) + x(t,s)$ for $t,s \in [a,b]_\mathbb{T}$. Then for each fixed $s \in [a,b]_\mathbb{T}$, $v(\cdot,s)$ is a solution of $M_c x = 0$ and satisfies the second boundary condition in (8.29). If h is continuous, then

$$x(t) := \int_a^b \frac{G(t,s)h(s)}{k_0(\alpha,s)} \nabla s$$

is the solution of the nonhomogeneous BVP (8.30) with $A = B = 0$.

Proof 8.6.6 *The existence and uniqueness of $u(t,s)$ is guaranteed by Theorem 8.6.1. Since for each fixed $s \in [a,b]_{\mathbb{T}}$, $u(\cdot,s)$ and $x(\cdot,s)$ are solutions of $M_c x = 0$, M_c in (8.1), we have for each fixed $s \in [a,b]_{\mathbb{T}}$ that $v(\cdot,s) = u(\cdot,s) + x(\cdot,s)$ is also a solution of $M_c x = 0$. It follows from (8.33) that $v(\cdot,s)$ satisfies the second boundary condition in (8.29) for each fixed $s \in [a,b]_{\mathbb{T}}$. First note that*

$$
\begin{aligned}
x(t) &= \int_a^b \frac{G(t,s)h(s)}{k_0(\alpha,s)} \nabla s \\
&= \int_a^t \frac{G(t,s)h(s)}{k_0(\alpha,s)} \nabla s + \int_t^b \frac{G(t,s)h(s)}{k_0(\alpha,s)} \nabla s \\
&= \int_a^t \frac{v(t,s)h(s)}{k_0(\alpha,s)} \nabla s + \int_t^b \frac{u(t,s)h(s)}{k_0(\alpha,s)} \nabla s \\
&= \int_a^t [u(t,s)+x(t,s)]\frac{h(s)}{k_0(\alpha,s)} \nabla s + \int_t^b \frac{u(t,s)h(s)}{k_0(\alpha,s)} \nabla s \\
&= \int_a^b \frac{u(t,s)h(s)}{k_0(\alpha,s)} \nabla s + \int_a^t \frac{x(t,s)h(s)}{k_0(\alpha,s)} \nabla s \\
&= \int_a^b \frac{u(t,s)h(s)}{k_0(\alpha,s)} \nabla s + w(t),
\end{aligned}
$$

where, by the variation of constants formula in Theorem 8.5.6, w is the solution of the IVP

$$ M_c w = h(t), \quad w(a) = D^\alpha w(a) = 0. \tag{8.34} $$

It follows that

$$ M_c x(t) = \int_a^b M_c u(t,s) \frac{h(s)}{k_0(\alpha,s)} \nabla s + M_c w(t) = M_c w(t) = h(t). $$

Hence x is a solution of the nonhomogeneous equation $M_c x = h(t)$. It remains to show that x satisfies the two boundary conditions in (8.29). Now

$$ \xi x(a) - \beta D^\alpha x(a) = \int_a^b [\xi u(a,s) - \beta D^\alpha u(a,s)]\frac{h(s)}{k_0(\alpha,s)} \nabla s = 0, $$

since for each fixed s, $u(\cdot,s)$ satisfies the first boundary condition in (8.29) and w satisfies (8.34). Hence x satisfies the first boundary condition in (8.29). From earlier in this proof,

$$
\begin{aligned}
x(t) &= \int_a^b u(t,s)\frac{h(s)}{k_0(\alpha,s)} \nabla s + \int_a^t x(t,s)\frac{h(s)}{k_0(\alpha,s)} \nabla s \\
&= \int_a^b v(t,s)\frac{h(s)}{k_0(\alpha,s)} \nabla s - \int_t^b x(t,s)\frac{h(s)}{k_0(\alpha,s)} \nabla s \\
&= \int_a^b v(t,s)\frac{h(s)}{k_0(\alpha,s)} \nabla s + \int_b^t x(t,s)\frac{h(s)}{k_0(\alpha,s)} \nabla s \\
&= \int_a^b v(t,s)\frac{h(s)}{k_0(\alpha,s)} \nabla s + y(t),
\end{aligned}
$$

where, by the variation of constants formula in Theorem 8.5.6, y solves

$$M_c y = h(t), \quad y(b) = D^\alpha y(b) = 0. \tag{8.35}$$

Then

$$\gamma x(b) + \delta D^\alpha x(b) \quad = \quad \int_a^b [\gamma v(b,s) + \delta D^\alpha v(b,s)] \frac{h(s)}{k_0(\alpha,s)} \nabla s = 0,$$

since for each fixed s, $v(\cdot,s)$ satisfies the second boundary condition in (8.29) and y satisfies (8.35). Hence x satisfies the second boundary condition in (8.29). The uniqueness of solutions of the nonhomogeneous BVP (8.30) (with $A = B = 0$) follows from Theorem 8.6.1. This completes the proof. □

Instead of the Cauchy function approach as above, in the next theorem we find another form of the Green function for the BVP (8.29); of particular interest is a symmetry condition on the square $[a,b]_\mathbb{T} \times [a,b]_\mathbb{T}$ satisfied by the Green function.

Theorem 8.6.7 *Assume that the BVP (8.29) has only the trivial solution. Let ϕ be the solution of the IVP*

$$M_c\phi = 0, \quad \phi(a) = \beta, \quad D^\alpha\phi(a) = \xi,$$

and let ψ be the solution of

$$M_c\psi = 0, \quad \psi(b) = \delta, \quad D^\alpha\psi(b) = -\gamma.$$

Then the Green function for the BVP (8.29) is given by

$$G(t,s) = \frac{1}{p^\rho(s)\,[\phi(s)D^\alpha\psi^\rho(s) - \psi(s)D^\alpha\phi^\rho(s)]} \begin{cases} \phi(t)\psi(s) & : a \le t \le s \le b, \\ \psi(t)\phi(s) & : a \le s \le t \le b. \end{cases} \tag{8.36}$$

Proof 8.6.8 *Let ϕ and ψ be as stated in the theorem. We use Theorem 8.6.5 to prove that G defined by (8.36) is the Green function for the BVP (8.29). Note that*

$$\xi\phi(a) - \beta D^\alpha\phi(a) = \xi\beta - \beta\xi = 0$$

and

$$\gamma\psi(b) + \delta D^\alpha\psi(b) = \gamma\delta - \delta\gamma = 0.$$

Hence ϕ and ψ satisfy the first and second boundary conditions in (8.29), respectively. Let

$$u(t,s) := \frac{\phi(t)\psi(s)}{p^\rho(s)\,[\phi(s)D^\alpha\psi^\rho(s) - \psi(s)D^\alpha\phi^\rho(s)]}$$

and

$$v(t,s) := \frac{\psi(t)\phi(s)}{p^\rho(s)\,[\phi(s)D^\alpha\psi^\rho(s) - \psi(s)D^\alpha\phi^\rho(s)]}$$

for $t \in [a,b]_\mathbb{T}$, $s \in [a,b]_\mathbb{T}$. Note that for each fixed $s \in [a,b]_\mathbb{T}$, $u(\cdot,s)$ and $v(\cdot,s)$ are solutions of $M_c x = 0$, M_c in (8.1), on $[a,b]_\mathbb{T}$. Also for each fixed $s \in [a,b]_\mathbb{T}$,

$$\xi u(a,s) - \beta D^\alpha u(a,s) = \frac{\psi(s)[\xi\phi(a) - \beta D^\alpha\phi(a)]}{p^\rho(s)\,[\phi(s)D^\alpha\psi^\rho(s) - \psi(s)D^\alpha\phi^\rho(s)]} = 0$$

and

$$\gamma v(b,s) + \delta D^\alpha v(b,s) = \frac{\phi(s)[\gamma \psi(b) + \delta D^\alpha \psi(b)]}{p^\rho(s)[\phi(s)D^\alpha \psi^\rho(s) - \psi(s)D^\alpha \phi^\rho(s)]} = 0.$$

Hence for each fixed $s \in [a,b]_\mathbb{T}$, $u(\cdot,s)$ and $v(\cdot,s)$ satisfy the first and second boundary conditions in (8.29), respectively. Let

$$\chi(t,s) := v(t,s) - u(t,s) = \frac{\psi(t)\phi(s) - \phi(t)\psi(s)}{p^\rho(s)[\phi(s)D^\alpha \psi^\rho(s) - \psi(s)D^\alpha \phi^\rho(s)]}.$$

It follows that for each fixed s, $\chi(\cdot,s)$ is a solution of $M_c x = 0$, $\chi(s,s) = 0$, and

$$D^\alpha \chi(\rho(s),s) = \frac{\phi(s)D^\alpha \psi^\rho(s) - \psi(s)D^\alpha \phi^\rho(s)}{p^\rho(s)[\phi(s)D^\alpha \psi^\rho(s) - \psi(s)D^\alpha \phi^\rho(s)]} = \frac{1}{p^\rho(s)}.$$

Consequently, $\chi(t,s) = x(t,s)$ is the Cauchy function for $M_c x = 0$, and we have

$$v(t,s) = u(t,s) + x(t,s).$$

It remains to prove that for each fixed s, $u(\cdot,s)$ satisfies (8.33). To see this, consider

$$\begin{aligned} \gamma u(b,s) + \delta D^\alpha u(b,s) &= \gamma v(b,s) + \delta D^\alpha v(b,s) - [\gamma x(b,s) + \delta D^\alpha x(b,s)] \\ &= -\gamma x(b,s) - \delta D^\alpha x(b,s). \end{aligned}$$

Hence by Theorem 8.6.5, $G(t,s)$ defined by (8.36) is the Green function for (8.29). This completes the proof. □

8.6.1 Conjugate Problem and Disconjugacy

In this subsection we examine Theorem 8.6.5 and Theorem 8.6.7 in more detail, in particular for the special case where the boundary conditions (8.33) are conjugate (also known as Dirichlet) boundary conditions.

Corollary 8.6.9 (Green Function for the Conjugate Problem) *Assume the BVP*

$$M_c x = 0, \quad x(a) = x(b) = 0, \tag{8.37}$$

has only the trivial solution. Let $x(t,s)$ be the Cauchy function for $M_c x = 0$, M_c in (8.1). For each fixed $s \in \mathscr{I}$, let $u(\cdot,s)$ be the unique solution of the BVP

$$M_c u(\cdot,s) = 0, \quad u(a,s) = 0, \quad u(b,s) = -x(b,s).$$

Then

$$G(t,s) = \begin{cases} u(t,s) & : t \leq s, \\ v(t,s) & : t \geq s, \end{cases}$$

where $v(t,s) = u(t,s) + x(t,s)$, is the Green function for the BVP (8.37). Moreover, for each fixed $s \in [a,b]_\mathbb{T}$, $v(\cdot,s)$ is a solution of $M_c x = 0$ and $v(b,s) = 0$.

Proof 8.6.10 *This corollary follows from Theorem 8.6.5 with $\xi = \gamma = 1$ and $\beta = \delta = 0$.* □

Corollary 8.6.11 *The Green function for the BVP* (8.37) *with* $q \equiv 0$ *is given by*

$$
G(t,s) = \frac{-1}{\int_a^b \frac{\widehat{E}_0(\tau,a)}{p(\tau)} \Delta_{\alpha,b}\tau}
\begin{cases}
\int_a^t \frac{\widehat{E}_0(\tau,a)}{p(\tau)} \Delta_{\alpha,t}\tau \int_s^b \frac{\widehat{E}_0(\tau,\rho(s))}{p(\tau)} \Delta_{\alpha,b}\tau & : a \le t \le s \le b, \\
\int_a^s \frac{\widehat{E}_0(\tau,a)}{p(\tau)} \Delta_{\alpha,t}\tau \int_t^b \frac{\widehat{E}_0(\tau,\rho(s))}{p(\tau)} \Delta_{\alpha,b}\tau & : a \le s \le t \le b.
\end{cases}
$$

Proof 8.6.12 *It is easy to check that the BVP*

$$
\widehat{D}_\alpha[pD^\alpha x](t) = 0, \qquad x(a) = x(b) = 0,
$$

has only the trivial solution from a modification of Example 8.6.3. By Example 8.5.3, the Cauchy function for

$$
\widehat{D}_\alpha[pD^\alpha x](t) = 0 \tag{8.38}
$$

is given by

$$
x(t,s) = \int_s^t \frac{\widehat{E}_0(\tau,\rho(s))}{p(\tau)} \Delta_{\alpha,t}\tau.
$$

By Corollary 8.6.9, $u(\cdot,s)$ from the statement of Corollary 8.6.9 solves (8.38) *for each fixed $s \in [a,b]_\mathbb{T}$ and satisfies*

$$
u(a,s) = 0 \quad and \quad u(b,s) = -x(b,s) = -\int_s^b \frac{\widehat{E}_0(\tau,\rho(s))}{p(\tau)} \Delta_{\alpha,b}\tau. \tag{8.39}
$$

Since

$$
x_1(t) = E_0(t,a) \quad and \quad x_2(t) = \int_a^t \frac{\widehat{E}_0(\tau,a)}{p(\tau)} \Delta_{\alpha,t}\tau
$$

are solutions of (8.38),

$$
u(t,s) = \xi(s)E_0(t,a) + \beta(s)\int_a^t \frac{\widehat{E}_0(\tau,a)}{p(\tau)} \Delta_{\alpha,t}\tau.
$$

Using the boundary conditions (8.39), *it can be shown that*

$$
u(t,s) = \frac{-1}{\int_a^b \frac{\widehat{E}_0(\tau,a)}{p(\tau)} \Delta_{\alpha,b}\tau} \int_a^t \frac{\widehat{E}_0(\tau,a)}{p(\tau)} \Delta_{\alpha,t}\tau \int_s^b \frac{\widehat{E}_0(\tau,\rho(s))}{p(\tau)} \Delta_{\alpha,b}\tau.
$$

Hence $G(t,s)$ has the desired form for $t \le s$. By Corollary 8.6.9 for $t \ge s$,

$$
G(t,s) = x(t,s) + u(t,s).
$$

Therefore for $t \ge s$,

$$
G(t,s) = \int_s^t \frac{\widehat{E}_0(\tau,\rho(s))}{p(\tau)} \Delta_{\alpha,t}\tau - \frac{\int_a^t \frac{\widehat{E}_0(\tau,a)}{p(\tau)} \Delta_{\alpha,t}\tau \int_s^b \frac{\widehat{E}_0(\tau,\rho(s))}{p(\tau)} \Delta_{\alpha,b}\tau}{\int_a^b \frac{\widehat{E}_0(\tau,a)}{p(\tau)} \Delta_{\alpha,b}\tau}
$$

$$= \frac{\int_s^t \frac{\widehat{E}_0(\tau,\rho(s))}{p(\tau)}\Delta_{\alpha,t}\tau \int_a^b \frac{\widehat{E}_0(\tau,a)}{p(\tau)}\Delta_{\alpha,b}\tau - \int_a^t \frac{\widehat{E}_0(\tau,a)}{p(\tau)}\Delta_{\alpha,t}\tau \int_s^b \frac{\widehat{E}_0(\tau,\rho(s))}{p(\tau)}\Delta_{\alpha,b}\tau}{\int_a^b \frac{\widehat{E}_0(\tau,a)}{p(\tau)}\Delta_{\alpha,b}\tau}$$

$$= \frac{\int_s^t \frac{\widehat{E}_0(\tau,\rho(s))}{p(\tau)}\Delta_{\alpha,t}\tau \int_a^b \frac{\widehat{E}_0(\tau,a)}{p(\tau)}\Delta_{\alpha,b}\tau - \int_a^t \frac{\widehat{E}_0(\tau,\rho(s))}{p(\tau)}\Delta_{\alpha,t}\tau \int_s^b \frac{\widehat{E}_0(\tau,a)}{p(\tau)}\Delta_{\alpha,b}\tau}{\int_a^b \frac{\widehat{E}_0(\tau,a)}{p(\tau)}\Delta_{\alpha,b}\tau}$$

$$= \frac{\int_s^t \frac{\widehat{E}_0(\tau,\rho(s))}{p(\tau)}\Delta_{\alpha,t}\tau \int_a^s \frac{\widehat{E}_0(\tau,a)}{p(\tau)}\Delta_{\alpha,b}\tau - \int_a^s \frac{\widehat{E}_0(\tau,\rho(s))}{p(\tau)}\Delta_{\alpha,t}\tau \int_s^b \frac{\widehat{E}_0(\tau,a)}{p(\tau)}\Delta_{\alpha,b}\tau}{\int_a^b \frac{\widehat{E}_0(\tau,a)}{p(\tau)}\Delta_{\alpha,b}\tau}$$

$$= \frac{\int_s^t \frac{\widehat{E}_0(\tau,a)}{p(\tau)}\Delta_{\alpha,b}\tau \int_a^s \frac{\widehat{E}_0(\tau,\rho(s))}{p(\tau)}\Delta_{\alpha,t}\tau - \int_a^s \frac{\widehat{E}_0(\tau,\rho(s))}{p(\tau)}\Delta_{\alpha,t}\tau \int_s^b \frac{\widehat{E}_0(\tau,a)}{p(\tau)}\Delta_{\alpha,b}\tau}{\int_a^b \frac{\widehat{E}_0(\tau,a)}{p(\tau)}\Delta_{\alpha,b}\tau}$$

$$= \int_a^s \frac{\widehat{E}_0(\tau,\rho(s))}{p(\tau)}\Delta_{\alpha,t}\tau \left(\frac{-\int_t^b \frac{\widehat{E}_0(\tau,a)}{p(\tau)}\Delta_{\alpha,b}\tau}{\int_a^b \frac{\widehat{E}_0(\tau,a)}{p(\tau)}\Delta_{\alpha,b}\tau} \right)$$

$$= -\frac{\int_a^s \frac{\widehat{E}_0(\tau,a)}{p(\tau)}\Delta_{\alpha,t}\tau \int_t^b \frac{\widehat{E}_0(\tau,\rho(s))}{p(\tau)}\Delta_{\alpha,b}\tau}{\int_a^b \frac{\widehat{E}_0(\tau,a)}{p(\tau)}\Delta_{\alpha,b}\tau}$$

which is the desired result. □

The following corollary follows immediately from Corollary 8.6.11 by taking

$$p^{-1}(\tau) = \widehat{E}_0(a,\tau)D^{\alpha}\left[\frac{\tau}{E_0(b,\tau)}\right]. \tag{8.40}$$

Corollary 8.6.13 *Let $a,b \in \mathbb{T}$ with $a < b$, and let p be given via (8.40). Then the Green function for the conjugate BVP*

$$\widehat{D}_{\alpha}\left[pD^{\alpha}x\right](t) = 0, \quad x(a) = x(b) = 0$$

is given by

$$G(t,s) = \frac{-\widehat{E}_0(a,\rho(s))}{(b-a)E_0(b,t)} \begin{cases} (t-a)(b-s) & : a \le t \le s \le b, \\ (s-a)(b-t) & : a \le s \le t \le b. \end{cases}$$

Proof 8.6.14 *By Corollary 8.6.11, the Green function for the BVP (8.37) with $q \equiv 0$ is given by*

$$G(t,s) = \frac{-1}{\int_a^b \frac{\widehat{E}_0(\tau,a)}{p(\tau)}\Delta_{\alpha,b}\tau} \begin{cases} \int_a^t \frac{\widehat{E}_0(\tau,a)}{p(\tau)}\Delta_{\alpha,t}\tau \int_s^b \frac{\widehat{E}_0(\tau,\rho(s))}{p(\tau)}\Delta_{\alpha,b}\tau & : a \le t \le s \le b, \\ \int_a^s \frac{\widehat{E}_0(\tau,a)}{p(\tau)}\Delta_{\alpha,t}\tau \int_t^b \frac{\widehat{E}_0(\tau,\rho(s))}{p(\tau)}\Delta_{\alpha,b}\tau & : a \le s \le t \le b, \end{cases}$$

which can be rewritten as

$$G(t,s) = \frac{-1}{\int_a^b \frac{\widehat{E}_0(\tau,a)}{p(\tau)}\Delta_{\alpha,b}\tau} \begin{cases} \int_a^t \frac{\widehat{E}_0(\tau,\rho(s))}{p(\tau)}\Delta_{\alpha,t}\tau \int_s^b \frac{\widehat{E}_0(\tau,a)}{p(\tau)}\Delta_{\alpha,b}\tau & : a \le t \le s \le b, \\ \int_a^s \frac{\widehat{E}_0(\tau,\rho(s))}{p(\tau)}\Delta_{\alpha,t}\tau \int_t^b \frac{\widehat{E}_0(\tau,a)}{p(\tau)}\Delta_{\alpha,b}\tau & : a \le s \le t \le b \end{cases}$$

by the semi-group property of the exponential function. Using (8.40) and the fundamental theorem of calculus, we see that

$$\int_a^b \frac{\widehat{E}_0(\tau,a)}{p(\tau)} \Delta_{\alpha,b}\tau \;=\; \int_a^b \widehat{E}_0(\tau,a)\widehat{E}_0(a,\tau)D^\alpha\left[\frac{\tau}{E_0(b,\tau)}\right]\Delta_{\alpha,b}\tau$$

$$= \;\frac{b}{E_0(b,b)} - \frac{a}{E_0(b,a)}E_0(b,a)$$

$$= \; b-a;$$

similarly

$$\int_s^b \frac{\widehat{E}_0(\tau,a)}{p(\tau)} \Delta_{\alpha,b}\tau = b - s$$

and

$$\int_t^b \frac{\widehat{E}_0(\tau,a)}{p(\tau)} \Delta_{\alpha,b}\tau = b - t.$$

Also,

$$\int_a^t \frac{\widehat{E}_0(\tau,\rho(s))}{p(\tau)} \Delta_{\alpha,t}\tau \;=\; \int_a^t \widehat{E}_0(\tau,\rho(s))\widehat{E}_0(a,\tau)D^\alpha\left[\frac{\tau}{E_0(b,\tau)}\right]\Delta_{\alpha,t}\tau$$

$$= \;\widehat{E}_0(a,\rho(s)) \int_a^t D^\alpha\left[\frac{\tau}{E_0(b,\tau)}\right]\Delta_{\alpha,t}\tau$$

$$= \;\widehat{E}_0(a,\rho(s)) \left(\frac{t}{E_0(b,t)} - \frac{a}{E_0(b,a)}E_0(t,a)\right)$$

$$= \;\frac{\widehat{E}_0(a,\rho(s))}{E_0(b,t)}(t-a),$$

and

$$\int_a^s \frac{\widehat{E}_0(\tau,\rho(s))}{p(\tau)} \Delta_{\alpha,t}\tau \;=\; \widehat{E}_0(a,\rho(s)) \int_a^s D^\alpha\left[\frac{\tau}{E_0(b,\tau)}\right]\Delta_{\alpha,t}\tau$$

$$= \;\widehat{E}_0(a,\rho(s))E_0(t,s) \int_a^s D^\alpha\left[\frac{\tau}{E_0(b,\tau)}\right]\Delta_{\alpha,s}\tau$$

$$= \;\widehat{E}_0(a,\rho(s))E_0(t,s) \left(\frac{s}{E_0(b,s)} - \frac{a}{E_0(b,a)}E_0(s,a)\right)$$

$$= \;\frac{\widehat{E}_0(a,\rho(s))}{E_0(b,t)}(s-a).$$

This concludes the proof. □

The following example follows immediately from Corollary 8.6.11 by taking $\mathbb{T} = \mathbb{R}$ and $p(\tau) \equiv 1$.

Example 8.6.15 *Let* $\mathbb{T} = \mathbb{R}$ *and* $d_\alpha\tau := \dfrac{d\tau}{k_0(\alpha,\tau)}$. *Then the Green function for the BVP*

$$D^\alpha D^\alpha x = 0, \quad x(a) = x(b) = 0$$

is given by

$$G(t,s) = \frac{-E_0(t,s)}{\int_a^b 1 d_\alpha \tau} \begin{cases} \int_a^t 1 d_\alpha \tau \int_s^b 1 d_\alpha \tau & : a \leq t \leq s \leq b, \\ \int_a^s 1 d_\alpha \tau \int_t^b 1 d_\alpha \tau & : a \leq s \leq t \leq b. \end{cases}$$

For example, if $k_1(t) = (1-\alpha)(\omega t)^\alpha$ and $k_0(t) = \alpha(\omega t)^{1-\alpha}$ for $\alpha \in (0,1]$ and $\omega, t \in (0,\infty)$, then

$$\int_a^t 1 d_\alpha \tau = \frac{t^\alpha - a^\alpha}{\alpha^2 \omega^{1-\alpha}} \quad \text{and} \quad E_0(t,s) = e^{-\left(\frac{1-\alpha}{2\alpha^2}\right)\omega^{2\alpha-1}(t^{2\alpha}-s^{2\alpha})}$$

for $a,s \in (0,\infty)$, and

$$G(t,s) = \frac{-e^{-\left(\frac{1-\alpha}{2\alpha^2}\right)\omega^{2\alpha-1}(t^{2\alpha}-s^{2\alpha})}}{\alpha^2 \omega^{1-\alpha}(b^\alpha - a^\alpha)} \begin{cases} (t^\alpha - a^\alpha)(b^\alpha - s^\alpha) & : a \leq t \leq s \leq b, \\ (s^\alpha - a^\alpha)(b^\alpha - t^\alpha) & : a \leq s \leq t \leq b. \end{cases}$$

The proofs of the following two theorems are similar to their classical ($\alpha = 1$) counterparts and thus are omitted.

Theorem 8.6.16 *Let M_c be as in (8.1). If $p > 0$ and $M_c x = 0$ is disconjugate on $[a,b]_\mathbb{T}$, then the Green function for the conjugate BVP (8.37) exists and satisfies*

$$G(t,s) < 0 \quad \text{for} \quad t,s \in (a,b)_\mathbb{T}.$$

Theorem 8.6.17 (Comparison Theorem for Conjugate BVPs) *Let M_c be as in (8.1). Let $p > 0$ and assume that $M_c x = 0$ is disconjugate on $[a,b]_\mathbb{T}$. If $u,v \in \mathbb{D}$ satisfy*

$$M_c u(t) \leq M_c v(t) \text{ for all } t \in [a,b]_\mathbb{T}, \quad u(a) \geq v(a), \quad u(b) \geq v(b),$$

then

$$u(t) \geq v(t) \quad \text{for all} \quad t \in [a,b]_\mathbb{T}.$$

8.6.2 Right Focal Problem

Similar to the subsection above on the conjugate boundary conditions and disconjugacy, here we examine Theorem 8.6.5 and Theorem 8.6.7 for the special case where the boundary conditions (8.33) are right focal boundary conditions, namely the boundary value problem

$$M_c x = 0, \quad x(a) = D^\alpha x(b) = 0. \tag{8.41}$$

Corollary 8.6.18 (Green Function for Focal BVPs) *Assume that the BVP (8.41) has only the trivial solution. For each fixed $s \in [a,b]_\mathbb{T}$, let $u(\cdot,s)$ be the solution of the BVP*

$$M_c u(\cdot,s) = 0, \quad u(a,s) = 0, \quad D^\alpha u(b,s) = -D^\alpha x(b,s),$$

where $x(t,s)$ is the Cauchy function of $M_c x = 0$. Then

$$G(t,s) = \begin{cases} u(t,s) & : a \leq t \leq s \leq b \\ u(t,s) + x(t,s) & : a \leq s \leq t \leq b \end{cases}$$

is the Green function for the right focal BVP (8.41).

Proof 8.6.19 *This follows from Theorem 8.6.5 with $\xi = \delta = 1$ and $\beta = \gamma = 0$.* □

Corollary 8.6.20 *The Green function for the focal BVP (8.41) with $q \equiv 0$ is given by*

$$G(t,s) = \begin{cases} -\int_a^t \dfrac{\widehat{E}_0(\tau,\rho(s))}{p(\tau)}\Delta_{\alpha,t}\tau & : a \leq t \leq s \leq b, \\[3mm] -\int_a^s \dfrac{\widehat{E}_0(\tau,\rho(s))}{p(\tau)}\Delta_{\alpha,t}\tau & : a \leq s \leq t \leq b. \end{cases}$$

Proof 8.6.21 *It is easy to see that (8.41) with $q \equiv 0$ has only the trivial solution. Hence we can apply Corollary 8.6.18 to find the focal Green function $G(t,s)$. For $t \leq s$, $G(t,s) = u(t,s)$, where for each fixed s, $u(\cdot,s)$ solves the BVP*

$$M_c u(\cdot,s) = 0, \quad u(a,s) = 0, \quad D^\alpha u(b,s) = -D^\alpha x(b,s),$$

and where $x(t,s)$ is the Cauchy function for (8.38). Since $M_c u(\cdot,s) = 0$ and $u(a,s) = 0$, we have that $u(\cdot,s)$ must have the form

$$u(t,s) = A(s)\int_a^t \frac{\widehat{E}_0(\tau,a)}{p(\tau)}\Delta_{\alpha,t}\tau$$

for some function A. The boundary condition at b is

$$D^\alpha u(b,s) = -D^\alpha x(b,s)$$

implies that

$$\frac{A(s)}{p(b)}\widehat{E}_0(b,a) = \frac{-\widehat{E}_0(b,\rho(s))}{p(b)},$$

which yields

$$A(s) = -\widehat{E}_0(a,\rho(s)).$$

Overall, we get that

$$u(t,s) = -\int_a^t \frac{\widehat{E}_0(\tau,\rho(s))}{p(\tau)}\Delta_{\alpha,t}\tau,$$

which is the desired expression for $G(t,s)$ if $t \leq s$. If $t \geq s$, then

$$\begin{aligned} G(t,s) &= u(t,s) + x(t,s) \\ &= -\int_a^t \frac{\widehat{E}_0(\tau,\rho(s))}{p(\tau)}\Delta_{\alpha,t}\tau + \int_s^t \frac{\widehat{E}_0(\tau,\rho(s))}{p(\tau)}\Delta_{\alpha,t}\tau \\ &= -\int_a^s \frac{\widehat{E}_0(\tau,\rho(s))}{p(\tau)}\Delta_{\alpha,t}\tau. \end{aligned}$$

This completes the proof. □

Example 8.6.22 *Let* $\mathbb{T} = \mathbb{R}$ *and* $d_\alpha \tau = \dfrac{d\tau}{k_0(\alpha, \tau)}$. *Then the Green function for the focal BVP (8.41) with* $p(t) \equiv 1$ *and* $q \equiv 0$ *is given by*

$$G(t,s) = -E_0(t,s) \begin{cases} \displaystyle\int_{a_s}^t 1 d_\alpha \tau & : a \le t \le s \le b, \\ \displaystyle\int_a^t 1 d_\alpha \tau & : a \le s \le t \le b. \end{cases}$$

For example, if $k_1(t) = (1-\alpha)(\omega t)^\alpha$ *and* $k_0(t) = \alpha(\omega t)^{1-\alpha}$ *for* $\alpha \in (0,1]$ *and* $\omega, t \in (0,\infty)$, *then*

$$\int_a^t 1 d_\alpha \tau = \frac{t^\alpha - a^\alpha}{\alpha^2 \omega^{1-\alpha}} \quad \text{and} \quad e_0(t,s) = e^{-\left(\frac{1-\alpha}{2\alpha^2}\right)\omega^{2\alpha-1}(t^{2\alpha}-s^{2\alpha})}$$

for $a, s \in (0,\infty)$, *and*

$$G(t,s) = \frac{-e^{-\left(\frac{1-\alpha}{2\alpha^2}\right)\omega^{2\alpha-1}(t^{2\alpha}-s^{2\alpha})}}{\alpha^2 \omega^{1-\alpha}} \begin{cases} t^\alpha - a^\alpha & : t \le s, \\ s^\alpha - a^\alpha & : t \ge s. \end{cases}$$

Corollary 8.6.23 *Let* $a, b \in \mathbb{T}$ *with* $a < b$, *and let* p *be given via (8.40). Then the Green function for the BVP*

$$\widehat{D}_\alpha [pD^\alpha x](t) = 0, \quad x(a) = D^\alpha x(b) = 0$$

is given by

$$G(t,s) = \frac{-\widehat{E}_0(a,\rho(s))}{E_0(b,t)} \begin{cases} t - a & : a \le t \le s \le b, \\ s - a & : a \le s \le t \le b. \end{cases}$$

8.6.3 Periodic Problem

Finally we consider periodic boundary conditions. In traditional ($\alpha = 1$ and $\mathbb{T} = \mathbb{R}$) calculus, the geometry of periodicity means returning to the same values in the sense that they lie along the same horizontal straight line (geodesic with slope zero). Considering the context of the derivative (8.1), we investigate the periodic BVP

$$M_c x = 0, \quad x(a) = E_0(a,b)x(b), \quad D^\alpha x(a) = E_0(a,b)D^\alpha x(b). \tag{8.42}$$

Theorem 8.6.24 *Let* M_c *be as in (8.1). Assume that the homogeneous periodic BVP (8.42) has only the trivial solution. Then for* $t \in (a,b)_\mathbb{T}$, *the nonhomogeneous BVP*

$$M_c x = h(t), \quad x(a) - E_0(a,b)x(b) = A, \quad D^\alpha x(a) - E_0(a,b)D^\alpha x(b) = B, \tag{8.43}$$

where A and B are given constants and h is continuous, has a unique solution.

Proof 8.6.25 *Let* x_1 *and* x_2 *be linearly independent solutions of* $M_c x = 0$. *Then*

$$x(t) = c_1 x_1(t) + c_2 x_2(t)$$

is a general solution of $M_c x = 0$. Note that x satisfies the boundary conditions in (8.42) if and only if c_1 and c_2 are constants satisfying

$$M \begin{pmatrix} c_1 \\ c_2 \end{pmatrix} = 0$$

with

$$M = \begin{pmatrix} x_1(a) - E_0(a,b)x_1(b) & x_2(a) - E_0(a,b)x_2(b) \\ D^\alpha x_1(a) - E_0(a,b)D^\alpha x_1(b) & D^\alpha x_2(a) - E_0(a,b)D^\alpha x_2(b) \end{pmatrix}.$$

Since we are assuming that (8.42) has only the trivial solution, it follows that

$$c_1 = c_2 = 0$$

is the unique solution of the above linear system. Hence

$$\det M \neq 0. \tag{8.44}$$

Now we show that (8.43) has a unique solution. Let u_0 be a fixed solution of $M_c u = h(t)$. Then a general solution of $M_c u = h(t)$ is given by

$$u(t) = a_1 x_1(t) + a_2 x_2(t) + u_0(t).$$

It follows that u satisfies the boundary conditions in (8.43) if and only if a_1 and a_2 are constants satisfying the system of equations

$$M \begin{pmatrix} a_1 \\ a_2 \end{pmatrix} = \begin{pmatrix} A - u_0(a) + E_0(a,b)u_0(b) \\ B - D^\alpha u_0(a) + E_0(a,b)D^\alpha u_0(b) \end{pmatrix}.$$

This system has a unique solution because of (8.44), and hence (8.43) has a unique solution. This completes the proof. □

Theorem 8.6.26 (Green Function for Periodic BVPs) *Assume that the homogeneous BVP (8.42) has only the trivial solution. For each fixed $s \in \mathcal{I}$, let $u(\cdot, s)$ be the solution of the BVP*

$$\begin{cases} M_c u(\cdot, s) = 0 \\ u(a,s) = E_0(a,b)\left[u(b,s) + x(b,s)\right] \\ D^\alpha u(a,s) = E_0(a,b)\left[D^\alpha u(b,s) + D^\alpha x(b,s)\right], \end{cases} \tag{8.45}$$

where $x(t,s)$ is the Cauchy function for $M_c x = 0$, M_c as in (8.1). Define

$$G(t,s) := \begin{cases} u(t,s) & : t \leq s, \\ v(t,s) & : t \geq s, \end{cases} \tag{8.46}$$

where $v(t,s) := u(t,s) + x(t,s)$. If h is continuous, then

$$x(t) := \int_a^b G(t,s) \frac{h(s)}{k_0(\alpha,s)} \nabla s$$

is the unique solution of the nonhomogeneous periodic BVP (8.43) with $A = B = 0$. Furthermore, for each fixed $s \in [a,b]_\mathbb{T}$, $v(\cdot, s)$ is a solution of $M_c x = 0$, and

$$u(a,s) = E_0(a,b)v(b,s), \quad D^\alpha u(a,s) = E_0(a,b)D^\alpha v(b,s).$$

Proof 8.6.27 *The existence and uniqueness of $u(t,s)$ is guaranteed by Theorem 8.6.24. Since $v(t,s) = u(t,s) + x(t,s)$, we have for each fixed s that $v(\cdot,s)$ is a solution of $M_c x = 0$. Using the boundary conditions in (8.45), it is easy to see that for each fixed s, $u(a,s) = E_0(a,b)v(b,s)$ and $D^\alpha u(a,s) = E_0(a,b)D^\alpha v(b,s)$. Let G be as in (8.46) and notice that*

$$
\begin{aligned}
x(t) &= \int_a^b G(t,s)\frac{h(s)}{k_0(\alpha,s)}\nabla s \\
&= \int_a^b u(t,s)\frac{h(s)}{k_0(\alpha,s)}\nabla s + \int_a^t x(t,s)\frac{h(s)}{k_0(\alpha,s)}\nabla s \\
&= \int_a^b u(t,s)\frac{h(s)}{k_0(\alpha,s)}\nabla s + z(t),
\end{aligned}
$$

where, by the variation of constants formula in Theorem 8.5.6, z solves

$$
M_c z = h(t), \quad z(a) = D^\alpha z(a) = 0.
$$

Hence

$$
M_c x(t) = \int_a^b M_c u(t,s)\frac{h(s)}{k_0(\alpha,s)}\nabla s + M_c z(t) = M_c z(t) = h(t).
$$

Thus x is a solution of $M_c x = h(t)$. Note that

$$
\begin{aligned}
x(a) &= \int_a^b u(a,s)\frac{h(s)}{k_0(\alpha,s)}\nabla s + z(a) = \int_a^b E_0(a,b)v(b,s)\frac{h(s)}{k_0(\alpha,s)}\nabla s \\
&= \int_a^b E_0(a,b)G(b,s)\frac{h(s)}{k_0(\alpha,s)}\nabla s = E_0(a,b)x(b),
\end{aligned}
$$

and

$$
\begin{aligned}
D^\alpha x(a) &= \int_a^b D^\alpha u(a,s)\frac{h(s)}{k_0(\alpha,s)}\nabla s + D^\alpha z(a) \\
&= \int_a^b E_0(a,b)D^\alpha v(b,s)\frac{h(s)}{k_0(\alpha,s)}\nabla s \\
&= \int_a^b E_0(a,b)D^\alpha G(b,s)\frac{h(s)}{k_0(\alpha,s)}\nabla s \\
&= E_0(a,b)D^\alpha x(b).
\end{aligned}
$$

Hence, x satisfies the periodic boundary conditions in (8.42). This completes the proof. □

CHAPTER 9

The Conformable Laplace Transform

Suppose that $\mathbb{T}$ is a time scale with forward jump operator and delta differentiation operator σ and Δ, respectively. Let $\alpha \in (0,1]$, k_0 and k_1 satisfy $(A1)$, $k_0 \in \mathscr{C}^1_{rd}(\mathbb{T})$ and $|E_0(\infty,0)| < \infty$.

9.1 DEFINITION AND PROPERTIES

Let $h \in \mathscr{C}^1_{rd}(\mathbb{T})$ and $g \in \mathscr{R}_c$ be such that

$$zh^\sigma E_g^\sigma(\cdot,0) = -gE_g(\cdot,0)$$

$$D^\alpha h - zhh^\sigma + (z-k_1)h^\sigma - k_1 h = 0 \tag{9.1}$$

$$h(0) = 1$$

for $z \in \mathscr{H}_c(h)$, where $\mathscr{H}_c(h)$ consists of all complex numbers $z \in \mathscr{R}_c$ for which $z - k_1 \in \mathscr{R}_c$ and

$$k_0 + h^\sigma z(\mu - k_1) \neq 0.$$

Remark 9.1.1 *Note that there exists a unique $h \in \mathscr{C}^1_{rd}(\mathbb{T})$ that satisfies the second equation and the third equation of the system (9.1). Hence, there exists a unique $g \in \mathscr{R}_c$ that satisfies the first equation of (9.1).*

Remark 9.1.2 *If $\alpha = 1$, then $h = 1$ and $g = \ominus z$.*

Remark 9.1.3 *Note that from the first equation (9.1), we get*

$$zh^\sigma \left(1 + \mu \frac{g-k_1}{k_0}\right) = -g.$$

Definition 9.1.4 *Assume that $f : \mathbb{T} \to \mathbb{C}$ is regulated. Then the conformable Laplace transform of f is defined by*

$$\mathscr{L}_c(f)(z) = \int_0^\infty f(t)h^\sigma(t)E_g^\sigma(t,0)\Delta_{\alpha,\infty}t$$

for $z \in \mathcal{D}_c(f)$, where $\mathcal{D}_c(f)$ consists of all complex numbers $z \in \mathcal{H}_c(h)$ for which the improper integral exists.

Theorem 9.1.5 *Let $f, g : \mathbb{T} \to \mathbb{C}$ be regulated functions, $a, b \in \mathbb{C}$. Then*

$$\mathscr{L}_c(af + bg)(z) = a\mathscr{L}_c(f)(z) + b\mathscr{L}_c(g)(z),$$

$z \in \mathcal{D}_c(f) \bigcap \mathcal{D}_c(g)$.

Proof 9.1.6 *We have*

$$
\begin{aligned}
\mathscr{L}_c(af+bg)(z) &= \int_0^\infty (af(t)+bg(t))h^\sigma(t)E_g^\sigma(t,0)\Delta_{\alpha,\infty}t \\
&= \int_0^\infty af(t)h^\sigma(t)E_g^\sigma(t,0)\Delta_{\alpha,\infty}t \\
&\quad + \int_0^\infty bg(t)h^\sigma(t)E_g^\sigma(t,0)\Delta_{\alpha,\infty}t \\
&= a\int_0^\infty f(t)h^\sigma(t)E_g^\sigma(t,0)\Delta_{\alpha,\infty}t \\
&\quad + b\int_0^\infty g(t)h^\sigma(t)E_g^\sigma(t,0)\Delta_{\alpha,\infty}t \\
&= a\mathscr{L}_c(f)(z) + b\mathscr{L}_c(g)(z), \quad z \in \mathcal{D}_c(f) \bigcap \mathcal{D}_c(g).
\end{aligned}
$$

This completes the proof. □

Example 9.1.7 *We will compute*

$$\mathscr{L}_c(1)(z), \quad z \in \mathcal{D}_c(1).$$

We have

$$
\begin{aligned}
\mathscr{L}_c(1)(z) &= \int_0^\infty h^\sigma(t)E_g^\sigma(t,0)\Delta_{\alpha,\infty}t \\
&= -\frac{1}{z}\int_0^\infty gE_g(t,0)\Delta_{\alpha,\infty}t \\
&= -\frac{1}{z}\int_0^\infty D^\alpha E_g(t,0)\Delta_{\alpha,\infty}t \\
&= -\frac{1}{z}\left(E_g(\infty,0) - E_g(0,0)E_0(\infty,0)\right) \\
&= \frac{1}{z}E_0(\infty,0)
\end{aligned}
$$

for all $z \in \mathcal{D}_c(1)$ for which

$$E_g(\infty,0) = 0.$$

This ends the example.

Theorem 9.1.8 *Let $n \in \mathbb{N}$, $f : \mathbb{T} \to \mathbb{C}$ be such that $(D^{\alpha})^k f$, $k \in \{0, 1, \dots, n\}$, are regulated. Then*

$$\mathscr{L}_c\left((D^{\alpha})^n f\right)(z) = z^n \mathscr{L}_c(f)(z)$$

$$-E_0(\infty, 0)\left(f(0)z^{n-1} + D^{\alpha}f(0)z^{n-2} + \cdots \right. \tag{9.2}$$

$$\left. + (D^{\alpha})^{n-1} f(0)\right)$$

for any

$$z \in \mathscr{D}_c(f) \bigcap \mathscr{D}_c\left(D^{\alpha}(f)\right) \bigcap \cdots \bigcap \mathscr{D}_c\left((D^{\alpha})^n f\right) \tag{9.3}$$

for which

$$\lim_{t \to \infty}\left((D^{\alpha})^k f(t)h(t)E_g(t, 0)\right) = 0, \quad k \in \{0, \dots, n-1\}. \tag{9.4}$$

Proof 9.1.9 *We will use the principle of the mathematical induction.*

1. *Let $n = 1$. Using integration by parts and the definition for h and g, we get*

$$\mathscr{L}_c\left(D^{\alpha}f\right)(z) = \int_0^{\infty} D^{\alpha}f(t)h^{\sigma}(t)E_g^{\sigma}(t, 0)\Delta_{\alpha, \infty}t$$

$$= \lim_{t \to \infty}\left(f(t)h(t)E_g(t, 0)\right) - f(0)h(0)E_g(0, 0)E_0(\infty, 0)$$

$$-\int_0^{\infty}\left(f(t)D^{\alpha}\left(hE_g(\cdot, 0)\right)(t) - k_1(\alpha, t)f(t)h^{\sigma}(t)E_g^{\sigma}(t, 0)\right)\Delta_{\alpha, \infty}t$$

$$= -f(0)E_0(\infty, 0)$$

$$-\int_0^{\infty}\left(f(t)\left(D^{\alpha}h(t)E_g^{\sigma}(t, 0) + h(t)D^{\alpha}E_g(t, 0)\right.\right.$$

$$\left. -k_1(\alpha, t)h(t)E_g^{\sigma}(t, 0)\right)$$

$$\left. -k_1(\alpha, t)f(t)h^{\sigma}(t)E_g^{\sigma}(t, 0)\right)\Delta_{\alpha, \infty}t$$

$$= -f(0)E_0(\infty, 0)$$

$$-\int_0^{\infty}\left(f(t)D^{\alpha}h(t)E_g^{\sigma}(t, 0) + f(t)h(t)g(t)E_g(t, 0)\right.$$

$$-k_1(\alpha, t)f(t)h(t)E_g^{\sigma}(t, 0)$$

$$\left. -k_1(\alpha,t)f(t)h^\sigma(t)E_g^\sigma(t,0) \right) \Delta_{\alpha,\infty}t$$

$$= -f(0)E_0(\infty,0)$$

$$-\int_0^\infty f(t)\left(D^\alpha h(t) - zh(t)h^\sigma(t) - k_1(\alpha,t)h(t) \right.$$

$$\left. -k_1(\alpha,t)h^\sigma(t) \right) E_g^\sigma(t,0)\Delta_{\alpha,\infty}t$$

$$= -f(0)E_0(\infty,0)+z\int_0^\infty f(t)h^\sigma(t)E_g^\sigma(t,0)\Delta_{\alpha,\infty}t$$

$$= -f(0)E_0(\infty,0)+z\mathscr{L}_c(f)(z)$$

for any $z \in \mathscr{D}_c(f)$ for which

$$\lim_{t\to\infty}\left(f(t)h(t)E_g(t,0)\right)=0.$$

2. *Assume that (9.2) holds for any z that satisfies (9.3) and (9.4).*

3. *We will prove that*

$$\mathscr{L}_c\left((D^\alpha)^{n+1}f\right)(z) = z^{n+1}\mathscr{L}_c(f)(z)$$

$$-E_0(\infty,0)\left(f(0)z^n + D^\alpha f(0)z^{n-1} + \cdots \right.$$

$$\left. + (D^\alpha)^n f(0) \right)$$

for any

$$z \in \mathscr{D}_c(f)\bigcap\mathscr{D}_c\left(D^\alpha(f)\right)\bigcap\cdots\bigcap\mathscr{D}_c\left((D^\alpha)^{n+1}f\right) \qquad (9.5)$$

for which

$$\lim_{t\to\infty}\left((D^\alpha)^k f(t)h(t)E_g(t,0)\right)=0, \quad k\in\{0,\ldots,n\}. \qquad (9.6)$$

Really, using (9.2), we get

$$\mathscr{L}_c\left((D^\alpha)^{n+1}f\right)(z) = \mathscr{L}_c\left(D^\alpha\left((D^\alpha)^n f\right)\right)$$

$$= z\mathscr{L}_c\left((D^\alpha)^n f\right)(z) - (D^\alpha)^n f(0)E_0(\infty,0)$$

$$= z^{n+1}\mathscr{L}_c(f)(z)$$

$$-E_0(\infty,0)\left(f(0)z^n + D^\alpha f(0)z^{n-1}\right.$$

$$\left. +\cdots + (D^\alpha)^{n-1} f(0)z\right)$$

$$-(D^\alpha)^n f(0)E_0(\infty,0)$$

$$= z^{n+1}\mathscr{L}_c(f)(z)$$

$$-E_0(\infty,0)\left(f(0)z^n + D^\alpha f(0)z^{n-1}\right.$$

$$\left. +\cdots + (D^\alpha)^{n-1} f(0)z + (D^\alpha)^n f(0)\right)$$

for any z which satisfies (9.5) and (9.6). This completes the proof. □

Theorem 9.1.10 *Assume that $f : \mathbb{T} \to \mathbb{C}$ is regulated. If*

$$F(t) = \int_0^t f(s)\Delta_{\alpha,t}s, \quad t \in \mathbb{T},$$

then

$$\mathscr{L}_c(F)(z) = \frac{1}{z}\mathscr{L}_c(f)(z)$$

for those $z \in \mathscr{D}_c(F)\bigcap\mathscr{D}_c(f)$ for which

$$\lim_{t\to\infty}\left(h(t)F(t)E_g(t,0)\right) = 0. \tag{9.7}$$

Proof 9.1.11 *Note that*

$$D^\alpha F(t) = f(t), \quad t \in \mathbb{T}.$$

Hence, by Theorem 9.1.8, we get

$$\mathscr{L}_c(f)(z) = \mathscr{L}_c(D^\alpha F(t))$$

$$= -F(0)E_0(\infty,0) + z\mathscr{L}_c(F)(z)$$

$$= z\mathscr{L}_c(F)(z)$$

for those $z \in \mathscr{D}_c(F)\bigcap\mathscr{D}_c(f)$ for which (9.7) holds. From here,

$$\mathscr{L}_c(F)(z) = \frac{1}{z}\mathscr{L}_c(f)(z)$$

for those $z \in \mathscr{D}_c(F)\bigcap\mathscr{D}_c(f)$ for which (9.7) holds. This completes the proof. □

Theorem 9.1.12 *We have*

$$\mathscr{L}_c\left(h_n(\cdot,0)\right)(z) = \frac{E_0(\infty,0)}{z^{n+1}}, \quad n \in \mathbb{N}_0, \tag{9.8}$$

for those $z \in \mathscr{D}_c(h_n(\cdot,0))$ for which

$$\lim_{t\to\infty}\left(h_n(t,0)h(t)E_g(t,0)\right) = 0. \tag{9.9}$$

Proof 9.1.13 *We will use the principle of the mathematical induction.*

1. *Let $n = 0$. Then*

$$h_0(t,0) = E_0(t,0), \quad t \in \mathbb{T}.$$

Hence, and considering Theorem 9.1.8, we get

$$\begin{aligned}
0 &= \mathscr{L}_c\left(D^\alpha E_0(\cdot,0)\right)(z) \\[2mm]
&= z\mathscr{L}_c(E_0(\cdot,0))(z) - E_0(0,0)E_0(\infty,0) \\[2mm]
&= z\mathscr{L}_c(E_0(\cdot,0))(z) - E_0(\infty,0)
\end{aligned}$$

for those $z \in \mathscr{D}_c(E_0(\cdot,0))$ for which

$$\lim_{t\to\infty}\left(h(t)h_0(t,0)E_g(t,0)\right) = 0. \tag{9.10}$$

Hence,

$$\mathscr{L}_c(h_0(\cdot,0))(z) = \frac{E_0(\infty,0)}{z}$$

for those $z \in \mathscr{D}_c(h_0(\cdot,0))$ for which (9.10) holds.

2. *Assume (9.8) for some $n \in \mathbb{N}$ and for those $z \in \mathscr{D}_c(h_n(\cdot,0))$ for which (9.9) holds.*

3. *We will prove that*

$$\mathscr{L}_c\left(h_{n+1}(\cdot,0)\right)(z) = \frac{E_0(\infty,0)}{z^{n+2}} \tag{9.11}$$

for those $z \in \mathscr{D}_c\left(h_{n+1}(\cdot,0)\right)$ for which

$$\lim_{t\to\infty}\left(h_{n+1}(t,0)h(t)E_g(t,0)\right) = 0. \tag{9.12}$$

Really, using that

$$D^\alpha h_{n+1}(t,0) = h_n(t,0), \quad t \in \mathbb{T},$$

and using Theorem 9.1.8, we get

$$\begin{aligned}
\frac{E_0(\infty,0)}{z^{n+1}} &= \mathscr{L}_c\left(h_n(\cdot,0)\right)(z) \\[2mm]
&= \mathscr{L}_c\left(D^\alpha h_{n+1}(\cdot,0)\right)(z)
\end{aligned}$$

$$= z\mathscr{L}_c\left(h_{n+1}(\cdot,0)\right)(z)$$

$$-h_{n+1}(0,0)E_0(\infty,0)$$

$$= z\mathscr{L}_c\left(h_{n+1}(\cdot,0)\right)(z)$$

for those $z \in \mathscr{D}_c\left(h_{n+1}(\cdot,0)\right)$ for which (9.12) holds. Hence, we get (9.11) for those $z \in \mathscr{D}_c\left(h_{n+1}(\cdot,0)\right)$ for which (9.12) holds. This completes the proof. □

Example 9.1.14 *Let $f \in \mathscr{R}_c$ be a constant. We will compute*

$$\mathscr{L}_c\left(E_f(\cdot,0)\right)(z).$$

Note that

$$D^{\alpha}E_f(t,0) = fE_f(t,0), \quad t \in \mathbb{T}^{\kappa}.$$

Then, using Theorem 9.1.8, we get

$$f\mathscr{L}_c\left(E_f(\cdot,0)\right)(z) = \mathscr{L}_c\left(fE_f(\cdot,0)\right)$$

$$= \mathscr{L}_c\left(D^{\alpha}E_f(\cdot,0)\right)(z)$$

$$= z\mathscr{L}_c\left(E_f(\cdot,0)\right)(z)$$

$$-E_f(0,0)E_0(\infty,0)$$

for those $z \in \mathscr{D}_c(E_f(\cdot,0))$ for which

$$\lim_{t\to\infty}\left(E_f(t,0)h(t)E_g(t,0)\right) = 0.$$

Hence,

$$(z-f)\mathscr{L}_c\left(E_f(\cdot,0)\right)(z) = E_0(\infty,0)$$

or

$$\mathscr{L}_c(E_f(\cdot,0)) = \frac{E_0(\infty,0)}{z-f}$$

for those $z \in \mathscr{D}_c(E_f(\cdot,0))$ for which

$$\lim_{t\to\infty}\left(E_f(t,0)h(t)E_g(t,0)\right) = 0.$$

Example 9.1.15 *Let $f \in \mathscr{R}_c$ be a constant. We will compute*

$$\mathscr{L}_c\left(Cosh_f(\cdot,0)\right)(z).$$

We have

$$
\begin{aligned}
\mathscr{L}_c\left(Cosh_c(\cdot,0)\right)(z) &= \mathscr{L}_c\left(\frac{E_f(\cdot,0)+E_{-f}(\cdot,0)}{2}\right)(z) \\
&= \frac{1}{2}\mathscr{L}_c\left(E_f(\cdot,0)\right)(z) + \frac{1}{2}\mathscr{L}_c\left(E_{-f}(\cdot,0)\right)(z) \\
&= \frac{E_0(\infty,0)}{2(z-f)} + \frac{E_0(\infty,0)}{2(z+f)} \\
&= \frac{zE_0(\infty,0)}{z^2-f^2}
\end{aligned}
$$

for those $z \in \mathscr{D}_c(E_f(\cdot,0)) \bigcap \mathscr{D}_c(E_{-f}(\cdot,0))$ *for which*

$$
\lim_{t\to\infty}\left(E_f(t,0)h(t)E_g(t,0)\right) = \lim_{t\to\infty}\left(E_{-f}(t,0)h(t)E_g(t,0)\right) = 0.
$$

Example 9.1.16 *Let* $f \in \mathscr{R}_c$ *be a constant. We will compute*

$$
\mathscr{L}_c\left(Sinh_f(\cdot,0)\right)(z).
$$

We have

$$
\begin{aligned}
\mathscr{L}_c\left(Sinh_c(\cdot,0)\right)(z) &= \mathscr{L}_c\left(\frac{E_f(\cdot,0)-E_{-f}(\cdot,0)}{2}\right)(z) \\
&= \frac{1}{2}\mathscr{L}_c\left(E_f(\cdot,0)\right)(z) - \frac{1}{2}\mathscr{L}_c\left(E_{-f}(\cdot,0)\right)(z) \\
&= \frac{E_0(\infty,0)}{2(z-f)} - \frac{E_0(\infty,0)}{2(z+f)} \\
&= \frac{fE_0(\infty,0)}{z^2-f^2}
\end{aligned}
$$

for those $z \in \mathscr{D}_c(E_f(\cdot,0)) \bigcap \mathscr{D}_c(E_{-f}(\cdot,0))$ *for which*

$$
\lim_{t\to\infty}\left(E_f(t,0)h(t)E_g(t,0)\right) = \lim_{t\to\infty}\left(E_{-f}(t,0)h(t)E_g(t,0)\right) = 0.
$$

Exercise 9.1.17 *Let* $f,g \in \mathscr{R}_c$ *be constants. Under "suitable" assumptions find the following.*

1. $\mathscr{L}_c\left(E_f(\cdot,0)+Ch_{fg}(\cdot,0)\right)(z)$,

2. $\mathscr{L}_c\left(2E_f(\cdot,0)-3E_{f+g}(\cdot,0)+E_{f-g}(\cdot,0)\right)(z)$,

3. $\mathscr{L}_c\left(Sh_{fg}(\cdot,0)-3E_{f^2}(\cdot,0)\right)(z)$,

4. $\mathscr{L}_c\left(Cos_f(\cdot,0)\right)(z)$,

5. $\mathscr{L}_c\left(Sin_f(\cdot,0)\right)(z)$.

9.2 DECAY OF THE EXPONENTIAL FUNCTION

For $\lambda \in \mathscr{R}(\mathbb{T})$, define

$$m_\lambda(t,s) = \int_s^t \frac{1}{1+\mu(\tau)\lambda} \Delta\tau, \quad s,t \in \mathbb{T}.$$

Theorem 9.2.1 *Let $s \in \mathbb{T}$ and $\lambda \in \mathscr{R}^+([s,\infty))$. Then*

1. *$m_\lambda^{\Delta_t}(t,s) \geq 0$ for any $t \in [s,\infty)$.*

2. *$\lim_{t\to\infty} m_\lambda(t,s) = \infty$.*

Proof 9.2.2   *1. Since $\lambda \in \mathscr{R}^+([s,\infty))$, we have*

$$1+\mu(t)\lambda > 0, \quad t \in [s,\infty).$$

Hence,

$$
\begin{aligned}
m_\lambda^{\Delta_t}(t,s) &= \frac{1}{1+\mu(t)\lambda} \\[2mm]
&> 0, \quad t \in [s,\infty).
\end{aligned}
$$

2. We consider the following two cases.

  (a) Let $\sup\mu(\mathbb{T}) < \infty$. Then

$$
\begin{aligned}
m_\lambda(t,s) &= \int_s^t \frac{1}{1+\lambda\mu(\tau)} \Delta\tau \\[2mm]
&\geq \int_s^t \frac{1}{1+\mu(\tau)|\lambda|} \Delta\tau \\[2mm]
&\geq \frac{t-s}{1+|\lambda|\sup\mu(\mathbb{T})} \\[2mm]
&\to \infty, \quad as \quad t \to \infty.
\end{aligned}
$$

  (b) Let $\sup\mu(\mathbb{T}) = \infty$. Since $\lambda \in \mathscr{R}^+([s,\infty))$, we conclude that $\lambda \geq 0$.

    i. Let $\lambda = 0$. Then

$$
\begin{aligned}
m_\lambda(t,s) &= t-s \\[2mm]
&\to \infty, \quad as \quad t \to \infty.
\end{aligned}
$$

    ii. Let $\lambda > 0$. Take a sequence $\{\xi_k\}_{k\in\mathbb{N}} \subset [s,\infty)$ such that the sequence $\{\mu(\xi_k)\}_{k\in\mathbb{N}}$ is increasing and divergent. Hence,

$$m_\lambda(t,s) = \int_s^t \frac{1}{1+\lambda\mu(\tau)} \Delta\tau$$

$$\geq \sum_{\substack{\sigma(\xi_k) \leq t \\ \xi_k \geq s}} \int_{\xi_k}^{\sigma(\xi_k)} \frac{1}{1 + \lambda \mu(\tau)} \Delta \tau$$

$$= \sum_{\substack{\sigma(\xi_k) \leq t \\ \xi_k \geq s}} \frac{\mu(\xi_k)}{1 + \lambda \mu(\xi_k)}.$$

Since

$$\lim_{k \to \infty} \frac{\mu(\xi_k)}{1 + \lambda \mu(\xi_k)} = \frac{1}{\lambda},$$

we get

$$\lim_{t \to \infty} m_\lambda(t,s) \geq \lim_{t \to \infty} \sum_{\substack{\sigma(\xi_k) \leq t \\ \xi_k \geq s}} \frac{\mu(\xi_k)}{1 + \lambda \mu(\xi_k)}$$

$$= \infty.$$

This completes the proof. □

Theorem 9.2.3 *Let $s \in \mathbb{T}$ and $f \in \mathscr{R}^+([s,\infty))$. Then*

$$0 < e_f(t,s) < e^{\int_s^t f(\tau)\Delta \tau}, \quad t \in [s,\infty).$$

Proof 9.2.4 *Observe that*

$$\frac{\log(1 + \mu(\tau)f(\tau))}{\mu(\tau)} = \frac{\mu(\tau)f(\tau) - (\mu(\tau)f(\tau) - \log(1 + \mu(\tau)f(\tau)))}{\mu(\tau)}$$

$$\leq f(\tau), \quad \tau \in [s,t], \quad t \in [s,\infty).$$

Then

$$0 < e_f(t,s)$$

$$= e^{\int_s^t \frac{\log(1 + \mu(\tau)f(\tau))}{\mu(\tau)} \Delta \tau}$$

$$\leq e^{\int_s^t f(\tau)\Delta \tau}, \quad t \in [s,\infty).$$

This completes the proof. □

For $h > 0$ define

$$\mathbb{C}_h = \left\{ z \in \mathbb{C} : z \neq -\frac{1}{h} \right\} \quad \text{and} \quad \mathbb{R}_h = \mathbb{C}_h \bigcap \mathbb{R},$$

and $\mathbb{C}_0 = \mathbb{C}$, $\mathbb{R}_0 = \mathbb{R}$. For $h \geq 0$ define the functions $\Psi_h : \mathbb{C}_h \times \mathbb{R}_h \to \mathbb{R}$ and $\mathrm{Re}_h : \mathbb{C}_h \to \mathbb{R}$ as follows

$$
\Psi_h(z, \lambda) = \begin{cases} \dfrac{1}{h} \log \left| \dfrac{1 + h\lambda}{1 + hz} \right| & \text{if} \quad h > 0 \\[2ex] \lambda - \mathrm{Re}(z) & \text{if} \quad h = 0, \end{cases}
$$

$$
\mathrm{Re}_h(z) = \begin{cases} \dfrac{1}{h} (|1 + hz| - 1) & \text{if} \quad h > 0 \\[2ex] \mathrm{Re}(z) & \text{if} \quad h = 0. \end{cases}
$$

Theorem 9.2.5 *Let $h \geq 0$ and $\lambda \in \mathbb{R}_h$. Then:*

1. *$\Psi_h(z, \lambda) = \Psi_h(\mathrm{Re}_h(z), \lambda)$ for any $z \in \mathbb{C}_h$.*

2. *Suppose $1 + h\lambda > 0$ and $x_1, x_2 \in \mathbb{R}$, $1 + hx_1 > 0$, $1 + hx_2 > 0$. If $x_1 < x_2$, then*

$$
\Psi_h(x_1, \lambda) > \Psi_h(x_2, \lambda).
$$

Proof 9.2.6   *1. Let $h = 0$. Then*

$$
\Psi_h(z, \lambda) = \lambda - Re(z),
$$

$$
Re_h(z) = Re(z),
$$

$$
\Psi_h(Re_h(z), \lambda) = \lambda - Re(Re_h(z))
$$

$$
= \lambda - Re(z).
$$

Thus,
$$
\Psi_h(z, \lambda) = \Psi_h(Re_h(z), \lambda).
$$

Let $h > 0$. Then

$$
\begin{aligned}
\Psi_h(Re_h(z), \lambda) &= \frac{1}{h} \log \left| \frac{1 + h\lambda}{1 + hRe_h(z)} \right| \\[2ex]
&= \frac{1}{h} \log \left| \frac{1 + h\lambda}{1 + |1 + hz| - 1} \right| \\[2ex]
&= \frac{1}{h} \log \left| \frac{1 + h\lambda}{1 + hz} \right| \\[2ex]
&= \Psi_h(z, \lambda).
\end{aligned}
$$

2. *Let $h = 0$. Then*

$$\Psi_h(x_1, \lambda) \;=\; \lambda - x_1$$

$$> \;\; \lambda - x_2$$

$$= \;\; \Psi_h(x_2, \lambda).$$

Suppose that $h > 0$. Then

$$\Psi_h(x_1, \lambda) \;=\; \frac{1}{h} \log \left| \frac{1 + h\lambda}{1 + hx_1} \right|$$

$$= \;\; \frac{1}{h} \log \frac{1 + h\lambda}{1 + hx_1}$$

$$> \;\; \frac{1}{h} \log \frac{1 + h\lambda}{1 + hx_2}$$

$$= \;\; \Psi_h(x_2, \lambda).$$

This completes the proof. □

For $h \geq 0$ and $z \in \mathbb{C}_h$, define

$$\psi_h(z) = \begin{cases} \dfrac{1}{h} \log|1 + hz| & \text{if} \quad h > 0 \\[2ex] \text{Re}(z) & \text{if} \quad h = 0. \end{cases}$$

Define the minimal graininess function $\mu_* : \mathbb{T} \to [0, \infty)$ by

$$\mu_*(s) = \inf_{\tau \in [s, \infty)} \mu(\tau), \quad s \in \mathbb{T}.$$

For $h \geq 0$ and $\lambda \in \mathbb{R}$, define

$$\mathbb{C}_h(\lambda) = \{z \in \mathbb{C}_h : \text{Re}_h(z) > \lambda\}.$$

Theorem 9.2.7 *Let $s \in \mathbb{T}$ and $\lambda \in' \mathscr{R}^+([s, \infty))$. Then, for any $z \in \mathbb{C}_{\mu_*(s)}(\lambda)$, we have the following properties.*

1. *$|e_{\lambda \ominus z}(t, s)| \leq e_{\lambda \ominus \text{Re}_{\mu_*(s)}(s)}(t, s), \quad t \in [s, \infty)$.*

2. *$\lim_{t \to \infty} e_{\lambda \ominus \text{Re}_{\mu_*(s)}(z)}(t, s) = 0$.*

3. *$\lim_{t \to \infty} e_{\lambda \ominus z}(t, s) = 0$.*

Proof 9.2.8   *1. Observe that*

$$\psi_{\mu(\tau)}((\lambda \ominus z)(\tau)) \;=\; \begin{cases} \dfrac{1}{\mu(\tau)}\log\left|1+\mu(\tau)((\lambda \ominus z)(\tau))\right| & \text{if } \mu(\tau) > 0 \\[2ex] Re(\lambda - z) & \text{if } \mu(\tau) = 0 \end{cases}$$

$$=\; \begin{cases} \dfrac{1}{\mu(\tau)}\log\left|\dfrac{1+\mu(\tau)\lambda}{1+\mu(\tau)z}\right| & \text{if } \mu(\tau) > 0 \\[2ex] \lambda - Re(z) & \text{if } \mu(\tau) = 0 \end{cases}$$

$$=\; \Psi_{\mu(\tau)}(z, \lambda).$$

Then, using Theorem 9.2.5, we get

$$\left|e_{\lambda \ominus z}(t,s)\right| \;=\; e^{\int_s^t \psi_{\mu(\tau)}((\lambda \ominus z)(\tau))\Delta\tau}$$

$$=\; e^{\int_s^t \Psi_{\mu(\tau)}(z,\lambda)\Delta\tau}$$

$$=\; e^{\int_s^t \Psi_{\mu(\tau)}(Re_{\mu(\tau)}(z),\lambda)\Delta\tau}$$

$$\leq\; e^{\int_s^t \Psi_{\mu(s)}(Re_{\mu_*(\tau)}(z),\lambda)\Delta\tau}$$

$$=\; e_{\lambda \ominus Re_{\mu_*(s)}(z)}(t,s), \quad t \in [s,\infty).$$

2. We have

$$\left(\lambda \ominus Re_{\mu_*(s)}(z)\right)(t) \;=\; \frac{\lambda - Re_{\mu_*(s)}(z)}{1+\mu(t)Re_{\mu_*(s)}(z)}$$

$$=\; \frac{\lambda - Re_{\mu_*(s)}(z)}{1+\lambda\mu(t)+\mu(t)\left(Re_{\mu_*(s)}(z)-\lambda\right)}$$

and

$$1+\mu(t)\left(\lambda \ominus Re_{\mu_*(s)}(z)\right)(t) \;=\; 1+\frac{\mu(t)\left(\lambda - Re_{\mu_*(s)}(z)\right)}{1+\lambda\mu(t)+\mu(t)\left(Re_{\mu_*(s)}(z)-\lambda\right)}$$

$$=\; \frac{1+\lambda\mu(t)}{1+\lambda\mu(t)+\mu(t)\left(Re_{\mu_*(s)}(z)-\lambda\right)}.$$

Therefore

$$\lambda \ominus Re_{\mu_*(s)}(z) < 0 \quad \text{and} \quad \lambda \ominus Re_{\mu_*(s)}(z) \in \mathscr{R}^+([s,\infty)).$$

Hence, and considering Theorem 9.2.3, it follows that

$$e_{\lambda \ominus Re_{\mu_*(s)}(z)}(t,s) \;\leq\; e^{\int_s^t \left(\lambda \ominus Re_{\mu_*(s)}(z)\right)(\tau)\Delta\tau}$$

$$= e^{\left(\lambda - Re_{\mu_*(s)}(z)\right)\int_s^t \frac{\Delta\tau}{1+\mu(\tau)Re_{\mu_*(s)}(z)}}$$

$$= e^{-\left(Re_{\mu_*(s)}(z)-\lambda\right)m_{Re_{\mu_*(s)}(z)}(t,s)}, \quad t \in [s,\infty).$$

Since $z \in \mathbb{C}_{\mu_*(s)}(\lambda)$ and $Re_{\mu_*(s)}(z) \in \mathscr{R}^+([s,\infty))$, we have

$$\lim_{t\to\infty} m_{Re_{\mu_*(s)}(z)}(t,s) = \infty$$

and

$$\lim_{t\to\infty} e_{\lambda\ominus Re_{\mu_*(s)}(z)}(t,s) = 0.$$

3. *By 1) and 2), we get*

$$\left|\lim_{t\to\infty} e_{\lambda\ominus z}(t,s)\right| \leq \lim_{t\to\infty} e_{\lambda\ominus Re_{\mu_*(s)}(z)}(t,s)$$

$$= 0.$$

This completes the proof. □

9.3 CONVERGENCE OF THE CONFORMABLE LAPLACE TRANSFORM

Definition 9.3.1 *A function $f \in \mathscr{C}_{rd}(\mathbb{T})$ is said to be of conformable exponential order β on $[0,\infty)$ if there exist $\beta \in \mathscr{R}^+([0,\infty))$ and a constant $K > 0$ such that*

$$\left|f(t)h^\sigma(t)e_{\frac{g-k_1}{k_0}}(\sigma(t),0)e_{-\frac{k_1}{k_0}}(\infty,\sigma(t))\right| \leq K\left|e_{\beta\ominus z}(t,0)\right|k_0(\alpha,t)$$

for any $t \in [0,\infty)$ and $z \in \mathbb{C}$.

Theorem 9.3.2 *Let $f \in \mathscr{C}_{rd}([0,\infty))$ be of conformable exponential order β. Then the conformable Laplace transform $\mathscr{L}_c(f)$ exists on $\mathbb{C}_{\mu_*(0)}(\beta)$ and converges absolutely. In the case $\mu_*(0) > 0$, we have*

$$\lim_{|z|\to\infty} \mathscr{L}_c(f)(z) = 0.$$

Proof 9.3.3 *Using Theorem 9.2.7, we have*

$$|\mathscr{L}_c(f)(z)| = \lim_{t\to\infty}\left|\int_0^t f(\tau)h^\sigma(\tau)E_g^\sigma(\tau,0)\Delta_{\alpha,\infty}\tau\right|$$

$$= \lim_{t\to\infty}\left|\int_0^t f(\tau)h^\sigma(\tau)E_g^\sigma(\tau,0)\frac{E_0(\infty,\sigma(\tau))}{k_0(\alpha,\tau)}\Delta\tau\right|$$

$$= \lim_{t \to \infty} \left| \int_0^t f(\tau) h^\sigma(\tau) e_{\frac{g-k_1}{k_0}}(\sigma(\tau),0) \frac{e_{-\frac{k_1}{k_0}}(\infty,\sigma(\tau))}{k_0(\alpha,\tau)} \Delta\tau \right|$$

$$\le \lim_{t \to \infty} \int_0^t \left| f(\tau) h^\sigma(\tau) e_{\frac{g-k_1}{k_0}}(\sigma(\tau),0) \frac{e_{-\frac{k_1}{k_0}}(\infty,\sigma(\tau))}{k_0(\alpha,\tau)} \right| \Delta\tau$$

$$\le K \lim_{t \to \infty} \int_0^t \frac{|e_{\beta \ominus z}(\tau,0)|}{|1+\mu(\tau)z|} \Delta\tau$$

$$\le K \lim_{t \to \infty} \int_0^t \frac{e_{\beta - Re_{\mu_*(0)}}(z)(\tau,0)}{1+\mu(\tau)Re_{\mu_*(0)}(z)} \Delta\tau$$

$$= \frac{K}{\beta - Re_{\mu_*(0)}(z)} \lim_{t \to \infty} \int_0^t \frac{\beta - Re_{\mu_*(0)}(z)}{1+\mu(\tau)Re_{\mu_*(0)}(z)} e_{\beta \ominus Re_{\mu_*(0)}(z)}(\tau,0) \Delta\tau$$

$$= \frac{K}{\beta - Re_{\mu_*(0)}(z)} \lim_{t \to \infty} \int_0^t \left(\beta \ominus Re_{\mu_*(0)}(z)(z)\right) e_{\beta \ominus Re_{\mu_*(0)}(z)}(\tau,0) \Delta\tau$$

$$= \frac{K}{\beta - Re_{\mu_*(0)}(z)} \lim_{t \to \infty} \int_0^t e^\Delta_{\beta \ominus Re_{\mu_*(0)}(z)}(\tau,0) \Delta\tau$$

$$= \frac{K}{Re_{\mu_*(0)}(z) - \beta} \lim_{t \to \infty} \left(1 - e_{\beta \ominus Re_{\mu_*(0)}(z)}(t,0)\right)$$

$$= \frac{K}{Re_{\mu_*(0)}(z) - \beta}, \quad z \in \mathbb{C}_{\mu_*(0)}(\beta).$$

Let $\mu_*(0) > 0$. Note that $|z| \to \infty$ implies $Re_{\mu_*(0)}(z) \to \infty$ and

$$\left| \lim_{|z| \to \infty} \mathscr{L}_c(f) \right| \le \lim_{|z| \to \infty} \frac{K}{Re_{\mu_*(0)}(z) - \beta} = 0.$$

This completes the proof. □

Theorem 9.3.4 Let $f \in \mathscr{C}_{rd}([0,\infty))$ be of conformable exponential order β. Then the conformable Laplace transform $\mathscr{L}_c(f)$ converges uniformly on $\mathbb{C}_{\mu_*(0)}(\gamma)$ for any $\gamma > \beta$.

Proof 9.3.5 By the proof of Theorem 9.3.2, it follows that

$$|\mathscr{L}_c(f)(z)| \le \frac{K}{Re_{\mu_*(0)}(z) - \beta}$$

$$\le \frac{K}{\gamma - \beta}.$$

Hence, for any $\varepsilon > 0$ there exists an $r \in [0,\infty)$ such that

$$\left| \int_t^\infty f(\tau) h^\sigma(\tau) E_g^\sigma(\tau,0) \Delta_{\alpha,\infty}\tau \right| < \varepsilon$$

for any $t \in [r,\infty)$ and for any $z \in \mathbb{C}_{\mu_*(0)}(\gamma)$. This completes the proof. □

Theorem 9.3.6 (Inversion Formula) *Suppose* $0 < \mu_{\min} \leq \mu(t) \leq \mu_{\max} < \infty$ *and*

$$a_1 \leq |k_0(\alpha,t) - \mu(t)k_1(\alpha,t)| \leq b_1, \quad t \in \mathbb{T},$$

for some positive constants a_1 *and* b_1. *If* $f : \mathbb{T} \to \mathbb{R}$ *is regulated and*

$$\int_{a-i\infty}^{a+i\infty} |\mathcal{L}_c(f)(z)||dz| < \infty$$

and the poles of $\mathcal{L}_c(f)(z)$ *are regressive constants* $\{z_1, \ldots, z_n\}$ *of finite order, then*

$$f(t) = \frac{k_0(\alpha,t)}{E_0(\infty,\sigma(t))} \sum_{j=1}^{n} Res_{z=z_j}\left(\frac{1}{h(t)E_g(t,0)}\mathcal{L}_c(f)(z)\right), \quad t \in \mathbb{T},$$

for any $z \in \mathbb{C}$ *with* $Re(z) > a$.

Proof 9.3.7 *Firstly, we will find an expression for* $h(\cdot)E_g(\cdot,0)$. *We have*

$$D^\alpha\left(h(\cdot)E_g(\cdot,0)\right)(t) = D^\alpha h(t)E_g^\sigma(t,0) + h(t)g(t)E_g(t,0)$$

$$- k_1(\alpha,t)h(t)E_g^\sigma(t,0), \quad t \in \mathbb{T}^\kappa. \tag{9.13}$$

By the second equation of (9.1), *we obtain*

$$0 = D^\alpha h(t)E_g^\sigma(t,0) - zh(t)h^\sigma(t)E_g^\sigma(t,0)$$

$$+ (z - k_1(\alpha,t))h^\sigma(t)E_g^\sigma(t,0) - k_1(\alpha,t)h(t)E_g^\sigma(t,0), \quad t \in \mathbb{T}^\kappa. \tag{9.14}$$

By the first equation of (9.1), *we get*

$$zh(t)h^\sigma(t)E_g^\sigma(t,0) = -h(t)g(t)E_g(t,0), \quad t \in \mathbb{T}^\kappa.$$

Hence, employing (9.14) *and* (9.13), *we find*

$$0 = D^\alpha h(t)E_g^\sigma(t,0) + h(t)g(t)E_g(t,0)$$

$$+ (z - k_1(\alpha,t))h^\sigma(t)E_g^\sigma(t,0) - k_1(\alpha,t)h(t)E_g^\sigma(t,0)$$

$$= D^\alpha\left(h(\cdot)E_g(\cdot,0)\right)(t) + (z - k_1(\alpha,t))h^\sigma(t)E_g^\sigma(t,0), \quad t \in \mathbb{T}^\kappa,$$

or

$$D^\alpha\left(h(\cdot),E_g(\cdot,0)\right)(t) = -(z - k_1(\alpha,t))h^\sigma(t)E_g^\sigma(t,0), \quad t \in \mathbb{T}^\kappa.$$

Let

$$r(t) = \frac{(k_1(\alpha,t) - z)(k_0(\alpha,t) - \mu(t)k_1(\alpha,t))}{k_0(\alpha,t) - \mu(t)(k_1(\alpha,t) - z)},$$

$$\bar{z} = \frac{1}{k_0(\alpha,t)}(r(t) - k_1(\alpha,t)), \quad t \in \mathbb{T}.$$

Then

$$\bar{z} = \frac{1}{k_0(\alpha,t)}(r(t) - k_1(\alpha,t))$$

$$= \frac{1}{k_0(\alpha,t)}\left(\frac{(k_1(\alpha,t) - z)(k_0(\alpha,t) - \mu(t)k_1(\alpha,t))}{k_0(\alpha,t) - \mu(t)(k_1(\alpha,t) - z)} - k_1(\alpha,t)\right)$$

$$= \frac{1}{k_0(\alpha,t)(k_0(\alpha,t) - \mu(t)(k_1(\alpha,t) - z))}\left(k_0(\alpha,t)k_1(\alpha,t) - \mu(t)(k_1(\alpha,t))^2\right.$$

$$- zk_0(\alpha,t) + z\mu(t)k_1(\alpha,t) - k_0(\alpha,t)k_1(\alpha,t)$$

$$\left. + \mu(t)(k_1(\alpha,t))^2 - z\mu(t)k_1(\alpha,t)\right)$$

$$= -\frac{zk_0(\alpha,t)}{k_0(\alpha,t)(k_0(\alpha,t) - \mu(t)(k_1(\alpha,t) - z))}$$

$$= -\frac{z}{k_0(\alpha,t) - \mu(t)(k_1(\alpha,t) - z)},$$

$$1 + \mu(t)\bar{z} = 1 + \mu(t)\left(-\frac{z}{k_0(\alpha,t) - \mu(t)(k_1(\alpha,t) - z)}\right)$$

$$= \frac{k_0(\alpha,t) - \mu(t)k_1(\alpha,t) + z\mu(t) - z\mu(t)}{k_0(\alpha,t) - \mu(t)(k_1(\alpha,t) - z)}$$

$$= \frac{k_0(\alpha,t) - \mu(t)k_1(\alpha,t)}{k_0(\alpha,t) - \mu(t)(k_1(\alpha,t) - z)},$$

$$\ominus\bar{z} = -\frac{\bar{z}}{1 + \mu(t)\bar{z}}$$

$$= \frac{\frac{z}{k_0(\alpha,t) - \mu(t)(k_1(\alpha,t) - z)}}{\frac{k_0(\alpha,t) - \mu(t)k_1(\alpha,t)}{k_0(\alpha,t) - \mu(t)(k_1(\alpha,t) - z)}}$$

$$= \frac{z}{k_0(\alpha,t) - \mu(t)k_1(\alpha,t)}, \quad t \in \mathbb{T}.$$

Denote

$$\tilde{z} = \ominus\bar{z}, \quad t \in \mathbb{T}.$$

Therefore

$$h(t)E_g(t,0) = E_r(t,0)$$

$$= e_{\frac{r-k_1}{k_0}}(t,0)$$

$$= e_{\bar{z}}(t,0)$$

$$= e_{\ominus\bar{z}}(t,0), \quad t \in \mathbb{T}.$$

Hence,

$$\mathscr{L}_c(f)(z) = \int_0^\infty f(t)h^\sigma(t)E_g^\sigma(t,0)\frac{E_0(\infty,\sigma(t))}{k_0(\alpha,t)}\Delta t$$

$$= \int_0^\infty f(t)\frac{E_0(\infty,\sigma(t))}{k_0(\alpha,t)}e_{\ominus\tilde{z}}(t,0)\Delta t$$

$$= \mathscr{L}\left(f(\cdot)\frac{E_0(\infty,\sigma(\cdot))}{k_0(\alpha,\cdot)}\right)(\tilde{z}).$$

Hence, by the inverse formula for the Laplace transform on time scales, we get

$$f(t)\frac{E_0(\infty,\sigma(t))}{k_0(\alpha,t)} = \sum_{j=1}^n Res_{\tilde{z}=\tilde{z}_j}e_{\tilde{z}}(t,0)\mathscr{L}(f)(\tilde{z})$$

$$= \sum_{j=1}^n Res_{z=z_j}\frac{1}{h(t)E_g(t,0)}\mathscr{L}_c(f)(z), \quad t \in \mathbb{T},$$

for any $z \in \mathbb{C}$ with $Re(z) > a$, whereupon

$$f(t) = \frac{k_0(\alpha,t)}{E_0(\infty,\sigma(t))}\sum_{j=1}^n Res_{z=z_j}\left(\frac{1}{h(t)E_g(t,0)}\mathscr{L}_c(f)(z)\right), \quad t \in \mathbb{T},$$

for any $z \in \mathbb{C}$ with $Re(z) > a$. This completes the proof. □

Theorem 9.3.8 (Uniqueness of the Inverse) *If the functions $f,g : \mathbb{T} \to \mathbb{R}$ are regulated and have the same Laplace transform, then $f = g$ a.e.*

Proof 9.3.9 *We have*

$$\mathscr{L}_c(f)(z) = \mathscr{L}_c(g)(z).$$

By the proof of Theorem 9.3.6, it follows that

$$\mathscr{L}\left(f(\cdot)\frac{E_0(\infty,\sigma(\cdot))}{k_0(\alpha,\cdot)}\right)(\tilde{z}) = \mathscr{L}\left(g(\cdot)\frac{E_0(\infty,\sigma(\cdot))}{k_0(\alpha,\cdot)}\right)(\tilde{z}).$$

By the uniqueness of the inverse of the Laplace transform on time scales, it follows

$$f(\cdot)\frac{E_0(\infty,\sigma(\cdot))}{k_0(\alpha,\cdot)} = g(\cdot)\frac{E_0(\infty,\sigma(\cdot))}{k_0(\alpha,\cdot)} \quad a.e.,$$

from where $f = g$ a.e. This completes the proof. □

Definition 9.3.10 *Suppose $0 < \mu_{\min} \le \mu(t) \le \mu_{\max} < \infty$ and*

$$a_1 \le |k_0(\alpha,t) - \mu(t)k_1(\alpha,t)| \le b_1, \quad t \in \mathbb{T},$$

for some positive constants a_1 and b_1. If $f_1, f_2 : \mathbb{T} \to \mathbb{R}$ are regulated and

$$\int_{a-i\infty}^{a+i\infty} |\mathscr{L}_c(f_1)(z)\mathscr{L}_c(f_2)(z)||dz| < \infty$$

and the poles of $\mathscr{L}_c(f_1)(z)\mathscr{L}_c(f_2)(z)$ are regressive constants $\{z_1,\ldots,z_n\}$ of finite order, then we say that f_1 and f_2 is a conformable Laplace pair.

Definition 9.3.11 *Let $f_1, f_2 : \mathbb{T} \to \mathbb{R}$ be a conformable Laplace pair. Then we define the conformable convolution of f_1 and f_2 by*

$$(f_1 \star_c f_2)(t) = \frac{k_0(\alpha,t)}{E_0(\infty,\sigma(t))} \sum_{j=1}^{n} Res_{z=z_j} \left(\frac{1}{h(t)E_g(t,0)} \mathscr{L}_c(f_1)(z)\mathscr{L}_c(f_2)(z) \right), \quad t \in \mathbb{T},$$

for any $z \in \mathbb{C}$ with $Re(z) > a$.

If $f_1, f_2 : \mathbb{T} \to \mathbb{R}$ is a conformable Laplace pair, then

$$(f_1 \star_c f_2)(t) = (f_2 \star_c f_1)(t), \quad t \in \mathbb{T},$$

and

$$\mathscr{L}_c(f_1 \star_c f_2)(z) = \mathscr{L}_c(f_1)(z)\mathscr{L}_c(f_2)(z)$$

for any $z \in \mathbb{C}$ with $Re(z) > a$. This ends the example.

Exercise 9.3.12 *Let $f_1, f_2, f_3 : \mathbb{T} \to \mathbb{R}$ be regulated and f_1 and f_2, f_1 and f_3, f_2 and f_3 be conformable Laplace pairs such that the poles of $\mathscr{L}_c(f_1)(z)\mathscr{L}_c(f_2)(z)$, $\mathscr{L}_c(f_1)(z)\mathscr{L}_c(f_3)(z)$ and $\mathscr{L}_c(f_2)(z)\mathscr{L}_c(f_3)(z)$ are regressive constants $\{z_1,\ldots,z_n\}$ of finite order. Prove that*

$$(f_1 \star_c f_2) \star_c f_3 = f_1 \star_c (f_2 \star_c f_3).$$

Theorem 9.3.13 *Let $f_1, f_2 : \mathbb{T} \to \mathbb{R}$ be a conformable Laplace pair. If f_1 is conformable Δ-differentiable, then*

$$D^\alpha (f_1 \star_c f_2)(t) = (D^\alpha f_1 \star_c f_2)(t) + E_0(\infty,0)f_1(0)f_2.$$

If f_2 is conformable Δ-differentiable, then

$$D^\alpha (f_1 \star_c f_2)(t) = (f_1 \star_c D^\alpha f_2)(t) + E_0(\infty,0)f_1 f_2(0).$$

Proof 9.3.14 *Suppose that f_1 is conformable Δ-differentiable. Then*

$$\mathscr{L}_c \left(D^\alpha(f_1 \star_c f_2) \right)(z) = z\mathscr{L}_c(f_1 \star_c f_2)(z) - E_0(\infty)(f_1 \star_c f_2)(0)$$

$$= z\mathscr{L}_c(f_1)(z)\mathscr{L}_c(f_2)(z) - (E_1(\infty,0)f_1(0))\mathscr{L}_c(f_2)(z)$$

$$+ (E_0(\infty,0)f_1(0)) \mathscr{L}_c(f_2)(z)$$

$$= (z\mathscr{L}_c(f_1)(z) - E_0(\infty,0)f_1(0)) \mathscr{L}_c(f_2)(z)$$

$$+ E_0(\infty,0)f_1(0)\mathscr{L}_c(f_2)(z)$$

$$= \mathscr{L}_c(D^\alpha f_1)(z)\mathscr{L}_c(f_2)(z)$$

$$+ \mathscr{L}_c(E_0(\infty,0)f_1(0)f_2)(z)$$

$$= \mathscr{L}_c(D^\alpha f_1 \star_c f_2)(z)$$

$$+ \mathscr{L}_c(E_0(\infty,0)f_1(0)f_2)(z)$$

$$= \mathscr{L}_c(D^\alpha f_1 \star_c f_2 + E_0(\infty,0)f_1(0)f_2),$$

whereupon

$$D^\alpha(f_1 \star_c f_2)(t) = (D^\alpha f_1 \star_c f_2)(t) + E_0(\infty,0)f_1(0)f_2.$$

The case when f_2 is conformable Δ-differentiable we leave to the reader as an exercise. This completes the proof. □

9.4 APPLICATIONS TO IVPS

Consider the following IVP

$$(D^\alpha)^n y + a_{n-1}(D^\alpha)^{n-1}y + \cdots + a_1 D^\alpha y + a_0 y = f(t), \quad t > 0, \qquad (9.15)$$

$$(D^\alpha)^{n-1}y(0) = b_{n-1},$$

$$\vdots \qquad\qquad (9.16)$$

$$D^\alpha y(0) = b_1,$$

$$y(0) = b_0,$$

where $a_i, b_i \in \mathbb{C}$, $i \in \{1,\ldots,n-1\}$, $f: \mathbb{T} \to \mathbb{R}$ is regulated. Take the conformable Laplace transform of both sides of equation (9.15) and using the initial conditions (9.16), we get

$$\mathscr{L}_c(f)(z) = \mathscr{L}_c\left((D^\alpha)^n y + a_{n-1}(D^\alpha)^{n-1}y + \cdots + a_1 D^\alpha y + a_0 y\right)(z)$$

$$= \mathscr{L}_c((D^\alpha)^n y)(z) + a_{n-1}\mathscr{L}_c\left((D^\alpha)^{n-1}y\right)(z)$$
$$+ \cdots + a_1\mathscr{L}_c(D^\alpha y)(z) + a_0\mathscr{L}_c(y)(z)$$

$$= z^n \mathscr{L}_c(y)(z) - E_0(\infty, 0) \left(y(0)z^{n-1} + D^\alpha y(0)z^{n-2} + \cdots + (D^\alpha)^{n-1} y(0) \right)$$

$$+ a_{n-1} \left(z^{n-1} \mathscr{L}_c(y)(z) - E_0(\infty, 0) \right.$$

$$\left. \left(y(0)z^{n-2} + D^\alpha y(0)z^{n-3} + \cdots + (D^\alpha)^{n-2} y(0) \right) \right)$$

$$+ \cdots$$

$$+ a_1 \left(z \mathscr{L}_c(y)(z) - E_0(\infty, 0)y(0) \right)$$

$$+ a_0 \mathscr{L}_c(y)(z)$$

$$= \left(z^n + a_{n-1}z^{n-1} + \cdots + a_1 z + a_0 \right) \mathscr{L}_c(y)(z)$$

$$- E_0(\infty, 0) \left(b_0 z^{n-1} + b_1 z^{n-2} + \cdots + b_{n-2} z + b_{n-1} \right)$$

$$- E_0(\infty, 0) \left(b_0 z^{n-2} + b_1 z^{n-3} + \cdots + b_{n-3} z + b_{n-2} \right)$$

$$- \cdots$$

$$- E_0(\infty, 0)b_0$$

$$= \left(z^n + a_{n-1}z^{n-1} + \cdots + a_1 z + a_0 \right) \mathscr{L}_c(y)(z)$$

$$- E_0(\infty, 0) \left(b_0 z^{n-1} + (b_1 + b_2)z^{n-2} + (b_2 + b_1 + b_0)z^{n-3} \right.$$

$$+ \cdots + (b_{n-2} + b_{n-1} + \cdots + b_1 + b_0)$$

$$\left. + b_{n-1} + b_{n-2} + \cdots + b_1 + b_0 \right).$$

Let

$$l(z) = E_0(\infty, 0) \left(b_0 z^{n-1} + (b_1 + b_2)z^{n-2} + (b_2 + b_1 + b_0)z^{n-3} \right.$$

$$+ \cdots + (b_{n-2} + b_{n-1} + \cdots + b_1 + b_0)$$

$$\left. + b_{n-1} + b_{n-2} + \cdots + b_1 + b_0 \right).$$

Then

$$\left(z^n + a_{n-1}z^{n-1} + \cdots + a_1 z + a_0\right)\mathscr{L}_c(y)(z) = \mathscr{L}_c(f)(z) + l(z)$$

or

$$\mathscr{L}_c(y)(z)(y)(z) = \frac{1}{z^n + a_{n-1}z^{n-1} + \cdots + a_1 z + a_0}\left(\mathscr{L}_c(f)(z) + l(z)\right).$$

Hence,

$$y(t) = \mathscr{L}_c^{-1}\left(\frac{1}{z^n + a_{n-1}z^{n-1} + \cdots + a_1 z + a_0}\left(\mathscr{L}_c(f)(z) + l(z)\right)\right), \quad t \geq 0.$$

Example 9.4.1 *Consider the IVP*

$$(D^\alpha)^2 y + 3D^\alpha y + 2y = 1, \quad t > 0,$$

$$D^\alpha y(0) = 1,$$

$$y(0) = 0.$$

We take the conformable Laplace transform of both sides of the considered equation and using the initial conditions, we find

$$\begin{aligned}
\mathscr{L}_c(1)(z) &= \mathscr{L}_c\left((D^\alpha)^2 y + 3D^\alpha y + 2y\right)(z) \\[2mm]
&= \mathscr{L}_c\left((D^\alpha)^2 y\right)(z) + 3\mathscr{L}_c(D^\alpha y)(z) + 2\mathscr{L}_c(y)(z) \\[2mm]
&= z^2 \mathscr{L}_c(y)(z) - E_0(\infty,0) + 3z\mathscr{L}_c(y)(z) + 2\mathscr{L}_c(y)(z) \\[2mm]
&= (z^2 + 3z + 2)\mathscr{L}_c(y)(z) - E_0(\infty,0)
\end{aligned}$$

or

$$\frac{E_0(\infty,0)}{z} + E_0(\infty,0) = (z^2 + 3z + 2)\mathscr{L}_c(y)(z),$$

or

$$\begin{aligned}
\mathscr{L}_c(y)(z) &= E_0(\infty,0)\left(\frac{1}{z(z+1)(z+2)} + \frac{1}{(z+1)(z+2)}\right) \\[2mm]
&= E_0(\infty,0)\left(\frac{1}{2z} - \frac{1}{z+1} + \frac{1}{2(z+2)} + \frac{1}{z+1} - \frac{1}{z+2}\right) \\[2mm]
&= E_0(\infty,0)\left(\frac{1}{2z} - \frac{1}{2(z+2)}\right).
\end{aligned}$$

Thus,

$$y(t) = \frac{1}{2} - \frac{1}{2}E_{-2}(t,0), \quad t \geq 0.$$

This ends the example.

Exercise 9.4.2 *Using the conformable Laplace transform, find a solution of the following IVP*

$$(D^\alpha)^2 y + 5D^\alpha y + 6y = 3 + 2E_1(t,0), \quad t > 0,$$

$$D^\alpha y(0) = -11,$$

$$y(0) = 1.$$

9.5 ADVANCED PRACTICAL PROBLEMS

Problem 9.5.1 *Let $f, g \in \mathscr{R}_c$ be constants. Under "suitable" assumptions find the following.*

1. $\mathscr{L}_c\left(C_{fg}(\cdot,0)\right)(z),$

2. $\mathscr{L}_c\left(S_{fg}(\cdot,0)\right)(z),$

3. $\mathscr{L}_c\left(Cos_f(\cdot,0) + 4S_{fg}(\cdot,0) - 5E_{f+g}(\cdot,0)\right)(z),$

4. $\mathscr{L}_c\left(Sin_f(\cdot,0) - 3Cos_f(\cdot,0)\right)(z),$

5. $\mathscr{L}_c\left(Sin_f(\cdot,0) - 4h_{11}(\cdot,0)\right)(z).$

Problem 9.5.2 *Let $\mathbb{T} = 3^{\mathbb{N}_0}$,*

$$k_1(\alpha,t) = (1-\alpha)t^\alpha, \quad k_0(\alpha,t) = \alpha t^{1-\alpha}, \quad \alpha \in (0,1], \quad t \in \mathbb{T},$$

$$u(t,s) = (s+t)^2 + t^4, \quad (t,s) \in \mathbb{T} \times \mathbb{T}.$$

Find

$$D_t^{\frac{1}{2}} u(t,s), \quad D_s^{\frac{1}{3}} u(t,s), \quad D_s^{\frac{1}{4}} u(t,s).$$

Problem 9.5.3 *Using the conformable Laplace transform, find solutions of the following IVPs:*

1.

$$(D^\alpha)^2 y - 7D^\alpha y + 12y = 2E_1(t,0) + 3Cosh_2(t,0), \quad t > 0,$$

$$D^\alpha y(0) = 2,$$

$$y(0) = -1.$$

2.

$$(D^\alpha)^2 y - 9y = E_2(t,0) - Sinh_2(t,0), \quad t > 0,$$

$$D^\alpha y(0) = 0,$$

$$y(0) = 1.$$

3.

$$(D^\alpha)^3 y - 6(D^\alpha)^2 y + 11 D^\alpha y - 6y = E_4(t,0) - Sinh_1(t,0) + 3Cosh_7(t,0), \quad t > 0,$$

$$(D^\alpha)^2 y(0) = 4,$$

$$D^\alpha y(0) = 1,$$

$$y(0) = 1.$$

Derivatives on Banach Spaces

A.1 REMAINDERS

Let X and Y be normed spaces. With $o(X,Y)$ we will denote the set of all maps $r : X \to Y$ for which there is some map $\alpha : X \to Y$ such that

1. $r(x) = \alpha(x)\|x\|$ for all $x \in X$,

2. $\alpha(0) = 0$,

3. α is continuous at 0.

Definition A.1.1 *The elements of $o(X,Y)$ will be called remainders.*

Exercise A.1.2 *Prove that $o(X,Y)$ is a vector space.*

Definition A.1.3 *Let $f : X \to Y$ be a function and $x_0 \in X$. We say that f is stable at x_0 if there are some $\varepsilon > 0$ and some $c > 0$ such that $\|x - x_0\| \leq \varepsilon$ implies*

$$\|f(x - x_0)\| \leq c\|x - x_0\|.$$

Example A.1.4 *Let $T : X \to Y$ be a linear bounded operator. Then*

$$\|T(x - 0)\| \;=\; \|T(x)\|$$

$$\leq \;\|T\|\|x\|, \quad x \in X.$$

Hence, T is stable at 0.

Theorem A.1.5 *Let X, Y, Z and W be normed spaces, $r \in o(X,Y)$, and assume $f : W \to X$ is stable at 0, $g : Y \to Z$ is stable at 0. Then $r \circ f \in o(W,Y)$ and $g \circ r \in o(X,Z)$.*

Proof A.1.6 *Since* $r \in o(X, Y)$*, then there is a map* $\alpha : X \to Y$ *such that*

$$r(x) = \alpha(x)\|x\|, \quad x \in X,$$

$\alpha(0) = 0$ *and* α *is continuous at* 0. *Define* $\beta : W \to Y$ *such that*

$$\beta(w) = \begin{cases} \dfrac{\|f(w)\|}{\|w\|}\alpha(f(w)) & \text{if} \quad w \neq 0, \\[2ex] 0 & \text{if} \quad w = 0, \end{cases}$$

$w \in W$. *Since* $f : W \to Z$ *is stable at* 0, *then there are constants* $\varepsilon > 0$ *and* $c > 0$ *such that* $\|w\| \leq \varepsilon$ *implies that*

$$\|f(w)\| \leq c\|w\|.$$

Hence,

$$\|f(0)\| = 0 \quad \text{and} \quad f(0) = 0.$$

Next, $\beta(0) = 0$ *and if* $w \neq 0$, $\|w\| \leq \varepsilon$, *we get*

$$\|\beta(w)\| = \frac{\|f(w)\|}{\|w\|}\|\alpha(f(w))\|$$

$$\leq c\|\alpha(f(w))\|.$$

From here, using that

$$f(w) \to 0 \quad \text{as} \quad w \to 0$$

and

$$\alpha(f(w)) \to 0 \quad \text{as} \quad w \to 0,$$

we get

$$\beta(w) \to 0 \quad \text{as} \quad w \to 0.$$

Therefore $\beta : W \to Y$ *is continuous at* 0. *Also,*

- *if* $w = 0$*, then*

$$\beta(0) = 0,$$

$$r \circ f(0) = \alpha(f(0))\|f(0)\|$$

$$= 0$$

$$= \beta(0).$$

- *if* $w \neq 0$*, then*

$$r \circ f(w) = \alpha(f(w))\|f(w)\|$$

$$= \frac{\|w\|\beta(w)}{\|f(w)\|}\|f(w)\|$$

$$= \|w\|\beta(w).$$

Therefore $r \circ f \in o(W, Y)$.

Since $g : Y \to Z$ *is stable at* 0, *then there are constants* $\varepsilon_1 > 0$ *and* $c_1 > 0$ *such that* $\|w\| \leq \varepsilon_1$ *implies*

$$\|g(w)\| \leq c_1\|w\|.$$

Define $\gamma : X \to Y$ *by*

$$\gamma(x) = \begin{cases} \dfrac{g(\|x\|\alpha(x))}{\|x\|} & \text{if } x \neq 0, \\ \\ 0 & \text{if } x = 0. \end{cases}$$

Then

$$g(\|x\|\alpha(x)) = \|x\|\gamma(x), \quad x \in X.$$

For $x \neq 0$, $x \in X$, *we have*

$$\|\gamma(x)\| = \frac{\|g(\|x\|\alpha(x))\|}{\|x\|}$$

$$\leq \frac{c_1\|x\|\alpha(x)}{\|x\|}$$

$$= c_1\alpha(x).$$

Then

$$\gamma(x) \to 0 \quad \text{as} \quad x \to 0, \quad x \in X.$$

Also,

$$g \circ r(x) = g(r(x))$$

$$= g(\alpha(x)\|x\|)$$

$$= \gamma(x)\|x\|, \quad x \in X.$$

This completes the proof. ☐

A.2 DEFINITION AND UNIQUENESS OF THE FRÉCHET DERIVATIVE

Suppose that X and Y are normed spaces, U is an open subset of X and $x_0 \in U$. With $\mathcal{L}(X, Y)$ we will denote the vector space of all linear bounded operators from X to Y.

Definition A.2.1 *We say that a function $f : X \to Y$ is Fréchet differentiable at x_0 if there is some $L \in \mathscr{L}(X,Y)$ and $r \in o(X,Y)$ such that*

$$f(x) = f(x_0) + L(x - x_0) + r(x - x_0), \quad x \in U.$$

The operator L will be called the Fréchet derivative of the function f at x_0. We will write $Df(x_0) = L$.

Suppose that $L_1, L_2 \in \mathscr{L}(X,Y)$ and $r_1, r_2 \in o(X,Y)$ are such that

$$f(x) = f(x_0) + L_1(x - x_0) + r_1(x - x_0),$$

$$f(x) = f(x_0) + L_2(x - x_0) + r_2(x - x_0), \quad x \in U.$$

Then

$$f(x_0) + L_1(x - x_0) + r_1(x - x_0) = f(x_0) + L_2(x - x_0) + r_2(x - x_0), \quad x \in U,$$

or

$$L_1(x - x_0) - L_2(x - x_0) = r_2(x - x_0) - r_1(x - x_0), \quad x \in U.$$

Also, let $\alpha_1, \alpha_2 : X \to Y$ be such that

$$r_1(x) = \|x\| \alpha_2(x), \quad r_2(x) = \|x\| \alpha_1(x),$$

$$\alpha_1(0) = \alpha_2(0) = 0,$$

α_1 and α_2 are continuous at 0. Then

$$L_1(x - x_0) - L_2(x - x_0) = \|x - x_0\| \alpha_1(x - x_0) - \|x - x_0\| \alpha_2(x - x_0)$$

$$= \|x - x_0\| (\alpha_1(x - x_0) - \alpha_2(x - x_0)), \quad x \in U.$$

Let $x \in X$ be arbitrarily chosen. Then there is some $h > 0$ such that for all $|t| \le h$ we have $x_0 + tx \in U$. Hence,

$$L_1(tx) - L_2(tx) = \|tx\| (\alpha_1(tx) - \alpha_2(tx))$$

or

$$t(L_1(x) - L_2(x)) = |t| \|x\| (\alpha_1(tx) - \alpha_2(tx)),$$

or

$$L_1(x) - L_2(x) = \operatorname{sign}(t) \|x\| (\alpha_1(tx) - \alpha_2(tx))$$

$$\to 0 \quad \text{as} \quad t \to 0.$$

Because $x \in X$ was arbitrarily chosen, we conclude that $L_1 = L_2$ and $r_1 = r_2$.

Definition A.2.2 *We denote by $\mathscr{C}^1(U,Y)$ the set of all functions $f : U \to Y$ that are Fréchet differentiable at each point of U and $Df : U \to \mathscr{L}(X,Y)$ is continuous. We denote by $\mathscr{C}^2(U,Y)$ the set of all functions $f \in \mathscr{C}^1(U,Y)$ such that $Df : U \to \mathscr{L}(X,Y)$ is Fréchet differentiable at each point of U and*

$$D(Df) : U \to \mathscr{L}(X,\mathscr{L}(X,Y))$$

is continuous.

Theorem A.2.3 *Let $f_1, f_2 : U \to Y$ be Fréchet differentiable at x_0 and $a,b \in \mathbb{R}$. Then $af_1 + bf_2$ is Fréchet differentiable at x_0.*

Proof A.2.4 *Let $r_1, r_2 \in o(X,Y)$ be such that*

$$f_1(x) \;=\; f_1(x_0) + Df_1(x_0)(x-x_0) + r_1(x-x_0),$$

$$f_2(x) \;=\; f_2(x_0) + Df_2(x_0)(x-x_0) + r_2(x-x_0), \quad x \in U.$$

Hence,

$$
\begin{aligned}
(af_1 + bf_2)(x) \;=\;& a\left(f_1(x_0) + Df_1(x_0)(x-x_0) + r_1(x-x_0)\right) \\
& + b\left(f_2(x_0) + Df_2(x_0)(x-x_0) + r_2(x-x_0)\right)
\end{aligned}
$$

$$
=\; af_1(x_0) + bf_2(x_0)
$$

$$
+ \left(aDf_1(x_0) + bDf_2(x_0)\right)(x-x_0)
$$

$$
+ \left(ar_1(x-x_0) + br_2(x-x_0)\right), \quad x \in U.
$$

Note that $ar_1 + br_2 \in o(X,Y)$. This completes the proof. □

Theorem A.2.5 *A function $f : U \to Y$ is Fréchet differentiable at x_0 if and only if there is some function $F : U \to \mathscr{L}(X,Y)$ that is continuous at x_0 and for which*

$$f(x) - f(x_0) = F(x)(x-x_0), \quad x \in U.$$

Proof A.2.6   *1. Suppose that there is a function $F : U \to \mathscr{L}(X,Y)$ that is continuous at x_0 and*

$$f(x) - f(x_0) = F(x)(x-x_0), \quad x \in U.$$

Then

$$
\begin{aligned}
f(x) - f(x_0) \;=\;& F(x)(x-x_0) - F(x_0)(x-x_0) + F(x_0)(x-x_0)
\end{aligned}
$$

$$
=\; F(x_0)(x-x_0) + r(x-x_0),
$$

where

$$r(x) = \begin{cases} (F(x+x_0) - F(x_0))(x) & for \quad x+x_0 \in U, \\ \\ 0 & for \quad x+x_0 \notin U. \end{cases}$$

Define

$$\alpha(x) = \begin{cases} \dfrac{(F(x+x_0) - F(x_0))(x)}{\|x\|} & for \quad x+x_0 \in U, \quad x \neq 0, \\ \\ 0 & for \quad x+x_0 \notin U, \\ \\ 0 & for \quad x = 0. \end{cases}$$

Then

$$r(x) = \alpha(x)\|x\|, \quad x \in X.$$

Let $\varepsilon > 0$ be arbitrarily chosen. Since $F : U \to \mathscr{L}(X,Y)$ is continuous at x_0, there exists some $\delta > 0$ for which $\|x\| < \delta$ implies

$$\| (F(x+x_0) - F(x_0))(x) \| \quad \leq \quad \|F(x+x_0) - F(x_0)\| \|x\|$$

$$< \quad \varepsilon \|x\|.$$

Therefore

$$|\alpha(x)| < \varepsilon$$

for $\|x\| < \delta$, i.e., α is continuous at 0. From here, we conclude that $r \in o(X,Y)$ and $F(x_0) = Df(x_0)$.

2. *Suppose that f is Fréchet differentiable at x_0. Then there is some $r \in o(X,Y)$ such that*

$$f(x) = f(x_0) + Df(x_0)(x - x_0) + r(x - x_0), \quad x \in U,$$

where $Df(x_0) \in \mathscr{L}(X,Y)$. Since $r \in o(X,Y)$, there is some $\alpha : X \to Y$ such that

$$r(x) \quad = \quad \alpha(x)\|x\|,$$

$$\alpha(0) \quad = \quad 0,$$

$$\alpha(x) \quad \to \quad 0 \quad as \quad x \to 0.$$

By the Hahn-Banach extension theorem, it follows that there is some $\lambda_x \in X^*$ such that

$$\lambda_x x = \|x\|$$

and

$$|\lambda_x v| \leq \|v\|, \quad v \in X.$$

Then

$$r(x) = (\lambda_x x)\alpha(x), \quad x \in X,$$

and

$$f(x) = f(x_0) + Df(x_0)(x - x_0) + \left(\lambda_{x-x_0}(x - x_0)\right)\alpha(x - x_0), \quad x \in U.$$

Let $F : U \to \mathscr{L}(X, Y)$ be defined as follows

$$F(x)(v) = Df(x_0)(v) + \left(\lambda_{x-x_0}v\right)\alpha(x - x_0), \quad x \in U, \quad v \in X.$$

We have

$$f(x) = f(x_0) + F(x)(x - x_0), \quad x \in U,$$

$$r(x - x_0) = \left(\lambda_{x-x_0}(x - x_0)\right)\alpha(x - x_0)$$

$$= f(x) - f(x_0) - Df(x_0)(x - x_0)$$

$$= F(x)(x - x_0) - Df(x_0)(x - x_0), \quad x \in U.$$

Note that

$$\|F(x)(v) - F(x_0)(v)\| = \|Df(x_0)(v) + \left(\lambda_{x-x_0}v\right)\alpha(x - x_0) - Df(x_0)(v)\|$$

$$= \|\left(\lambda_{x-x_0}v\right)\alpha(x - x_0)\|$$

$$= |\lambda_{x-x_0}v|\|\alpha(x - x_0)\|$$

$$\leq \|v\|\|\alpha(x - x_0)\|, \quad x \in U, \quad v \in X.$$

Then

$$\|F(x) - F(x_0)\| \leq \|\alpha(x - x_0)\|, \quad x \in U.$$

Consequently, F is continuous at x_0. This completes the proof. □

Theorem A.2.7 *Let Z be a normed space, assume $f : U \to Z$ is Fréchet differentiable at x_0, $g : f(U) \to Z$ is Fréchet differentiable at $f(x_0)$. Then $g \circ f : U \to Z$ is Fréchet differentiable at x_0 and*

$$D(g \circ f)(x_0) = Dg(f(x_0)) \circ Df(x_0).$$

Proof A.2.8 *Let*

$$y_0 = f(x_0),$$

$$L_1 = Df(x_0),$$

$$L_2 = Dg(y_0).$$

There exist $r_1 \in o(X,Y)$, $r_2 \in o(Y,Z)$ such that

$$f(x) = f(x_0) + L_1(x - x_0) + r_1(x - x_0), \quad x \in U,$$

$$g(y) = g(y_0) + L_2(y - y_0) + r_2(y - y_0), \quad y \in f(U).$$

Hence,

$$g(f(x)) = g(f(x_0)) + L_2(f(x) - y_0) + r_2(f(x) - y_0)$$

$$= g(y_0) + L_2(L_1(x - x_0) + r_1(x - x_0))$$

$$+ r_2(L_1(x - x_0) + r_1(x - x_0))$$

$$= g(y_0) + L_2(L_1(x - x_0)) + L_2(r_1(x - x_0))$$

$$+ r_2(L_1(x - x_0) + r_1(x - x_0)), \quad x \in U.$$

Define $r_3 : X \to Z$ as follows

$$r_3(x) = r_2(L_1(x) + r_1(x)), \quad x \in U.$$

Fix $c > \|L_1\|$ and we represent r_1 as follows

$$r_1(x) = \alpha_1(x)\|x\|, \quad x \in U.$$

We have that $\alpha_1 : X \to Y$, $\alpha_1(0) = 0$ and α_1 is continuous at 0. Then there exists some $\delta > 0$ such that if $\|x\| < \delta$, then

$$\|\alpha_1(x)\| < c - \|L_1\|.$$

Hence, if $\|x\| < \delta$, then

$$\|r_1(x)\| \leq (c - \|L_1\|)\|x\|.$$

Then, $\|x\| < \delta$ implies

$$\|L_1(x) + r_1(x)\| \leq \|L_1(x)\| + \|r_1(x)\|$$

$$\leq \|L_1\|\|x\| + (c - \|L_1\|)\|x\|$$

$$= c\|x\|.$$

Then $x \to L_1(x) + r_1(x)$ is stable at 0. Hence, by Theorem A.1.5, we get $r_3 \in o(X,Z)$. Define $r : X \to Z$ as follows

$$r = L_1 \circ r_1 + r_3.$$

We have $r \in o(X,Z)$ and

$$g \circ f(r) = g \circ f(x_0) + L_2 \circ L_1(x - x_0) + r(x - x_0), \quad x \in U.$$

Since $L_1 \in \mathscr{L}(X,Y)$, $L_2 \in \mathscr{L}(Y,Z)$, we have $L_2 \circ L_1 \in \mathscr{L}(X,Z)$. Therefore $g \circ f$ is Fréchet differentiable at x_0 and

$$L_2 \circ L_1 = Dg(y_0) \circ Df(x_0)$$

$$= Dg(f(x_0)) \circ Df(x_0).$$

This completes the proof. □

Theorem A.2.9 *Let $f_1, f_2 : U \to \mathbb{R}$ be Fréchet differentiable at x_0. Then $f_1 \cdot f_2$ is Fréchet differentiable at x_0 and*

$$D(f_1 \cdot f_2)(x_0) = f_2(x_0)Df_1(x_0) + f_1(x_0)Df_2(x_0).$$

Proof A.2.10 *Let $r_1, r_2 \in o(X, \mathbb{R})$ be such that*

$$f_1(x) = f_1(x_0) + Df_1(x_0)(x - x_0) + r_1(x - x_0),$$

$$f_2(x) = f_2(x_0) + Df_2(x_0)(x - x_0) + r_2(x - x_0), \quad x \in U.$$

Hence,

$$f_1(x)f_2(x) = (f_1(x_0) + Df_1(x_0)(x - x_0) + r_1(x - x_0))$$

$$\times (f_2(x_0) + Df_2(x_0)(x - x_0) + r_2(x - x_0))$$

$$= f_1(x_0)f_2(x_0) + f_1(x_0)Df_2(x_0)(x - x_0) + f_2(x_0)Df_1(x_0)(x - x_0)$$

$$+ f_1(x_0)r_2(x - x_0) + Df_1(x_0)(x - x_0)Df_2(x_0)(x - x_0)$$

$$+ Df_1(x_0)(x - x_0)r_2(x - x_0) + r_1(x - x_0)f_2(x_0)$$

$$+ Df_2(x_0)(x - x_0)r_1(x - x_0) + r_1(x - x_0)r_2(x - x_0), \quad x \in U.$$

Let $r : X \to \mathbb{R}$ be defined as follows

$$r(x) = f_1(x_0)r_2(x) + Df_1(x_0)xDf_2(x_0)x$$

$$+ Df_1(x_0)xr_2(x) + r_1(x)f_2(x_0)$$

$$+ Df_2(x_0)xr_1(x) + r_1(x)r_2(x), \quad x \in U.$$

Then

$$f_1(x)f_2(x) = f_1(x_0)f_2(x_0) + f_1(x_0)Df_2(x_0)(x - x_0)$$

$$+ f_2(x_0)Df_1(x_0)(x - x_0) + r(x - x_0), \quad x \in U.$$

Note that

$$|Df_1(x_0)xDf_2(x_0)x| \le \|Df_1(x_0)\|\|Df_2(x_0)\|\|x\|^2, \quad x \in U.$$

Define $\alpha : X \to \mathbb{R}$ as follows

$$\alpha(x) = \begin{cases} \dfrac{Df_1(x_0)xDf_2(x_0)x}{\|x\|}, & x \in U, \quad x \ne 0, \\[2mm] 0, & x = 0. \end{cases}$$

Then

$$Df_1(x_0)xDf_2(x_0)x = \alpha(x)\|x\|, \quad x \in U,$$

$$|\alpha(x)| = \frac{|Df_1(x_0)xDf_2(x_0)x|}{\|x\|}$$

$$\le \frac{\|Df_1(x_0)\|\|Df_2(x_0)\|\|x\|^2}{\|x\|}$$

$$= \|Df_1(x_0)\|\|Df_2(x_0)\|\|x\|, \quad x \in U, \quad x \ne 0.$$

Then

$$\alpha(x) \to 0 \quad as \quad x \to 0.$$

From here, $r \in o(X, \mathbb{R})$. This completes the proof. □

A.3 THE GÂTEAUX DERIVATIVE

Let X and Y be normed spaces and U be an open subset of X. Let also, $x_0 \in U$.

Definition A.3.1 *Let $f : U \to Y$. If there is some $T \in \mathscr{L}(X, Y)$ such that*

$$\lim_{t \to 0} \frac{f(x_0 + tv) - f(x_0)}{t} = Tv$$

for any $v \in X$, we say that f is Gâteaux differentiable at x_0. We write $f'(x_0) = T$. If f is Gâteaux differentiable at any point of U, then we say that f is Gâteaux differentiable on U.

Example A.3.2 *Let $f : \mathbb{R}^2 \to \mathbb{R}$ be defined as follows*

$$f(x_1,x_2) = \begin{cases} \dfrac{x_1^4}{x_1^6 + x_2^3} & for \quad (x_1,x_2) \neq (0,0), \\ 0 & for \quad (x_1,x_2) = (0,0). \end{cases}$$

Let $v = (v_1,v_2) \in \mathbb{R}^2$, $(v_1,v_2) \neq (0,0)$, be arbitrarily chosen. We have, for $t \neq 0$,

$$f(0+tv) = \frac{t^4 v_1^4}{t^6 v_1^6 + t^3 v_2^3}$$

$$= \frac{t v_1^4}{t^3 v_1^6 + v_2^3},$$

$$\lim_{t \to 0} \frac{f(0+tv) - f(0)}{t} = \lim_{t \to 0} \frac{t v_1^4}{t \left(t^3 v_1^6 + v_2^3\right)}$$

$$= \lim_{t \to 0} \frac{v_1^4}{t^3 v_1^6 + v_2^3}$$

$$= \frac{v_1^4}{v_2^3}.$$

Therefore

$$f'(0,0)(v_1,v_2) = \frac{v_1^4}{v_2^3}, \quad (v_1,v_2) \in \mathbb{R}^2, \quad (v_1,v_2) \neq (0,0).$$

This ends the example.

Theorem A.3.3 *If $f : U \to Y$ is Fréchet differentiable at x_0, then it is Gâteaux differentiable at x_0.*

Proof A.3.4 *Since $f : U \to Y$ is Fréchet differentiable at x_0, then there is some $r \in o(X,Y)$ such that*

$$f(x) = f(x_0) + Df(x_0)(x - x_0) + r(x - x_0), \quad x \in U,$$

and

$$r(x) = \alpha(x)\|x\|, \quad x \in X,$$

where $\alpha : X \to Y$, $\alpha(0) = 0$, α is continuous at 0. Then, for $v \in X$ and $t \in \mathbb{R}$, $|t|$ small enough, we have

$$\frac{f(x_0 + tv) - f(x_0)}{t} = \frac{Df(x_0)(tv) + r(tv)}{t}$$

$$= \frac{t Df(x_0)(v) + |t| \|v\| \alpha(tv)}{t}$$

$$= Df(x_0)(v) + sign(t) \|v\| \alpha(tv)$$

$$\rightarrow Df(x_0)(v) \quad as \quad t \rightarrow 0.$$

This completes the proof. □

A Chain Rule

B.1   MEASURE CHAINS

Let $\mathbb{T}$ be some set of real numbers.

Definition B.1.1 *A triple $(\mathbb{T}, \leq, \nu)$ is called a measure chain provided it satisfies the following axioms.*

(A1) *The relation "$\leq$" satisfies, for $r, s, t \in \mathbb{T}$,*

    1. $t \leq t$ (reflexive),

    2. if $t \leq r$ and $r \leq s$, then $t \leq s$ (transitive),

    3. if $t \leq r$ and $r \leq t$, then $t = r$ (antisymmetric),

    4. either $r \leq s$ or $s \leq r$ (total).

(A2) *Any nonvoid subset of $\mathbb{T}$ which is bounded above has a least upper bound, i.e., the measure chain $(T, \leq)$ is conditionally complete.*

(A3) *The mapping $\nu : \mathbb{T} \times \mathbb{T} \to \mathbb{R}$ has the following properties, for $r, s, t \in \mathbb{T}$.*

    1. $\nu(r, s) + \nu(s, t) = \nu(r, t)$ (cocycle property),

    2. if $r > s$, then $\nu(r, s) > 0$ (strong isotony),

    3. ν is continuous (continuity).

Example B.1.2 *Let $\mathbb{T}$ be any nonvoid closed subset of real numbers, "$\leq$" is the usual order relation between real numbers and*

$$\nu(r, s) = r - s, \quad r, s \in \mathbb{T}.$$

Definition B.1.3 *The forward jump operator σ and the backward jump operator ρ are defined as follows.*

$$\sigma(t) = \inf\{s \in \mathbb{T} : s > t\}, \quad \rho(t) = \sup\{s \in \mathbb{T} : s < t\},$$

328 ■ Conformable Dynamic Equations on Time Scales

where

$$\sigma(t) \;=\; t \quad if \quad t = \max \mathbb{T},$$

$$\rho(t) \;=\; t \quad if \quad t = \min \mathbb{T}.$$

The graininess function is defined as follows.

$$\mu(t) = \nu(\sigma(t),t), \quad t \in \mathbb{T}.$$

The notions left-scattered, left-dense, right-scattered, right-dense, isolated, and $\mathbb{T}^{\kappa}$ are defined as in the case of time scales.

Definition B.1.4 *Let X be a Banach space with a norm $\|\cdot\|$. We say that $f : \mathbb{T} \to X$ is differentiable at $t \in \mathbb{T}$ if there exists $f^{\Delta}(t) \in X$ such that for any $\varepsilon > 0$ there exists a neighborhood U of t such that*

$$\|f(\sigma(t)) - f(s) - f^{\Delta}(t)\nu(\sigma(t),s)\| \leq \varepsilon |\nu(\sigma(t),s)|$$

for all $s \in U$. In this case $f^{\Delta}(t)$ is said to be a derivative of f at t.

Theorem B.1.5 *We have*

$$\nu^{\Delta}(\cdot,t) = 1, \quad t \in \mathbb{T}.$$

Proof B.1.6 *Let $t \in \mathbb{T}$. Let also, $\varepsilon > 0$ be arbitrarily chosen and U is a neighborhood of t. Then*

$$\nu(\sigma(t),s) + \nu(s,t) \;=\; \nu(\sigma(t),t), \quad s \in \mathbb{T},$$

and

$$|\nu(\sigma(t),t) - \nu(s,t) - \nu(\sigma(t),s)| \;=\; |\nu(\sigma(t),t) - \nu(\sigma(t),t)|$$

$$=\; 0$$

$$\leq\; \varepsilon |\nu(\sigma(t),s)|,$$

for any $s \in U$. This completes the proof. □

As in the case of time scales, one can prove the following assertion.

Theorem B.1.7 *Let $f,g : \mathbb{T} \to X$ and $t \in \mathbb{T}$.*

1. *If $t \in \mathbb{T}^{\kappa}$, then f has at most one derivative at t.*

2. *If f is differentiable at t, then f is continuous at t.*

3. *If f is continuous at t and t is right-scattered, then f is differentiable at t and*

$$f^{\Delta}(t) = \frac{f(\sigma(t)) - f(t)}{\mu(t)}.$$

4. *If f and g are differentiable at $t \in \mathbb{T}^{\kappa}$ and $\alpha, \beta \in \mathbb{R}$, then $\alpha f + \beta g$ is differentiable at t and*

$$(\alpha f + \beta g)^{\Delta}(t) = \alpha f^{\Delta}(t) + \beta g^{\Delta}(t).$$

5. *If f and g are differentiable at $t \in \mathbb{T}^{\kappa}$ and "$\cdot$" is bilinear and continuous, then $f \cdot g$ is differentiable at t and*

$$(f \cdot g)^{\Delta}(t) = f^{\Delta}(t) \cdot g(t) + f(\sigma(t)) \cdot g^{\Delta}(t).$$

6. *If f and g are differentiable at $t \in \mathbb{T}^{\kappa}$ and g is algebraically invertible, then $f \cdot g^{-1}$ is differentiable at t with*

$$\left(f \cdot g^{-1}\right)^{\Delta}(t) = \left(f^{\Delta}(t) - \left(f \cdot g^{-1}\right)(t) \cdot g^{\Delta}(t)\right) \cdot g^{-1}(\sigma(t)).$$

B.2 PÖTZSCHE'S CHAIN RULE

Throughout this section we suppose that $(\mathbb{T}, \leq, \nu)$ is a measure chain with forward jump operator σ and graininess μ. Assume that X and Y are Banach spaces and we will write $\|\cdot\|$ for the norms of X and Y. For a function $f : \mathbb{T} \times X \to Y$ and $x_0 \in X$, we denote the delta derivative of $t \to f(t, x_0)$ by $\Delta_1 f(\cdot, x_0)$, and for a $t_0 \in \mathbb{T}$ we denote the Fréchet derivative of $x \to f(t_0, x)$ by $D_2 f(t_0, \cdot)$, provided these derivatives exist.

Theorem B.2.1 (Pötzsche's Chain Rule) *For some fixed $t_0 \in \mathbb{T}^{\kappa}$, let $g : \mathbb{T} \to X$, $f : \mathbb{T} \times X \to Y$ be functions such that g, $f(\cdot, g(t_0))$ are differentiable at t_0, and let $U \subseteq \mathbb{T}$ be a neighborhood of t_0 such that $f(t, \cdot)$ is differentiable for $t \in U \bigcup \{\sigma(t_0)\}$, $D_2 f(\sigma(t_0), \cdot)$ is continuous on the line segment*

$$\{g(t_0) + h\mu(t_0)g^{\Delta}(t_0) \in X : h \in [0,1]\}$$

and $D_2 f$ is continuous at $(t_0, g(t_0))$. Then the composition function $F : \mathbb{T} \to Y$, $F(t) = f(t, g(t))$ is differentiable at t_0 with derivative

$$F^{\Delta}(t_0) = \Delta_1 f(t_0, g(t_0))$$

$$+ \left(\int_0^1 D_2 f(\sigma(t_0), g(t_0) + h\mu(t_0)g^{\Delta}(t_0))dh\right)g^{\Delta}(t_0).$$

Proof B.2.2 *Let $U_0 \subseteq U$ be a neighborhood of t_0 such that*

$$\mu(t_0) \leq |\nu(t, \sigma(t_0))| \quad \text{for} \quad t \in U_0.$$

Let

$$\Phi(t, h) = D_2 f(t, g(t_0) + h(g(t) - g(t_0))), \quad t \in U_0, \quad h \in [0,1].$$

Note that there exists a constant $C > 0$ such that

$$\|\Phi(\sigma(t_0), h) - \Phi(t_0, h)\| \leq C|v(t, \sigma(t_0))| \quad \text{for} \quad t \in U_0, \quad h \in [0, 1].$$

Let $\varepsilon > 0$ be arbitrarily chosen. We choose $\varepsilon_1 > 0$, $\varepsilon_2 > 0$ small enough such that

$$\varepsilon_1 \left(1 + C \| \int_0^1 \Phi(\sigma(t_0), h) dh \| \right) + \varepsilon_2 \left(\varepsilon_1 + 2\|g^\Delta(t_0)\| \right) \leq \varepsilon.$$

Since g and $f(\cdot, g(t_0))$ are differentiable at t_0, there exists a neighborhood $U_1 \subseteq U_0$ of t_0 such that

$$\|g(t) - g(t_0)\| \leq \varepsilon_1,$$

$$\|g(t) - g(\sigma(t_0)) - v(t, \sigma(t_0))g^\Delta(t_0)\| \leq \varepsilon_1 |v(t, \sigma(t_0))|,$$

$$\|f(t, g(t_0)) - f(\sigma(t_0), g(t_0)) - v(t, \sigma(t_0))\Delta_1 f(t_0, g(t_0))\| \leq \varepsilon_1 |v(t, \sigma(t_0))|$$

for $t \in U_1$. Hence,

$$
\begin{aligned}
\|g(t) - g(t_0)\| &= \|g(t) - g(\sigma(t_0)) - v(t, \sigma(t_0))g^\Delta(t_0) + g^\Delta(t_0)v(t, \sigma(t_0)) \\
&\quad + g(\sigma(t_0)) - g(t_0)\| \\
&\leq \|g(t) - g(\sigma(t_0)) - v(t, \sigma(t_0))g^\Delta(t_0)\| \\
&\quad + \|g^\Delta(t_0)\||v(t, \sigma(t_0))| + \|g(\sigma(t_0)) - g(t_0)\| \\
&\leq \varepsilon_1 |v(t, \sigma(t_0))| + \|g^\Delta(t_0)\||v(t, \sigma(t_0))| \\
&\quad + \|g^\Delta(t_0)\|\mu(t_0) \\
&= \left(\varepsilon_1 + \|g^\Delta(t_0)\| \right) |v(t, \sigma(t_0))| + \|g^\Delta(t_0)\|\mu(t_0) \\
&\leq \left(\varepsilon_1 + 2\|g^\Delta(t_0)\| \right) |v(t, \sigma(t_0))|, \quad t \in U_1.
\end{aligned}
$$

Since g is continuous at t_0 and $D_2 f$ is continuous at $(t_0, g(t_0))$, there exists a neighborhood $U_2 \subseteq U$ of t_0 so that

$$\|\Phi(t, h) - \Phi(t_0, h)\| \leq \varepsilon_2 \quad \text{for} \quad t \in U_2, \quad h \in [0, 1].$$

Hence,

$$\|F(t) - F(\sigma(t_0)) - v(t, \sigma(t_0)) \left(\Delta_1 f(t_0, g(t_0)) + \int_0^1 \Phi(\sigma(t_0), h) dh g^\Delta(t_0) \right) \|$$

$$= \quad \| f(t,g(t)) - f(\sigma(t_0),g(\sigma(t_0))) - f(\sigma(t_0),g(t_0)) + f(\sigma(t_0),g(t_0))$$

$$-f(t,g(t_0)) + f(t,g(t_0))$$

$$-v(t,\sigma(t_0))\Delta_1 f(t_0,g(t_0))$$

$$-v(t,\sigma(t_0)) \int_0^1 \Phi(\sigma(t_0),h) dh g^\Delta(t_0)$$

$$- \int_0^1 \Phi(\sigma(t_0),h) dh (g(t) - g(t_0))$$

$$+ \int_0^1 \Phi(\sigma(t_0),h) dh (g(t) - g(t_0)) \|$$

$$\leq \quad \| f(t,g(t_0)) - f(\sigma(t_0),g(t_0)) - v(t,\sigma(t_0))\Delta_1 f(t_0,g(t_0)) \|$$

$$+ \| \int_0^1 \Phi(\sigma(t_0),h) dh \left(g(t) - g(t_0) - v(t,\sigma(t_0)) g^\Delta(t_0) \right) \|$$

$$+ \| f(t,g(t)) - f(t,g(t_0)) - (f(\sigma(t_0),g(\sigma(t_0))) - f(\sigma(t_0),g(t_0)))$$

$$- \int_0^1 \Phi(\sigma(t_0),h) dh (g(t) - g(t_0)) \|$$

$$\leq \quad \| f(t,g(t_0)) - f(\sigma(t_0),g(t_0)) - v(t,\sigma(t_0))\Delta_1 f(t_0,g(t_0)) \|$$

$$+ \| \int_0^1 \Phi(\sigma(t_0),h) dh \| \| g(t) - g(t_0) - v(t,\sigma(t_0)) g^\Delta(t_0) \|$$

$$+ \| \int_0^1 (\Phi(t,h) - \Phi(\sigma(t_0),h)) dh (g(t) - g(t_0)) \|$$

$$\leq \quad \| f(t,g(t_0)) - f(\sigma(t_0),g(t_0)) - v(t,\sigma(t_0))\Delta_1 f(t_0,g(t_0)) \|$$

$$+ \| \int_0^1 \Phi(\sigma(t_0),h) dh \| \| g(t) - g(t_0) - v(t,\sigma(t_0)) g^\Delta(t_0) \|$$

$$+ \| \int_0^1 (\Phi(t,h) - \Phi(t_0,h)) dh \| \| g(t) - g(t_0) \|$$

$$+ \| \int_0^1 (\Phi(t_0,h) - \Phi(\sigma(t_0),h)) dh \| \| g(t) - g(t_0) \|$$

$$\leq \quad \varepsilon_1 |v(t, \sigma(t_0))| + \varepsilon_1 |v(t, \sigma(t_0))| \left\| \int_0^1 \Phi(\sigma(t_0), h) dh \right\|$$

$$+ \varepsilon_2 \left(\varepsilon_1 + 2\|g^\Delta(t_0)\| \right) |v(t, \sigma(t_0))|$$

$$+ \varepsilon_1 C |v(t, \sigma(t_0))|$$

$$= \quad \left(\varepsilon_1 \left(1 + C + \left\| \int_0^1 \Phi(\sigma(t_0), h) dh \right\| \right) \right.$$

$$+ \varepsilon_2 \left(\varepsilon_1 + 2\|g^\Delta(t_0)\| \right) \bigg) |v(t, \sigma(t_0))|$$

$$\leq \quad \varepsilon |v(t, \sigma(t_0))|, \quad t \in U_1 \bigcap U_2.$$

This completes the proof. □

Bibliography

[1] D. R. Anderson, J. Bullock, L. Erbe, A. Peterson and H. N. Tran. Nabla Dynamic Equations on Time Scales, *Pan American Mathematical Journal*, 13:1 (2003) 1–47.

[2] D. R. Anderson and D. J. Ulness. Newly Defined Conformable Derivatives, *Advances in Dynamical Systems and Applications*, Vol. 10, Number 2, pp. 109-137, 2015.

[3] D. R. Anderson. Second-Order Self-Adjoint Differential Equations Using a Proportional-Derivative Controller, *Communications on Applied Nonlinear Analysis*, Vol. 24(2017), Number 1, 17-48.

[4] F. M. Atici and G. Sh. Guseinov. On Green's Functions and Positive Solutions for Boundary Value Problems on Time Scales, *J. Comput. Appl. Math.*, Vol. 141, Issues 1-2, (2002) 75–99.

[5] B. Bayour, A. Hammoudi and D. F. M. Torres. A Truly Conformable Calculus on Time Scales, *Global and Stochastic Analysis*, Vol. 5, No. 1, June (2018), 1-14.

[6] M. Bohner and A. Peterson. *Dynamic Equations on Time Scales: An Introduction with Applications*, Birkhäuser, Boston, 2001.

[7] M. Bohner and A. Peterson(editors). *Advances in Dynamic Equations on Time Scales*, Birkhäuser, Boston, 2003.

[8] M. Bohner and S. Georgiev. *Multivariable Dynamic Calculus on Time Scales*, Springer, 2017.

[9] S. Georgiev. *Integral Equations on Time Scales*, Springer 2016.

[10] M.R.S. Rahmat. *A New Definition of Conformable Fractional Derivative on Arbitrary Time Scales*, Advances in Difference Equations, 2019.

Index